T0075042

Communications and Control Engineering

For other titles published in this series, go to
www.springer.com/series/61

Series Editors

A. Isidori • J.H. van Schuppen • E.D. Sontag • M. Thoma • M. Krstic

Published titles include:

Stability and Stabilization of Infinite Dimensional Systems with Applications
Zheng-Hua Luo, Bao-Zhu Guo and Omer Morgul

Nonsmooth Mechanics (Second edition)
Bernard Brogliato

Nonlinear Control Systems II
Alberto Isidori

L_2-Gain and Passivity Techniques in Nonlinear Control
Arjan van der Schaft

Control of Linear Systems with Regulation and Input Constraints
Ali Saberi, Anton A. Stoorvogel and Peddapullaiah Sannuti

Robust and H_∞ Control
Ben M. Chen

Computer Controlled Systems
Efim N. Rosenwasser and Bernhard P. Lampe

Control of Complex and Uncertain Systems
Stanislav V. Emelyanov and Sergey K. Korovin

Robust Control Design Using H_∞ Methods
Ian R. Petersen, Valery A. Ugrinovski and Andrey V. Savkin

Model Reduction for Control System Design
Goro Obinata and Brian D.O. Anderson

Control Theory for Linear Systems
Harry L. Trentelman, Anton Stoorvogel and Malo Hautus

Functional Adaptive Control
Simon G. Fabri and Visakan Kadirkamanathan

Positive 1D and 2D Systems
Tadeusz Kaczorek

Identification and Control Using Volterra Models
Francis J. Doyle III, Ronald K. Pearson and Babatunde A. Ogunnaike

Non-linear Control for Underactuated Mechanical Systems
Isabelle Fantoni and Rogelio Lozano

Robust Control (Second edition)
Jürgen Ackermann

Flow Control by Feedback
Ole Morten Aamo and Miroslav Krstic

Learning and Generalization (Second edition)
Mathukumalli Vidyasagar

Constrained Control and Estimation
Graham C. Goodwin, Maria M. Seron and José A. De Doná

Randomized Algorithms for Analysis and Control of Uncertain Systems
Roberto Tempo, Giuseppe Calafiore and Fabrizio Dabbene

Switched Linear Systems
Zhendong Sun and Shuzhi S. Ge

Subspace Methods for System Identification
Tohru Katayama

Digital Control Systems
Ioan D. Landau and Gianluca Zito

Multivariable Computer-controlled Systems
Efim N. Rosenwasser and Bernhard P. Lampe

Dissipative Systems Analysis and Control (Second edition)
Bernard Brogliato, Rogelio Lozano, Bernhard Maschke and Olav Egeland

Algebraic Methods for Nonlinear Control Systems
Giuseppe Conte, Claude H. Moog and Anna M. Perdon

Polynomial and Rational Matrices
Tadeusz Kaczorek

Simulation-based Algorithms for Markov Decision Processes
Hyeong Soo Chang, Michael C. Fu, Jiaqiao Hu and Steven I. Marcus

Iterative Learning Control
Hyo-Sung Ahn, Kevin L. Moore and YangQuan Chen

Distributed Consensus in Multi-vehicle Cooperative Control
Wei Ren and Randal W. Beard

Control of Singular Systems with Random Abrupt Changes
El-Kébir Boukas

Nonlinear and Adaptive Control with Applications
Alessandro Astolfi, Dimitrios Karagiannis and Romeo Ortega

Stabilization, Optimal and Robust Control
Aziz Belmiloudi

Control of Nonlinear Dynamical Systems
Felix L. Chernous'ko, Igor M. Ananievski and Sergey A. Reshmin

Periodic Systems
Sergio Bittanti and Patrizio Colaneri

Discontinuous Systems
Yury V. Orlov

Constructions of Strict Lyapunov Functions
Michael Malisoff and Frédéric Mazenc

Controlling Chaos
Huaguang Zhang, Derong Liu and Zhiliang Wang

Stabilization of Navier-Stokes Flows
Viorel Barbu

Distributed Control of Multi-agent Networks
Wei Ren and Yongcan Cao

Zhendong Sun · Shuzhi Sam Ge

Stability Theory of Switched Dynamical Systems

Zhendong Sun
College Automation Science & Engineering
Center for Control and Optimization
South China University of Technology
Guangzhou 510640
People's Republic of China
zdsun@scut.edu.cn

Shuzhi Sam Ge
Robotics Institue and Institute of Intelligent Systems and Information Technology
University of Electronic Science and Technology of China
Chengdu
People's Republic of China
and
Department of Electrical and Computer Engineering
The National University of Singapore
Singapore
Singapore
elegesz@nus.edu.sg

ISSN 0178-5354
ISBN 978-0-85729-255-1 e-ISBN 978-0-85729-256-8
DOI 10.1007/978-0-85729-256-8
Springer London Dordrecht Heidelberg New York

British Library Cataloguing in Publication Data
A catalogue record for this book is available from the British Library

Cover design: eStudio Calamar S.L.

Printed on acid-free paper

Springer is part of Springer Science+Business Media (www.springer.com)

In memory of Professor Wei-Bing Gao

Preface

Switched control systems have attracted much interest from the control community not only because of their inherent complexity, but also due to the practical importance with a wide range of their applications in nature, engineering, and social sciences. Switched systems are necessary because various natural, social, and engineering systems cannot be described simply by a single model, and many systems exhibit switching between several models depending on various environments. Natural biological systems switch strategies in accordance to environmental changes for survival. Switched behaviors have also been exhibited in a number of social systems. To achieve an improved performance, switching has been extensively utilized/exploited in many engineering systems such as electronics, power systems, and traffic control, among others.

Theoretical investigation and examination of switched control systems are academically more challenging due to their rich, diverse, and complex dynamics. Switching makes those systems much more complicated than standard systems. Many more complicated behaviors/dynamics and fundamentally new properties, which standard systems do not have, have been demonstrated on switched systems. From the viewpoint of control system design, switching brings an additional degree of freedom in control system design. Switching laws, in addition to control laws, may be utilized to manipulate switched systems to achieve a better performance of a system. This can be seen as an added advantage for control design to attain certain control purposes.

On the one hand, switching could be induced by any unpredictable sudden change in system dynamics/structures, such as a sudden change of a system structure due to the failure of a component/subsystems, or the accidental activation of any subsystems. On the other hand, the switching is introduced artificially to effectively control highly complex nonlinear systems under the umbrella of the so-called hybrid control. In both cases, an essential feature is the interaction between the continuous system dynamics and the discrete switching dynamics. Such switched dynamical systems typically consist of sets of subsystems and switching signals that coordinate the switching among the subsystems.

In this book, we investigate the stability issues under various switching mechanisms. For a controlled switching, the switching signal is a design variable just as the control input in the conventional systems. It is measurable and can be freely assigned. In this case, the stability is in fact a kind of stabilization by stabilizing switching design. For an arbitrary switching, the switching signal is blind and uncontrolled, and the stability is in fact a kind of robustness against the switching perturbations. Besides the two extreme cases, the switching signal could be constrained in that it is neither controlled nor free arbitrarily. In other words, partial information is known about the switching mechanism. Typical constrained switchings include (i) autonomous switching, where the switching signal is generated autonomously with a preassigned state-space-partition-based switching mechanism; (ii) dwell-time switching, where the minimum duration on each subsystem is known and positive; and (iii) random switching with a known stochastic distribution. Switched systems under various constrained switching might behave in a rich, diverse and complex manner.

The objective of this book is to present in a systematic manner the stability theory of switched dynamical systems under different switching mechanisms. By bringing forward fresh new concepts, novel methods, and innovative tools into the exploration of various switching schemes, we are to provide a state-of-the-art and comprehensive systematic treatment of the stability issues for switched dynamical systems.

The book is organized in five chapters. Except for Chap. 1 that briefly introduces the problem formations and the organization of the book, subsequent chapters exploit several important topics in detail in a timely manner.

In Chap. 2, we focus on the guaranteed stability analysis of switched dynamical systems under arbitrary switching. As global uniform asymptotic stability is equivalent to the existence of a common Lyapunov function of the subsystems, the Lyapunov approach plays a dominant role in the stability analysis. For switched linear systems, due to the fact that quadratic Lyapunov candidates are insufficient for coping with stability, emphasis is laid on the sets of functions which are universal in the sense that each asymptotically stable system admits a Lyapunov function from the function set. Both piecewise linear functions and piecewise quadratic functions are proven to be universal, and their connections to algebraic stability criteria are also established. We also pay much attention to the algebraic theory of discrete-time switched linear systems, where the stability is elegantly characterized by the spectral radius of the matrix set, which generalizes the standard matrix spectral theory. While determining the spectral radius has been proven to be *NP*-hard, we introduce the homogeneous polynomials to serve as common Lyapunov functions and utilize the sum-of-squares technique and the semi-definite programming to approximate the spectral radius. Finally, the more subtle issue of marginal stability is carefully examined, and its connection to the common weak Lyapunov function is established. We reveal that marginal stability admits a block triangular decomposition with clear spectral information, and this leads to an invariant set viewpoint for characterizing marginal stability and marginal instability.

Chapter 3 presents stability theory for switched dynamical systems under constrained switching. There are three types of constrained switching addressed in this chapter. The first type of constrained switching is the random switching with a preassigned jump distribution. When the subsystems are linear and the switching is governed by a Markov process, the switched linear system is known to be a jump linear system. We introduce various stability concepts and their criteria, and establish the connections to the guaranteed stability criteria in Chap. 2. The second is the piecewise affine systems where the state space is partitioned into a set of polyhedral cells, each relating to a subsystem, and hence the switching is totally autonomous. The piecewise quadratic Lyapunov approach, the surface Lyapunov approach, and the transition graph approach are introduced. The pros and cons of the approaches are compared and discussed. The third type of constrained switching is the dwell-time switching, where the switching duration between any two consecutive switches admits a positive lower bound. We address both the stability analysis, where the dwell time is preassigned, and the stabilizing switching design, where the minimum or maximum dwell time is to be designed. The design captures the capability and the limitation of the switching mechanism.

Chapter 4 is devoted to the stabilizing switching design for switched dynamical systems under controlled switching. It is proven that a switched Lyapunov function exists if the system is globally asymptotically stabilizable. However, counterexamples exhibit that even stabilizable planar switched linear systems may not admit any convex switched Lyapunov function. To overcome the intrinsic difficulty, we introduce a class of nonconvex functions known as min functions that are piecewise quadratic and prove that each stabilizable switched linear system admits a min function as a switched Lyapunov function. To further address the stabilizability and robustness of switched linear system, we propose a pathwise state-feedback switching strategy, which accounts to concatenating a finite number of switching paths based on appropriate partitions of the state space. By aggregating the overall system into a discrete-time piecewise linear system, we are able to prove that the switching strategy exponentially stabilizes the original switched linear system whenever it is asymptotically stabilizable. We develop a computational procedure to calculate a stabilizing pathwise state-feedback switching law for an asymptotically stabilizable switched linear system. To further investigate the robustness of the pathwise state-feedback switching strategy, we define a (relative) distance between two switching signals and prove that the closed-loop system is robust against structural/unstructural/switching perturbations.

With the stability theory presented in the previous chapters, we further exploit its connections and implications to several fundamental control problems in Chap. 5. For absolute stability of Lur'e systems, an elegant connection to the guaranteed stability of switched linear systems is established. Utilizing this connection, computational algorithms are presented to verify absolute stability for planar Lur'e systems. Another implication of the guaranteed stability criteria is the consensus analysis of multiagent systems with dynamic neighbors, and exponential agreement is reached if the graph is always strongly connected. For an intelligent system with linear local controllers and a fuzzy rule, it is naturally converted into a piecewise linear system, and hence the stability analysis can be conducted by means of the stability criteria presented in Chap. 3. This brings a new design method and a fresh observation to the fuzzy control problem. For a SISO linear process with unknown parameters, an adaptive control framework is established based on appropriate partitions of the parameter space and proper stabilizing switching strategy among the local controllers which are designed to stabilize the system in a local sense. Finally, for controllable switched linear systems, a multilinear feedback design approach is proposed to tackle the stabilization problem. The main idea is to associate a set of candidate linear controllers with each subsystem, such that the extended switched system is stabilizable. By utilizing the pathwise state-feedback switching design diagram, the problem of stabilization is solved in a constructive manner.

The book is primarily intended for researchers and engineers in the system and control community. It can also serve as complementary reading for nonlinear system theory at the postgraduate level.

Acknowledgments

There is a beginning and an end in everything. For the completion of the book, we are in debts to many distinguished individuals in our community. First of all, we would like to thank Dazhong Zheng, Tsinghua University, for bringing our attention to this area fourteen years ago, and much of the results were rooted fundamentally in the numerous seminars, discussions, and well-rounded education at the Haidian district, Beijing, led by many leading scientists and academics including Hanfu Chen, Daizhan Cheng, Lei Guo, and Huashu Qin, Chinese Academy of Sciences; Lin Huang, Peking University; Zongji Chen, Weibing Gao, and Zhanlin Wang, Beijing University of Aeronautics of Astronautics; and, of course, Dazhong Zheng from Tsinghua University.

Blessed by many discussions and encounters with many distinguished individuals in society, the following people played an important role in our journey in the field of hybrid systems and switching control: Panos J. Antsaklis, University of Notre Dame, Tamer Basar, the University of Illinois at Urbana-Champaign, John Baillieul and Christos G. Cassandras of Boston University, Colin B. Besant and John C. Allwright of Imperial College, Robert R. Bitmead and Miroslav Krstic of University of California at San Diego, David Clements, the University of New South Wales, Xiren Cao of the Hong Kong University of Science and Technology, Graham C. Goodwin, the University of Newcastle, Chang Chieh Hang and Tong-Heng Lee of the National University of Singapore, David Hill of the Australian National University, Jie Huang, the Chinese University of Hong Kong, Petar V. Kokotovic, University of California at Santa Barbara, Frank F. Lewis of University of Texas at Arlington, Iven Mareels of the University of Melbourne, Ian Postlethwaite of Leicester University, Robert N. Shorten of the National University of Ireland at Maynooth, Mark W. Spong, University of Texas at Dallas, Roberto Tempo of Politecnico di Torino, and Xiaohua Xia of the University of Pretoria.

For the final completion of the book, we gratefully acknowledge the unreserved support, constructive comments, and fruitful discussions from Hai Lin, Rui Li, Qijin Liu, Yupeng Qiao, Cheng Xiang, and Bugong Xu.

Much appreciation goes to our individual former and current students Trung T. Han, Yuping Peng, Thanh L. Vu, Jun Wu, Jiandong Xiong, and Zhengui Xue for the time and effort in proofreading and providing numerous useful comments and suggestions to improve the readability of the book.

We are also grateful to Anthony Doyle, Senior Editor of Springer-Verlag, for his kind invitation in publishing the book. Special thanks go to Oliver Jackson, Editor of Springer-Verlag London, and Charlotte Cross, Editorial Assistant of Springer-Verlag London, for their reliable help and patience in the process of publishing the book.

This book is the crystallization of our ten-year long endeavors in the hybrid systems and the rich experience in convergence, divergence, and chaos with much joys, gratitude, and friendship in the long journey.

Financial Support

We acknowledge the financial support from the National Natural Science Foundation of China under Grants 60925013, 60736024, U0735003, and 60674042, from the National Basic Research Program of China (973 Program) under Grant 2011CB707005, and from the Program for New Century Excellent Talents in Universities of China under Grant NCET-07-0303.

Wushan, Guangzhou, China — Zhendong Sun
Shahe, Chengdu, China/Kent Ridge Crescent, Singapore — Shuzhi Sam Ge

Contents

List of Symbols

$a \stackrel{\text{def}}{=} b$	Defines a to be b
iff	If and only if
s.t.	Such that
w.r.t.	With respect to
a.s.	Almost surely
$\mathbf{R}$	The field of real numbers
$\mathbf{C}$	The field of complex numbers
$\mathbf{N}$	The set of integers
$\mathbf{R}^+$	The set of positive real numbers
$\mathbf{R}_+$	The set of nonnegative real numbers
$\mathbf{N}^+$	The set of natural numbers
$\mathbf{N}_+$	The set of nonnegative integers
$\mathbf{R}^n$	The set of n-dimensional real vectors
$\mathbf{R}^{n\times m}$	The set of $n \times m$-dimensional real matrices
z^*	The conjugate transpose of complex number z
$\sqrt{-1}$	The imaginary unit
$\lfloor a \rfloor$	The largest integer less than or equal to a
$\lceil a \rceil$	The smallest integer larger than or equal to a
I (I_n)	The identity matrix (of dimension $n \times n$)
x^T or A^T	The transpose of vector x or matrix A
A^{-1}	The inverse of matrix A
A^+	The Moore-Penrose pseudo-inverse of matrix A
$[A_1, A_2]$	The matrix commutator $A_1A_2 - A_2A_1$
$A \otimes B$	Kronecker product of matrices A and B
$A \oplus B$	Kronecker sum of matrices A and B
$P > 0$ $(P \geq 0)$	Matrix P is Hermitian and positive (semi-)definite
$P < 0$ $(P \leq 0)$	Matrix P is Hermitian and negative (semi-)definite
$x \succ y$ $(x \succeq y)$	x is entrywise greater than (not less than) y
$\det A$	The determinant of a matrix A
$\text{tr}(A)$	The trace of a matrix A
$\text{disc}(A)$	The discriminant of a matrix A

$\lambda(A)$	The set of eigenvalues of A
$\lambda_{\max}(A)$	The maximum eigenvalue of a real symmetric matrix A
$\lambda_{\min}(A)$	The minimum eigenvalue of a real symmetric matrix A
$\rho(\mathbf{A})$	The spectral radius of a matrix set $\mathbf{A}$
$x_i, x(i)$	The ith element of a vector x
$A(i,j), A_{ij}$	The ijth element of a matrix A
$\|p\|$	The length of a switching path p
$\|x\|$	The norm of a vector x
$\\|A\\|$	The induced norm of matrix A
$\|x\|_p$	The ℓ_p norm of vector x
$\\|A\\|_p$	The induced l_p-norm of a matrix A
$\mu_{\|\cdot\|}$	The matrix measure induced by a norm $\|\cdot\|$
$\mathrm{Im}\, A$	The image of an operator/matrix A
$E\xi^\delta$	The δth-order moment of a random variable ξ
$S = \{x, y, \ldots\}$	The set S with quantities x, y, etc.
μS	The set $\{\mu x : x \in S\}$
$\#S$	The cardinality of a set S
$\max S$	The maximum element of a set S
$\min S$	The minimum element of a set S
$\sup S$	The smallest number that is larger than or equal to each element of a set S
$\inf S$	The largest number that is smaller than or equal to each element of a set S
$S_1 - S_2$	The set $\{s \in S_1 : s \notin S_2\}$
$\arg\max S$	The index of the maximum element of an ordered set S
$\arg\min S$	The index of the minimum element of an ordered set S
$\emptyset$	The empty set
$\mathrm{meas}\, \Omega$	The Lebesgue measure of a set Ω in $\mathbf{R}^n$
Ω^o	The interior of a set Ω
$\mathrm{co}\, \Omega$	The convex hull of a set Ω
$\mathbf{B}_r$	The ball centered at the origin with radius r
$\mathbf{H}_r$	The sphere centered at the origin with radius r
$\mathrm{mod}(a, b)$	The remainder of a divided by b
$\mathrm{sgn}(\cdot)$	The signum function
$\mathrm{sat}(\cdot)$	The saturation function with unit limits
$\lim_{s\uparrow t} f(s)$	The limit from the left of a function $f(\cdot)$ at t
$\lim_{s\downarrow t} f(s)$	The limit from the right of a function $f(\cdot)$ at t
$\mathcal{C}^k$	The set of functions with continuous kth-order derivative
$\mathcal{D}^+ V$	The upper Dini derivative of a function V
class $\mathcal{K}$	The set of continuous and strictly increasing functions that vanish at zero
class $\mathcal{K}_\infty$	The set of unbounded class $\mathcal{K}$ functions
class $\mathcal{KL}$	The set of functions $\beta: \mathbf{R}_+ \times \mathbf{R}_+ \mapsto \mathbf{R}_+$ with $\beta(\cdot, t) \in \mathcal{K}\ \forall t \geq 0$ and $\lim_{t\to\infty} \beta(r, t) = 0\ \forall r \geq 0$
MF_Γ	The Minkowski function of a region Γ

M	The index set $\{1, 2, \ldots, m\}$ of the discrete state
$\mathcal{T}$	The time space
$\mathcal{T}_s$	The set $\{t \in \mathcal{T} : t \geq s\}$
$[a, b)$	The time interval $\{t \in \mathcal{T} : a \leq t < b\}$
σ	The switching signal of the switched system
$\hat{i}$	The constant switching signal $\sigma(t) = i \ \forall t$
$\mathcal{S}_{[a,b)}$	The set of well-defined switching paths over $[a, b)$
$\mathcal{S}_{[t_0,\infty)}$ or $\mathcal{S}$	The set of well-defined switching signals over $[t_0, \infty)$
Ψ	The infinitesimal matrix of a Markov process
$\phi(t; t_0, x_0, \sigma)$	The solution of the switched system
$\Phi(t_1, t_2, \sigma)$	The state transition matrix of the switched linear system

Chapter 1
Introduction

1.1 Switched Dynamical Systems

Generally speaking, a *switched dynamical system* is a dynamical system in which switching plays a nontrivial role. More specifically, a switched dynamical system is a two-level hybrid system with the lower level governed by a set of modes described by differential and/or difference equations and the upper level a coordinator that orchestrates the switching among the modes. Clearly, the system admits continuous states that take values from a vector space and discrete states that take values from a discrete index set. The interaction between the continuous and discrete states makes switched dynamical systems widely representative and complicatedly behaved.

A forced-free switched dynamical system is mathematically described by

$$x^+(t) = f_\sigma\big(x(t)\big), \tag{1.1}$$

where $x \in \mathbf{R}^n$ is the *continuous state*, σ is the *discrete state* taking values from an index set $M \overset{\text{def}}{=} \{1, \ldots, m\}$, and f_k, $k \in M$, are vector fields. x^+ denotes the derivative operator in continuous time (i.e., $x^+(t) = \frac{d}{dt}x(t)$) and the shift forward operator in discrete time (i.e., $x^+(t) = x(t+1)$).

It is clear that the continuous state space is the n-dimensional Euclidean space, and the discrete state space is the index set M with a finite number of elements. The time space is either the set of real numbers in continuous time or the set of integers in discrete time. According to the continuous or discrete nature of the time space, the switched system is said to be continuous-time or discrete-time. As there are m subsystems, the system is also termed as *m-form switched system*.

In the system description, each individual mode

$$x^+(t) = f_k\big(x(t)\big) \tag{1.2}$$

for $k \in M$ is said to be a *subsystem* of the switched system. The discrete dynamics represented by the switching coordinator is usually called the *supervisor*. The supervisor produces the discrete state σ, also known as the *switching signal* or *switching*

Z. Sun, S.S. Ge, *Stability Theory of Switched Dynamical Systems*,
Communications and Control Engineering,
DOI 10.1007/978-0-85729-256-8_1, © Springer-Verlag London Limited 2011

law. If $\sigma(t) = i$, then we say that the ith subsystem is *active at time* t. A characteristic of a switched system is that at a time instant there is one (and only one) active subsystem. Another characteristic is that the continuous state evolves continuously, that is, the state does not "jump" in an impulsive way.

As we treat continuous-time systems and discrete-time systems in a unified framework, the *time space*, denoted by $\mathcal{T}$, may either be the real set ($\mathcal{T} = \mathbf{R}$) or the integer set ($\mathcal{T} = \mathbf{N}$). For a real number s, let $\mathcal{T}_s$ be the set of times that are greater than or equal to s, i.e., $\mathcal{T}_s = \{t \in \mathcal{T} : t \geq s\}$. For two real numbers t_1 and t_2 with $t_1 < t_2$, the time interval $[t_1, t_2)$ should be understood as

$$[t_1, t_2) = \{t \in \mathcal{T} : t \geq t_1,\ t < t_2\}.$$

Other types of time intervals should be understood in a similar manner. The measure of $[t_1, t_2)$ is the length $t_2 - t_1$ in continuous time and the cardinality $\#[t_1, t_2)$ in discrete time. For notational convenience, we take any function defined over a discrete time set to be sufficiently smooth. For a piecewise continuous function χ defined over a time interval $[t_1, t_2)$ and for a time $t \in (t_1, t_2)$, define

$$\chi(t+) = \lim_{s \downarrow t} \chi(s), \qquad \chi(t-) = \lim_{s \uparrow t} \chi(s)$$

in continuous time and

$$\chi(t+) = \chi(t+1), \qquad \chi(t-) = \chi(t-1)$$

in discrete time.

As can be seen from the system description, when subsystems (1.2) for $k = 1, \ldots, m$ are given, the dynamical behavior of the switched system is decided by the switching signal. In the literature, a switching signal is also termed as a *switching path* or a *switching law*. For clarity, in this book we distinguish the terminologies as follows.

A switching path is a right-continuous function defined over a finite time interval taking values from the index set, M. Given a time interval $[t_0, t_f)$ with $-\infty < t_0 < t_f < +\infty$, a switching path p defined over the interval is denoted $p_{[t_0, t_f)}$. For a switching path $p_{[t_0, t_f)}$, time $t \in (t_0, t_f)$ is said to be a *jump time instant* if

$$\sigma(t-) \neq \sigma(t).$$

Suppose that the ordered sequence of jump instants in (t_0, t_f) is $t_1 < t_2 < t_3 < \cdots$. Then, the ordered sequence $t_0, t_1, t_2, \ldots$ is said to be the *switching time sequence* of σ over $[t_0, t_f)$. Similarly, the ordered discrete state sequence $\sigma(t_0), \sigma(t_1), \sigma(t_2), \ldots$ is said to be the *switching index sequence* of σ over $[t_0, t_f)$. The sequence of ordered pairs

$$(t_0, i_0), (t_1, i_1), \ldots, (t_s, i_s)$$

is said to be the *switching sequence* of σ over $[t_0, t_f)$. The switching path is said to be *well defined* if there are a finite number of jump times within the interval. The set of well-defined switching paths defined over $[t_0, t_f)$ is denoted by $\mathcal{S}_{[t_0, t_f)}$.

A switching signal is a function defined over an infinite time horizon taking values from M. Suppose that θ is a switching signal defined over $[t_0, +\infty)$, and $[s_1, s_2)$ is a finite-length subinterval of $[t_0, +\infty)$, then, a switching path $p_{[s_1,s_2)}$ is said to be a *subpath* of θ if $p(t) = \theta(t)$ for all $t \in [s_1, s_2)$. The notion of switching time/index sequences could be defined in the same way as for the switching path. A switching signal is said to be *well defined* if all its subpaths are well defined. We denote by $\theta_{[t_0,+\infty)}$ the switching signal θ defined over $[t_0, +\infty)$. The set of well-defined switching signals defined over $[t_0, +\infty)$ is denoted by $\mathcal{S}_{[t_0,+\infty)}$ or $\mathcal{S}$ in short when $t_0 = 0$.

Given a function pair $(x(\cdot), \theta(\cdot))$ over $[t_0, t_1)$, where $x: [t_0, t_1) \mapsto \mathbf{R}^n$ is absolutely continuous, and $\theta: [t_0, t_1) \mapsto M$ is piecewise constant. The pair $(x(\cdot), \theta(\cdot))$ is said to be a *solution* of system (1.1) over $[t_0, t_1)$ if for almost all $t \in [t_0, t_1)$, we have

$$x^+(t) = f_{\theta(t)}\big(x(t)\big).$$

The term "for almost all $t \in [t_0, t_1)$" means that "for all $t \in [t_0, t_1)$ except for possibly a zero-measure subset". The solution over other types of time intervals, such as $[t_0, t_1]$ and $[t_0, +\infty)$, should be understood in the same way.

Switched system (1.1) is said to be *(globally) well defined* if for any $\theta \in \mathcal{S}_{[0,+\infty)}$ and $x_0 \in \mathbf{R}^n$, there exists a unique absolutely continuous function x on $[0, +\infty)$ with $x(0) = x_0$ such that pair $(x(\cdot), \theta(\cdot))$ is a solution of system (1.1) over $[0, +\infty)$. It is clear that, when each subsystem satisfies the Lipschitz condition, i.e.,

$$\lim \sup_{x_1 \neq x_2} \big|f_k(x_1) - f_k(x_2)\big|/|x_1 - x_2| < +\infty, \quad k \in M,$$

then, the switched system is always well defined. Throughout the book, we assume that the Lipschitz condition holds for the subsystems, and thus the well-definedness of the switched system is guaranteed.

For the switched system, a switching law is a switching rule that generates a switching path or a switching signal for a set of initial configurations. Throughout the book, we consider switching laws in the form

$$\sigma(t) = \varphi\big(t, \sigma(t-), x(t)\big), \tag{1.3}$$

where φ is a piecewise constant function taking values from M. A function $x(\cdot)$ is said to be a *(continuous) state trajectory* of system (1.1) via the switching law (1.3) over $[t_0, t_1)$ if both equations (1.1) and (1.3) hold for almost all $t \in [t_0, t_1)$. The corresponding switching path/signal σ is said to be generated by the switching law (1.3) along $x(\cdot)$ for initial state x_0 over $[t_0, t_1)$.

A switching law is said to be *well defined* if it generates a well-defined switching signal for any initial state. As a result, for switched system (1.1), a well-defined switching law can be represented by the set $\{\theta^x : x \in \mathbf{R}^n\}$, where θ^x is the well-defined switching signal generated by the switching law with initial state x. Another implication is that the switched system admits a unique solution for any initial configuration when both the switched system and the switching law are well defined. For notational convenience, the continuous state trajectory will be denoted by $\phi(\cdot; t_0, x_0, \sigma)$ or $\phi(\cdot; x_0, \sigma)$ in short when $t_0 = 0$.

1.2 Stability and Stabilizability of Switched Systems

When the dynamics of the subsystems are known, the behaviors of the switched system are totally decided by the switching mechanism. A notable and attractive feature of switched systems is that the switched system might produce complex and diverse behaviors, even with simple and fixed subsystems. In this book, we aim to develop stability theory for switched dynamical systems under various switching mechanisms. To this end, we need to introduce the concepts of stability and stabilizability for switched dynamical systems.

Let $\Upsilon = \{\Lambda^x : x \in \mathbf{R}^n\}$, where Λ^x is a nonempty subset of $\mathcal{S}$, the set of well-defined switching signals. The set, called the *feasible set of switching signals*, assigns each initial state a set of switching signals. This set induces a *feasible set of continuous state trajectories* defined by $\{\Gamma_x : x \in \mathbf{R}^n\}$, where Γ_x is the set of state trajectories with initial state x and switching signals in Λ^x, i.e.,

$$\Gamma_x = \left\{\phi(\cdot; 0, x, \theta) : \theta \in \Lambda^x\right\}.$$

To proceed, we need to introduce some mathematical notation. A real-valued function $\alpha : \mathbf{R}_+ \mapsto \mathbf{R}_+$ is said to be of *class* $\mathcal{K}$ if it is continuous, strictly increasing, and $\alpha(0) = 0$. If in addition, α is unbounded, then it is said to be of *class* $\mathcal{K}_\infty$. A function $\beta : \mathbf{R}_+ \times \mathbf{R}_+ \mapsto \mathbf{R}_+$ is said to be of *class* $\mathcal{KL}$ if $\beta(\cdot, t)$ is of class $\mathcal{K}$ for each fixed $t \ge 0$ and $\lim_{t\to+\infty} \beta(r, t) = 0$ for each fixed $r \ge 0$.

Definition 1.1 (Stability) Suppose that $\Upsilon = \{\Lambda^x : x \in \mathbf{R}^n\}$ is a feasible set of switching signals. Switched system (1.1) is said to be

(1) *stable w.r.t.* Υ if there exist a class $\mathcal{K}$ function ζ and a positive real number δ such that

$$\left|\phi(t; 0, x_0, \theta)\right| \le \zeta\left(|x_0|\right) \quad \forall t \in [0, +\infty),\ x_0 \in \mathbf{B}_\delta,\ \theta \in \Lambda^{x_0}$$

(2) *asymptotically stable w.r.t.* Υ if there exists a class $\mathcal{KL}$ function ξ such that

$$\left|\phi(t; 0, x_0, \theta)\right| \le \xi\left(|x_0|, t\right) \quad \forall t \in [0, +\infty),\ x_0 \in \mathbf{R}^n,\ \theta \in \Lambda^{x_0}$$

and

(3) *exponentially stable w.r.t.* Υ if there exist positive real numbers α and β such that

$$\left|\phi(t; 0, x_0, \theta)\right| \le \beta e^{-\alpha t}|x_0| \quad \forall t \in [0, +\infty),\ x_0 \in \mathbf{R}^n,\ \theta \in \Lambda^{x_0}$$

Note that the stability is uniform w.r.t. switching signals and that asymptotic/exponential stability is defined in a global fashion.

Definition 1.2 (Stabilizability) Suppose that $\Upsilon = \{\Lambda^x : x \in \mathbf{R}^n\}$ is a feasible set of switching signals. Switched system (1.1) is said to be

(1) *stabilizable w.r.t.* Υ if there exist a class $\mathcal{K}$ function ζ, a positive real number δ, and a switching law $\{\theta^x : x \in \mathbf{R}^n\}$ with $\theta^x \in \Lambda^x$ such that

$$\left|\phi\left(t; 0, x_0, \theta^{x_0}\right)\right| \leq \zeta\left(|x_0|\right) \quad \forall t \in [0, +\infty),\ x_0 \in \mathbf{B}_\delta$$

(2) *asymptotically stabilizable w.r.t.* Υ if there exist a class $\mathcal{KL}$ function ξ and a switching law $\{\theta^x : x \in \mathbf{R}^n\}$ with $\theta^x \in \Lambda^x$ such that

$$\left|\phi\left(t; 0, x_0, \theta^{x_0}\right)\right| \leq \xi\left(|x_0|, t\right) \quad \forall t \in [0, +\infty),\ x_0 \in \mathbf{R}^n$$

and

(3) *exponentially stabilizable w.r.t.* Υ if there exist positive real numbers α and β and a switching law $\{\theta^x : x \in \mathbf{R}^n\}$ with $\theta^x \in \Lambda^x$ such that

$$\left|\phi\left(t; 0, x_0, \theta^{x_0}\right)\right| \leq \beta e^{-\alpha t}|x_0| \quad \forall t \in [0, +\infty),\ x_0 \in \mathbf{R}^n$$

When the subsystem dynamics are fixed, the stability property is totally determined by the feasible set of switching signals. In particular, if $\Upsilon_1 \supseteq \Upsilon_2$, then stability w.r.t. Υ_1 implies stability w.r.t. Υ_2, and stabilizability w.r.t. Υ_2 implies stabilizability w.r.t. Υ_1.

1.2.1 Guaranteed Stability Under Arbitrary Switching

When the supervisory mechanism is totally blind, that is, switching among the subsystems can occur in an arbitrary way, then the stability is said to be *guaranteed stability*. The feasible set of switching signals is given by

$$\Upsilon_{\text{as}} = \left\{\Lambda^x : x \in \mathbf{R}^n\right\}, \quad \Lambda^x = \mathcal{S},\ \forall x \in \mathbf{R}^n. \tag{1.4}$$

As the feasible set is the largest one among all the feasible sets, guaranteed stability is the most strict among various stability notions. In fact, as will be revealed later, it is robust against any arbitrary switching mechanism. In particular, guaranteed stability implies stability of each subsystem. The converse is not necessarily true, as exhibited by the following example.

Example 1.1 Suppose that we have two planar linear subsystems

$$\dot{x} = \begin{bmatrix} 0 & 1 \\ -1 & -1 \end{bmatrix} x$$

and

$$\dot{x} = \begin{bmatrix} 0 & 1 \\ -1 - 3a & -1 - a \end{bmatrix} x,$$

where a is a nonnegative real parameter. Clearly, both subsystems are exponentially stable and in a companion form. When $a = 0$, the two subsystems coincide, and

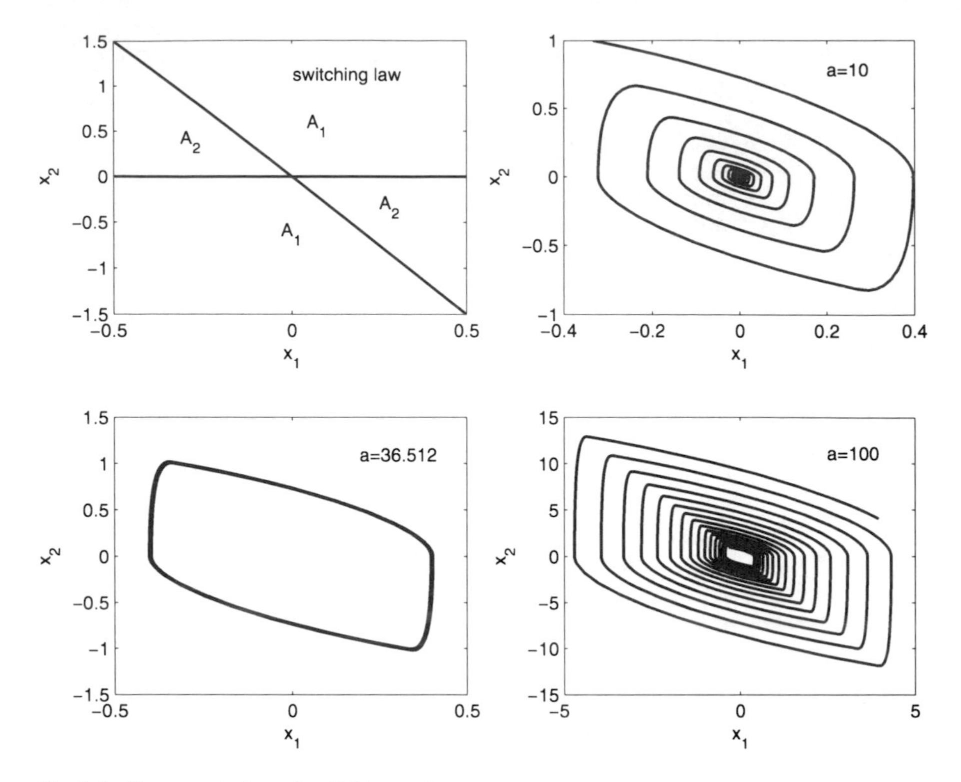

Fig. 1.1 Phase portraits and switching surfaces

the switched system is guaranteed exponentially stable. When $a \approx 36.512$, the switched system is guaranteed marginally stable (i.e., stable but not asymptotically stable). When $a > 36.512$, the switched system is not guaranteed stable. Figure 1.1 depicts the phase portraits (all initiated from $x_0 = [-1/3, 1]^T$) of the switched system under the most destabilizing switching law that is shown in the upper left, which produces worst state trajectories that diverge as fast as possible.

1.2.2 Dwell-Time Stability

A switching signal is said to be with *dwell time* τ if $t_{i+1} - t_i \geq \tau$ for any two consecutive jump times t_i and t_{i+1}. Let $\mathcal{S}_\tau$ be the set of well-defined switching signals with dwell time τ. It is a clear that $\mathcal{S} = \mathcal{S}_0 \supseteq \mathcal{S}_{\tau_1} \supseteq \mathcal{S}_{\tau_2}$ for $0 \leq \tau_1 \leq \tau_2$, and the subset relations are strict when $0 < \tau_1 < \tau_2$.

The set of arbitrary switching signals, $\mathcal{S}$, allows the supervisor to switch arbitrarily fast even without a uniform (positive) dwell time between the switching times.

For instance, the switching signal

$$\theta(t) = \begin{cases} 1 & \text{if } t \in [k, k + \frac{1}{k+2}),\ k = 0, 1, 2, \ldots, \\ 2 & \text{otherwise} \end{cases}$$

is well defined, but the length of the switching duration $[k, k + \frac{1}{k+2})$ approaches zero as k approaches infinity. It is clear that this switching signal belongs to $\mathcal{S}_0$, but it does not belong to any $\mathcal{S}_\tau$ when $\tau > 0$.

Fix a nonnegative τ. Let the feasible set of switching signals be

$$\Upsilon_\tau = \{\Lambda^x : x \in \mathbf{R}^n\}, \quad \Lambda^x = \mathcal{S}_\tau, \ \forall x \in \mathbf{R}^n.$$

The stability of the switched system w.r.t. Υ_τ is termed to be τ-*dwell-time stability*. It is clear that a necessary condition for τ-dwell-time stability is that each subsystem is stable. The converse is partly true in that, if each subsystem is exponentially stable, then the switched system is also τ-dwell-time exponentially stable for sufficiently large τ. Indeed, exponential stability of the subsystems implies the existence of a time $T > 0$ such that

$$|\phi_i(t; x_0)| \le \frac{1}{2}|x_0| \quad \forall t \in \mathcal{T}_T, \ x_0 \in \mathbf{R}^n,$$

where $\phi_i(\cdot; x_0)$ denotes the state trajectory of the ith subsystem with $x(0) = x_0$. It can be seen that the switched system is T-dwell-time exponentially stable. However, the assertion is not necessarily true for marginal or asymptotic stability. For instance, for the planar switched linear system with two marginally stable subsystems

$$\dot{x} = \begin{bmatrix} 0 & 1 \\ -2 & 0 \end{bmatrix} x$$

and

$$\dot{x} = \begin{bmatrix} 0 & 1 \\ -1/2 & 0 \end{bmatrix} x,$$

the switched system is unstable if we take the first subsystem when the state is in the second and fourth quadrants, and take the second subsystem otherwise. The lower left picture of Fig. 1.2 depicts a sample phase portrait of the switched system under the switching law. As the orbits for each subsystem are periodic, by incorporating one or more periods into each switching duration, the state keeps diverging, see the lower right picture of Fig. 1.2, where a period is incorporated in each switching duration. This implies that, for any $\tau > 0$, the feasible set Υ_τ contains destabilizing switching signals.

According to the above analysis, for dwell-time stability, the problem is to find a dwell time τ as small as possible such that the switched system is τ-dwell-time stable.

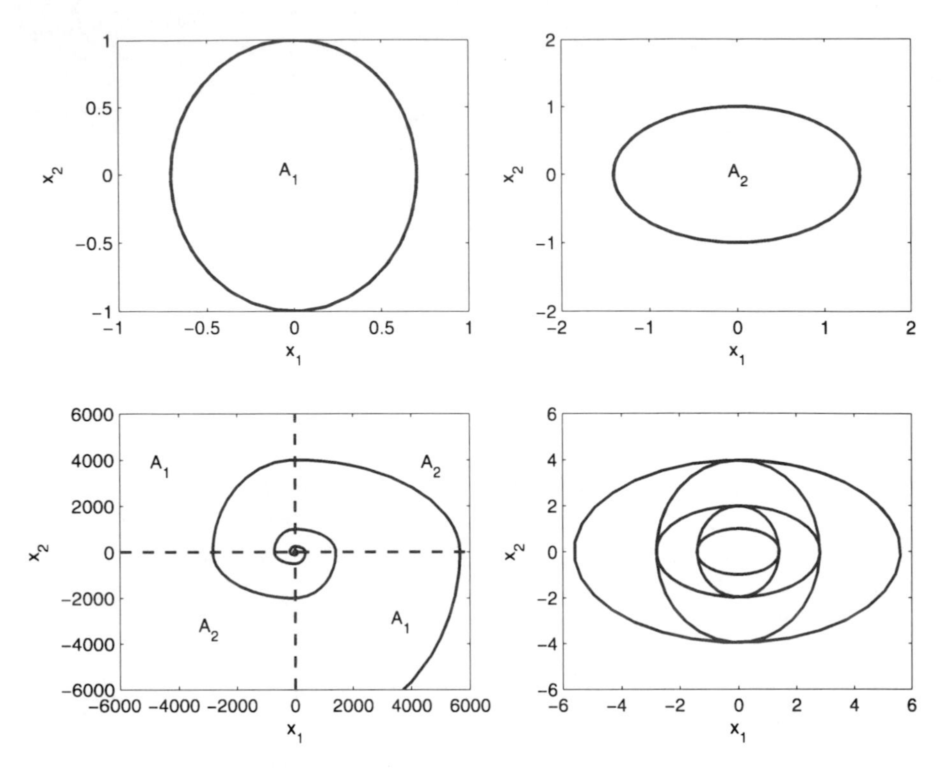

Fig. 1.2 Phase portraits of the switched marginally stable system

1.2.3 Autonomous Stability Under State-Driven Switching

Let $\gamma : \mathbf{R}^n \mapsto M$ be a piecewise constant function mapping the continuous state space into the discrete state space. Its inverse map, denoted γ^{-1}, is defined so that $\gamma(\gamma^{-1}(i)) = i$ for every $i \in \operatorname{Im}\gamma$, the image set of γ. Map γ induces a switching law given by

$$\sigma(t) = \gamma\big(x(t)\big), \tag{1.5}$$

which is state driven. When the switching law is well defined, it generates a feasible set of switching signals, denoted Υ^γ. The switched system is said to be *γ-autonomously stable* if it is stable w.r.t. the feasible set Υ^γ.

For a switched system with a state-driven switching, switches occur autonomously as the continuous state evolves. For an initial state, the switching law might generate a unique switching signal when it is well defined, or a deadlock might happen that prevents the switching from well-defined. Therefore, well-definedness is an important issue to be addressed. Another difficulty is the need to extend the solution concept to cope with possible sliding modes. While we will not go into details of the issues in the book, these indicate that the system behavior might be quite complex.

When each $\gamma^{-1}(i)$ is a convex polyhedron (or the empty set) and each subsystem is affine, the switched system is known to be a *piecewise affine system*, which is usually described by

$$x^{+}(t) = A_i x(t) + b_i, \quad C_i x(t) \succeq d_i, \tag{1.6}$$

where A_i, b_i, C_i, and d_i are constant matrices or vectors of appropriate dimensions, respectively, and $\{x : C_i x \succeq d_i\} = \gamma^{-1}(i)$, $i \in M$. When all b_i's vanish, the system is known as a *piecewise linear system*. If, in addition, each $\gamma^{-1}(i)$ is a cone, then the system is a *conewise linear system*.

It is clear that the stability properties of a piecewise affine system might not be consistent with that of the subsystems. One difficulty for stability analysis is to make clear the transition relationship among the polyhedral cells. As an example, consider the discrete-time planar four-form piecewise linear system given by

$$x(t+1) = A_i x(t), \quad x(t) \in \Omega_i, \quad i = 1, \ldots, 4,$$

where Ω_i is the ith quadrant for $i = 1, \ldots, 4$, and

$$A_1 = \begin{bmatrix} -0.5 & -1 \\ -1 & -0.5 \end{bmatrix}, \qquad A_2 = \begin{bmatrix} -0.3 & 0.2 \\ -0.2 & 0.3 \end{bmatrix}$$

$$A_3 = \begin{bmatrix} -1 & -0.5 \\ 0.5 & 0.5 \end{bmatrix}, \qquad A_4 = \begin{bmatrix} -0.8 & 0.5 \\ 0.5 & -0.6 \end{bmatrix}.$$

It can be seen that any (nonorigin) state in the first quadrant will enter into the third quadrant, then the fourth, then the second, and then the first once again. As a result, the discrete state is periodic with period $1, 3, 4, 2$. While not all the subsystems are stable, the transition matrix over a period, $A_2 A_4 A_3 A_1$, is Schur stable. Therefore, the piecewise linear system is asymptotically stable. Figure 1.3 shows the motions of the discrete and continuous states. The continuous state converges to the origin very slowly.

1.2.4 Stochastic Stabilities Under Random Switching

A random switching signal is a time-driven switching signal that fluctuates irregularly but obeys a distribution stochastically. Even when all the subsystems are deterministic, a random switching signal makes the switched system random in nature, and the stability notions have to be defined in a stochastic manner. A well-known feasible set of random switching signals is the *Markov jump* where switches between different subsystems are governed by a finite-state Markov process/chain. When the subsystems are linear and the switching is Markov jump, the switched system is known to be a *(Markovian) jump linear system*.

The stability properties of jump linear systems are much involved as there exist several stochastic stability concepts that differ in conservativeness as well as ease

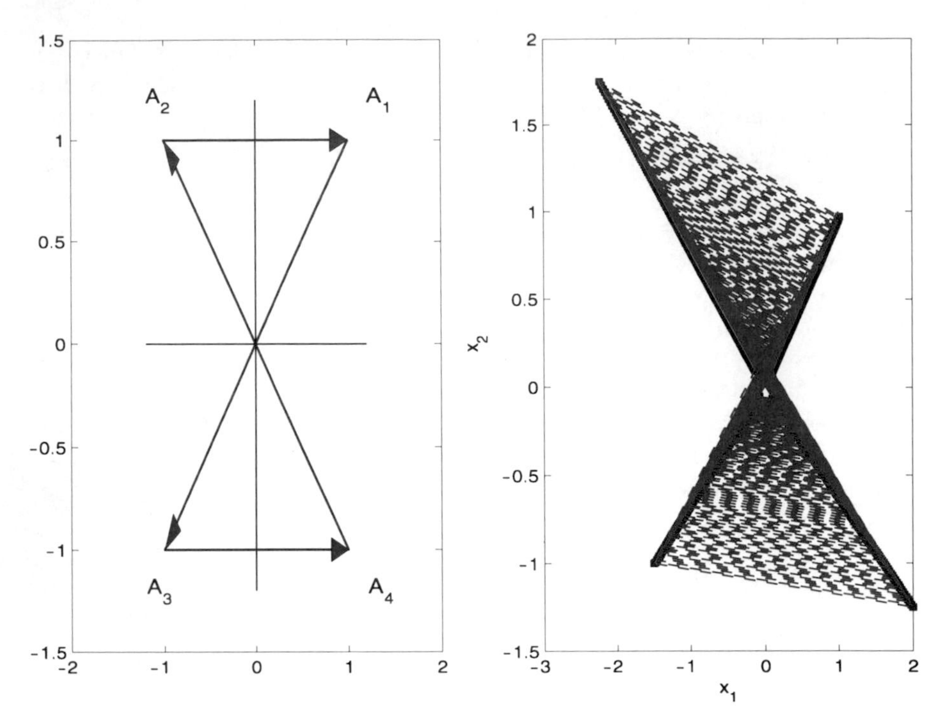

Fig. 1.3 Motions of discrete state (*left*) and continuous state (*right*)

of testability. The most important stability notions are mean-square stability and almost sure stability. While the detailed definitions will be presented in Sect. 3.2, *mean square stability* is the asymptotic convergence to zero of the second moment of the state norm, and *almost sure stability* means that the sample trajectory of the state converges to zero with probability one. To exhibit the difference between the stabilities, examine the simple scalar two-form jump linear system where $A_1 = -1$ and $A_2 = 0.5$, and the switching signal is a Markov stochastic process with the stationary transition probability

$$\Pr\big(\sigma(t+h) = j|\sigma(t) = i\big) = h + o(h), \quad i, j = 1, 2,\ i \neq j.$$

The state solution can be expressed by

$$x(t) = e^{A_1 t_1 + A_2 t_2} x_0 = e^{-t_1 + 0.5 t_2} x_0,$$

where t_1 and t_2 are the lengths of durations over $[0, t]$ at the first and second subsystems, respectively. Applying the law of large numbers, we have

$$\lim_{t\to+\infty} \frac{t_1}{t} = \lim_{t\to+\infty} \frac{t_2}{t} = \frac{1}{2} \quad \text{a.s.}$$

It follows that

$$\lim_{t\to+\infty}\left|x(t)\right|=0 \quad \text{a.s.}$$

On the other hand, it can be seen that

$$E\left\{x^2(t)\right\}=\sum_{i=1}^{2}E\left\{x^2(t)1_{\{\sigma(t)=i\}}\right\},$$

where $1_{\{\cdot\}}$ is the Dirac measure. Denote $Z_i(t)=E\{x^2(t)1_{\{\sigma(t)=i\}}\}$, $i=1,2$; simple calculation yields

$$\begin{aligned}
\frac{d}{dt}Z_1(t)&=-3Z_1(t)+Z_2(t),\\
\frac{d}{dt}Z_2(t)&=Z_1(t),
\end{aligned}$$

which is unstable. Therefore, the square moment is diverging with probability one, and the system is not mean square stable.

1.2.5 Stabilizing Switching Design

By Definition 1.2, a switched system is (asymptotically/exponentially) stabilizable if there is a feasible switching law that steers the system (asymptotically/exponentially) stable. Such a switching law is said to be a *stabilizing switching law* for the switched system. It is clear that, apart from formulating stabilizability criteria that characterize the existence of a stabilizing switching law, a more important issue is to explicitly calculate a stabilizing switching law. The latter is known to be the problem of stabilizing switching design.

When the set of feasible switching signals is given by Υ_{as} as defined in (1.4), the choice of a switching law is unconstrained, and the designer could take any switching mechanism into account. Otherwise, the switching design is constrained. In particular, if the switching law has to be initial-state-independent, i.e., $\theta^x=\theta^y$ for all x and y, then the stabilizability is said to be consistent stabilizability; if the switching law has to be time-independent, i.e., $y=\phi(s;0,x,\theta^x)\Longrightarrow\theta^y(t)=\theta^x(s+t)$ $\forall t\in\mathcal{T}_0$, then the stabilizability is said to be *spatial stabilizability*; and if the switching law has to be with a preassigned dwell time, then the stabilizability is said to be *dwell-time stabilizability*. As a comparison among the stabilizabilities, we examine the planar two-form continuous-time switched linear system with subsystems

$$A_1=\begin{bmatrix}2&0\\0&-1\end{bmatrix},\qquad A_2=\begin{bmatrix}1&1\\-1&1\end{bmatrix}.$$

While both subsystems are unstable, the switching law

$$\sigma(t) = \begin{cases} 1 & \text{if } x_1(t) = 0, \\ 2 & \text{otherwise} \end{cases} \tag{1.7}$$

steers the switched system exponentially stable. Therefore, the switched system is spatially stabilizable. On the other hand, the system is not consistently stabilizable, due to the fact that the sum of the subsystem eigenvalues is greater than zero (cf. [216]). As for dwell-time stabilizability, it can be seen that, for any preassigned dwell time τ, there is a stabilizing switching law that admits the dwell time. Indeed, this can be achieved by modifying the switching law to be

$$\sigma(t) = \begin{cases} 1 & \text{if } t \in \mathcal{T}_\tau \ \& \ x_1(t) = 0, \\ 2 & \text{otherwise.} \end{cases} \tag{1.8}$$

Note that the switching laws are not practically implementable due to the singular switching condition $x_1(t) = 0$. This brings the problem of robust switching design, which will be addressed in Chap. 4.

1.3 Organization of the Book

The book contains five chapters. Besides this short Introduction, there are four major chapters, which are briefly summarized as follows.

Chapter 2 focuses on stability analysis of switched dynamical systems under arbitrary switching. For general switched nonlinear systems, we establish the equivalence between marginal/asymptotic stability and the existence of a common weak/strong Lyapunov function. This reveals that the common Lyapunov approach is with full capacity in characterizing the guaranteed stability. For switched linear systems where the subsystems are linear, we further investigate the classes of functions that are universal for the stability. While quadratic functions are not universal as for stability of linear systems, we show that either piecewise linear/quadratic functions or norms are universal, which exhibit that the problem of stability is "convex" in nature. Furthermore, the well-known stability notions, such as attractivity, asymptotic stability, and exponential stability, are shown to be equivalent to each other. The equivalences, as in the linear stability theory, demonstrate that the stability is relatively simple in classification. Nevertheless, the verification of stability is by no means simple. Indeed, if we connect the absolute stability of a Lur'e system, which is notoriously difficult and still largely open though numerous achievements have been made during the past seventy years, with the guaranteed stability of the extreme switched linear system, then it can be seen that the former can be seen as a special case of the latter. For discrete-time switched linear systems, it has been made clear that the largest Lyapunov exponent is equal to the joint/generalized spectral radius, and asymptotic stability implies that the spectral radius is less than one. While the characteristic is insightful and elegant, the computation of the spectral radius is usually very hard. For this, we introduce a sum-of-squares technique approximating the spectral radius using high-order homogeneous polynomials as common Lyapunov functions. For a continuous-time system, we prove that asymptotic stability is equivalent to the negativeness of the least measure of the (subsystem) matrix set, which could be seen as a generalization of the largest real part of a single matrix. We also discuss the possibility of converting a switched system into a quasi-normal form by means of a common and possibly nonsquare coordinate transformation.

In Chap. 3, we investigate stabilities of switched linear systems under constrained switching. In the first part, we focus on stochastic stabilities of jump linear systems. Several stability notions are introduced including moment stability, stochastic stability, and almost sure stability. It is revealed that mean square asymptotic/exponential stability is equivalent to stochastic stability, which amounts to the feasibility of a coupled Lyapunov equation. A verifiable sufficient and necessary condition is also presented for mean square stability. For almost sure exponential stability, a sufficient and necessary condition is the norm contractivity of the state transition matrix within a finite number of switches. Though intractable, this exhibits that almost sure stability could be seen as stability of the averaged system.

In the second part of Chap. 3, we address the autonomous stability of piecewise affine systems. The piecewise quadratic Lyapunov approach, the surface Lyapunov approach, and the transition graph approach are introduced. The piecewise quadratic Lyapunov approach could provide less conservative exponential stability criteria at the cost of high computational burden. The surface Lyapunov approach, on the other hand, could reduce the computational burden by searching the Lyapunov functions over the switching surfaces instead of the total state space. It yields tractable asymptotic stability criteria for continuous-time piecewise linear systems with three or less switching surfaces. The transition graph approach provides an effective method for attractivity analysis with the aid of graphic decomposition, which is applicable to both discrete-time and continuous-time systems. We also address stabilities of conewise linear systems and present a constructive procedure for stability verification.

The third part of Chap. 3 is devoted to dwell-time stability and stabilizability. We show that dwell-time stability implies the existence of a dwell-time Lyapunov function. With the help of the sum-of-squares technique, we present a homogeneous polynomial Lyapunov function approach for calculating an upper bound of the least stable dwell time. For dwell-time stabilizability, the concept of ϵ-robust dwell time is introduced to keep a balance between the quality of the switching signal and the performance of the continuous state. A new combined switching strategy is also developed to further enlarge the dwell time without deteriorating the system performance.

The object of Chap. 4 is to develop a design scheme for the problem of stabilization by means of switching. We establish that a switched nonlinear system is asymptotically stabilizable iff it admits a smooth Lyapunov function. However, even for a planar switched linear system, a convex Lyapunov function does not necessarily exist. For this, the set of composite quadratic Lyapunov functions is examined from the viewpoint of optimal switching with quadratic cost indices, and we prove that the function set is universal in the sense that each asymptotically stabilizable system admits a Lyapunov function in the set. To further address the stabilization problem in a constructive manner, we propose a new switching mechanism named pathwise state-feedback switching, which concatenates a set of pre-assigned switching paths in a state-feedback way. We develop a computational procedure for calculating a stabilizing path-wise state-feedback switching law for an asymptotically stabilizable switched linear system. To address the robustness of the switching law, we define a (relative) distance between two switching signals and prove that the closed-loop system is robust against structural/unstructural/switching perturbations.

The last chapter, Chap. 5, provides a broader view of switched dynamical systems by connecting them with several well-known schemes in the system and control literature. First, for a planar two-form switched linear system, a phase-plane-based stability criterion is derived, and it is applied to the absolute stability of planar Lur'e systems. Second, we introduce a supervisory switching scheme for adaptive control of linear processes with large-scale parameter uncertainties. The scheme consists of a multicontroller, a multiestimator, a monitoring signal generator, and a switching logic. When properly designed, all the signals in the closed-loop system are bounded, while the output is asymptotically convergent. Third, we conduct stability analysis for Takagi–Sugeno (T–S) fuzzy systems. To this end, we introduce a framework of piecewise switched linear systems that combine and extend both piecewise affine systems and switched linear systems. A stability criterion for the system is presented with the help of the piecewise quadratic Lyapunov method. By aggregating a T–S fuzzy system into a piecewise switched linear system, a stability criterion is ready to obtain. Fourth, we examine the problem of consensus for a class of multiagent systems with proximity graphs and linear control protocols. We show that the problem is in fact a special stability problem for a piecewise linear system, which could be addressed by means of graph transition analysis. Finally, for controllable switched linear systems, a multilinear feedback design approach is proposed to tackle the stabilization problem. The main idea is to associate a set of candidate linear controllers with each subsystem so that the extended switched system is stabilizable. By utilizing the path-wise state-feedback switching design diagram, the problem of stabilization is solved in a constructive manner.

1.4 Notes and References

Switching is a phenomenon that widely exists in many real-world processes where logic elements interact with continuous dynamics. The study of switched dynamical systems takes root in modeling, control, and optimization for complex dynamical systems. First, switched dynamical systems could represent or approximately represent many practical systems, such as power electronics, gene networks, and intelligent softwares, to list a few. Second, switching control is a powerful design methodology widely used in dealing with highly complex systems including automotive control, flight control, and networked control. Third, even for systems which could be addressed by conventional methods/tools, the switching scheme provides an alternative diagram that could achieve the optimization in a better manner. Fourth, switched dynamical systems form a framework that extends both the conventional nonlinear (continuous) systems and discrete event systems. Therefore, it is an integration of the two types of dynamical systems.

The study of switched systems could be traced back to the 1960s when a series of works were devoted to optimality of switched system with either stochastic or autonomous switching [68, 240, 241, 262]. In the 1990s, a more general scheme of hybrid systems received increasing attention from both the control and computer communities. While a commonly acceptable definition of a hybrid system is still missing, it has been widely recognized that interaction between continuous dynamics and discrete dynamics is an important feature of a hybrid system. As a simple yet typical hybrid system, a switched system gradually became a major focus that has been attracting many researchers' interest. Since then, remarkable progress has been made in problems ranging from controllability/observability [49, 55, 67, 98, 105, 122, 236, 266] to stability/stabilizability [54, 60, 88, 104, 142, 147, 159, 164, 232, 259, 281] and optimality [20, 24, 123, 141, 200, 201, 220, 270, 279]. Meanwhile, some new problems that are unique to switched systems emerged, for instance, the problem of slow switching [66, 219, 227, 261, 263], the problem of robustness w.r.t. switching perturbations [225, 235], and the problem of bumpless transfer switching [86], to list a few. These problems are closely related to the issue of high-quality switching, which is a key in achieving good performance by means of switching.

For application-oriented switching control and design, the reader is referred to [2, 52, 65, 97, 155, 186, 209, 246, 283] for a short list. For more systematic reviews of recent development on switched systems, the reader is referred to the monographs [145, 146, 180, 234] and the survey papers [91, 151, 161, 208, 233, 245].

Chapter 2
Arbitrary Switching

2.1 Preliminaries

For switched dynamical systems, the switching may be induced by unpredictable change of system dynamics, such as a sudden change of system structure due to the failure of a component. In these cases, in order to keep the system working, it is necessary for the system to be stable under arbitrary switching. That is to say, the system should be stable under any possible switching law. Therefore, the stability is absolute or guaranteed regardless of the switching law. As the term "absolute stability" has been used to describe the global asymptotic stability for Lur'e systems with sector-bounded nonlinearities, here we use the term "guaranteed stability" to describe the stability of switched systems under arbitrary switching.

In this chapter, we consider the switched dynamical system given by

$$x^+(t) = f\bigl(x(t), \sigma(t)\bigr), \tag{2.1}$$

where $x(t) \in \mathbf{R}^n$ is the continuous state, $\sigma(t) \in M \stackrel{\text{def}}{=} \{1, \ldots, m\}$ is the discrete state, and $f : \mathbf{R}^n \times M \mapsto \mathbf{R}^n$ is a vector field with $f(\cdot, i)$ Lipschitz continuous for any $i \in M$.

It is clear that system (2.1) includes both continuous evolution and discrete elements. To emphasize the hybrid nature of the system, let us define new vector fields $f_i : \mathbf{R}^n \mapsto \mathbf{R}^n$ by

$$f_i(x) = f(x, i), \quad i \in M.$$

Then, the system can be rewritten as

$$x^+(t) = f_{\sigma(t)}\bigl(x(t)\bigr). \tag{2.2}$$

The issue of this chapter is the guaranteed stability analysis of switched dynamical system (2.2). For this, we assume that

Z. Sun, S.S. Ge, *Stability Theory of Switched Dynamical Systems*,
Communications and Control Engineering,
DOI 10.1007/978-0-85729-256-8_2, © Springer-Verlag London Limited 2011

(1) $f_i(0) = 0$ for all $i \in M$, which implies that the origin is an equilibrium.
(2) The system is globally Lipschitz continuous, that is, there exists a positive constant L such that

$$\left|f_i(x) - f_i(y)\right| \leq L|x - y| \quad \forall x, y \in \mathbf{R}^n, \ i \in M, \tag{2.3}$$

which guarantees the well-definedness of the switched system.

For clarity, we denote by $\phi(t; t_0, x_0, \sigma)$ the continuous state motion of system (2.2) at time t with initial condition $x(t_0) = x_0$ and switching path σ. By abuse of notation, we also use $\phi(t; x_0, \sigma)$ to denote the solution when $t_0 = 0$. The state evolution can be explicitly expressed in terms of the vector fields $f_i, i \in M$. Indeed, for any initial condition $x(t_0) = x_0$ and time $t > t_0$, in discrete time we have

$$\phi(t; t_0, x_0, \sigma) = f_{\sigma(t-1)} \circ \cdots \circ f_{\sigma(t_0+1)} \circ f_{\sigma(t_0)}(x_0), \tag{2.4}$$

where $\circ$ denotes the composition of functions, that is, $f_1 \circ f_2(x) \stackrel{\text{def}}{=} f_1(f_2(x))$. For continuous-time switched systems, the state evolution is

$$\phi(t; t_0, x_0, \sigma) = \Phi_{t-t_s}^{f_{i_s}} \circ \Phi_{t_s - t_{s-1}}^{f_{i_{s-1}}} \circ \cdots \circ \Phi_{t_2-t_1}^{f_{i_1}} \circ \Phi_{t_1-t_0}^{f_{i_0}}(x_0), \tag{2.5}$$

where $\Phi_t^f(x_0)$ denotes the value at t of the integral curve of f passing through $x(0) = x_0$, and $(t_0, i_0), \ldots, (t_s, i_s)$ is the switching sequence of σ in $[t_0, t)$. However, as the analytic expression of the curve $\Phi_t^f(x_0)$ is not available in general, the expression in (2.5) does not provide a sound basis for further analysis.

To present stability definitions for switched systems, we need more notation. Let $d(x, y)$ denote the Euclidean distance between vectors x and y. For a set $\Omega \subset \mathbf{R}^n$ and a vector $x \in \mathbf{R}^n$, let $|x|_\Omega = \inf_{y \in \Omega} d(x, y)$, and the normal norm $|x|_{\{0\}}$ is denoted by $|x|$ in short. For a set $\Omega \subset \mathbf{R}^n$ and a positive real number τ, let $\mathbf{B}(\Omega, \tau)$ be the τ-neighborhood of Ω, that is,

$$\mathbf{B}(\Omega, \tau) = \left\{x \in \mathbf{R}^n : |x|_\Omega \leq \tau\right\}.$$

Similarly, let $\mathbf{H}(\Omega, \tau)$ be the τ-sphere of Ω, that is,

$$\mathbf{H}(\Omega, \tau) = \left\{x \in \mathbf{R}^n : |x|_\Omega = \tau\right\}.$$

In particular, the closed ball $\mathbf{B}(\{0\}, \tau)$ will be denoted by $\mathbf{B}_\tau$ in short, and the sphere $\mathbf{H}(\{0\}, \tau)$ by $\mathbf{H}_\tau$ in short.

Definition 2.1 The origin equilibrium for system (2.2) is said to be

(1) *guaranteed globally attractive* if

$$\lim_{t \to +\infty} \left|\phi(t; x, \sigma)\right| = 0 \quad \forall x \in \mathbf{R}^n, \ \sigma \in \mathcal{S}$$

(2) *guaranteed globally uniformly attractive* if for any $\delta > 0$ and $\epsilon > 0$, there exists $T > 0$ such that

$$\left|\phi(t; x, \sigma)\right| < \epsilon \quad \forall t \in \mathcal{T}_T,\ |x| \le \delta,\ \sigma \in \mathcal{S}$$

(3) *guaranteed stable* if for any $\epsilon > 0$ and $\sigma \in \mathcal{S}$, there exists $\delta > 0$ such that

$$\left|\phi(t; x, \sigma)\right| \le \epsilon \quad \forall t \in \mathcal{T}_0,\ |x| \le \delta$$

(4) *guaranteed uniformly stable* if there exist $\delta > 0$ and a class $\mathcal{K}$ function γ such that

$$\left|\phi(t; x, \sigma)\right| \le \gamma\big(|x|\big) \quad \forall t \in \mathcal{T}_0,\ |x| \le \delta,\ \sigma \in \mathcal{S}$$

(5) *guaranteed globally asymptotically stable* if it is both guaranteed stable and guaranteed globally attractive
(6) *guaranteed globally uniformly asymptotically stable* if it is both guaranteed uniformly stable and guaranteed globally uniformly attractive
(7) *guaranteed globally exponentially stable* if for any $\sigma \in \mathcal{S}$, there exist $\alpha > 0$ and $\beta > 0$ such that

$$\left|\phi(t; x, \sigma)\right| \le \beta e^{-\alpha t} |x| \quad \forall t \in \mathcal{T}_0,\ x \in \mathbf{R}^n$$

and
(8) *guaranteed globally uniformly exponentially stable* if there exist $\alpha > 0$ and $\beta > 0$ such that

$$\left|\phi(t; x, \sigma)\right| \le \beta e^{-\alpha t} |x| \quad \forall t \in \mathcal{T}_0,\ x \in \mathbf{R}^n,\ \sigma \in \mathcal{S}$$

Note that the uniformity is referred to the switching signals rather than the initial time. By abuse of notation, we say that the system is guaranteed stable/attractive if the origin equilibrium is guaranteed stable/attractive. As we focus on the guaranteed stabilities with possible global attractivity in this chapter, the restrictive words "guaranteed" and "global" will be dropped for short in the sequel. Note also that the uniform asymptotic/exponential stability is consistent with that defined in Definition 1.1.

2.2 Switched Nonlinear Systems

In this section, we investigate the stability issue for switched nonlinear system

$$x^+(t) = f_{\sigma(t)}\big(x(t)\big) \tag{2.6}$$

under arbitrary switching. We assume that each vector field f_i is continuously differentiable.

2.2.1 Common Lyapunov Functions

The direct Lyapunov method provides a rigorous approach for studying stability of dynamical systems. Here, we review the preliminaries of Lyapunov stability theory.

A continuous function $V(x)\colon \mathbf{R}^n \mapsto \mathbf{R}$ with $V(0)=0$ is:

- *positive definite* ($V(x) \succ 0$) if $V(x) > 0\ \forall x \in \mathbf{R}^n - \{0\}$
- *positive semi-definite* ($V(x) \succeq 0$) if $V(x) \geq 0\ \forall x \in \mathbf{R}^n$
- *radially unbounded* if there exists a class $\mathcal{K}_\infty$ function $\alpha(\cdot)$ such that $V(x) \geq \alpha(|x|)\ \forall x \in \mathbf{R}^n$

Definition 2.2 Let Ω be a neighborhood of the origin. A function $V\colon \Omega \mapsto \mathbf{R}$ is said to be a *common weak Lyapunov function* (CWLF) for switched system (2.6) if

(1) it is lower semi-continuous in Ω
(2) it admits class $\mathcal{K}$ bounds, that is, there are class $\mathcal{K}$ functions α_1 and α_2 such that

$$\alpha_1\big(|x|\big) \leq V(x) \leq \alpha_2\big(|x|\big) \quad \forall x \in \Omega$$

and
(3) the upper Dini derivative of V along each vector f_i is nonpositive; that is, for all $x \in \Omega$ and $i \in M$, we have

$$\mathcal{D}^+ V(x)|_{f_i} \stackrel{\text{def}}{=} \limsup_{\tau \to 0^+} \frac{V(\phi(\tau; 0, x, \hat{i})) - V(x)}{\tau} \leq 0$$

in continuous time, where $\hat{i}$ stands for the constant switching signal $\sigma(t) = i\ \forall t$, and

$$\mathcal{D}^+ V(x)|_{f_i} \stackrel{\text{def}}{=} V\big(f_i(x)\big) - V(x) \leq 0$$

in discrete time

Remark 2.3 Note that we do not require that the common weak Lyapunov function is continuous. As a matter of fact, even a uniformly stable nonlinear system $\dot{x} = f(x)$ with f being sufficiently smooth may not admit any continuous weak Lyapunov function [14].

Definition 2.4 A function $V\colon \mathbf{R}^n \mapsto \mathbf{R}$ is said to be a *common* (*strong*) *Lyapunov function* (CLF) for switched system (2.6) if

(1) it is continuous everywhere and continuously differentiable except possibly at the origin
(2) it admits class $\mathcal{K}_\infty$ bounds, that is, there are class $\mathcal{K}_\infty$ functions α_1 and α_2 such that

$$\alpha_1\big(|x|\big) \leq V(x) \leq \alpha_2\big(|x|\big) \quad \forall x \in \mathbf{R}^n,$$

and

(3) there is a class $\mathcal{K}$ function $\alpha_3 : \mathbf{R}^n \mapsto \mathbf{R}_+$ such that

$$\mathcal{D}^+ V(x)|_{f_i} \leq -\alpha_3\big(|x|\big) \quad \forall x \in \mathbf{R}^n,\ i \in M \tag{2.7}$$

Remark 2.5 In continuous time, it can be seen that

$$\mathcal{D}^+ V(x)|_{f_i} = \limsup_{\tau \to 0^+} \frac{V(x + f_i(x)\tau) - V(x)}{\tau} \quad \forall x \in \mathbf{R}^n,\ i \in M, \tag{2.8}$$

due to the local Lipschitz continuity of V. For a continuously differentiable function V, its Dini derivative coincides with the (Lie) derivative of V along the vector field

$$\mathcal{D}^+ V(x)|_f = \frac{d}{dt} V(x) \stackrel{\text{def}}{=} L_f V(x) = \frac{\partial}{\partial x} V(x) f(x).$$

Suppose that system (2.6) admits a common weak Lyapunov function V. Then for any state trajectory $x(t) = \phi(t; x_0, \sigma)$ in Ω, we have $V(x(t)) \leq V(x_0)$ for all $t \geq 0$. For any $\epsilon > 0$, choose δ such that

$$\mathbf{B}_\delta \subset \Omega, \quad \big\{x : V(x) \leq \delta\big\} \subset \mathbf{B}_\epsilon.$$

Then, we have $|x(t)| \leq \epsilon$ for all $t \geq 0$ if $x_0 \in \mathbf{B}_\delta$. This shows that the system is uniformly stable.

Next, suppose that system (2.6) admits a common Lyapunov function V. We are to show that the system is uniformly asymptotically stable. For this, we assume that the system is in continuous time, and the discrete-time case can be treated in a similar way. Fix an initial state $x_0 \neq 0$ and a switching signal σ, and denote $x(t) = x(t; x_0, \sigma)$. It follows from Definition 2.4 that

$$\limsup_{\tau \to 0^+} \frac{V(x(t + \tau)) - V(x)}{\tau} \leq -\alpha_4\big(V\big(x(t)\big)\big) \quad \forall t \in \mathcal{T}_0, \tag{2.9}$$

where $\alpha_4 \stackrel{\text{def}}{=} \alpha_3 \circ \alpha_2^{-1}$. Define the function $\eta : \mathbf{R}^+ \mapsto \mathbf{R}$ by

$$\eta(t) = \begin{cases} -\int_1^t \frac{1}{\min(\tau, \alpha_4(\tau))}\, d\tau, & t \in (0, 1), \\ -\int_1^t \frac{1}{\alpha_4(\tau)}\, d\tau, & t \geq 1. \end{cases}$$

It is clear that η is strictly decreasing, differentiable, and $\lim_{t \downarrow 0} \eta(t) = +\infty$. From (2.9) it can be seen that

$$\begin{aligned} \eta\big(V\big(x(t)\big)\big) - \eta\big(V(x_0)\big) &= \int_0^t \dot{\eta}\big(V\big(x(\tau)\big)\big)\, dV\big(x(s)\big) \\ &\geq \int_0^t 1\, ds = t \quad \forall t \geq 0. \end{aligned} \tag{2.10}$$

Define the function $\pi : \mathbf{R}_+ \times \mathbf{R}_+ \mapsto \mathbf{R}_+$ by

$$\pi(s,t) = \begin{cases} 0, & s = 0, \\ \eta^{-1}(\eta(s) + t), & s > 0, \end{cases}$$

which can be verified to be a $\mathcal{KL}$-function. As η is strictly decreasing, it follows from (2.10) that

$$V\big(x(t)\big) \le \pi\big(V(x_0), t\big) \quad \forall t \ge 0.$$

Define the other $\mathcal{KL}$-function β by

$$\beta(s,t) = \alpha_1^{-1}\big(\pi\big(\alpha_2(s), t\big)\big), \quad s, t \in \mathbf{R}_+.$$

It can be seen that

$$\big|x(t)\big| \le \beta\big(|x_0|, t\big) \quad \forall t \ge 0. \tag{2.11}$$

This, together with the fact that the origin is an equilibrium of the switched system, implies that the system is uniformly asymptotically stable. Indeed, to achieve uniform stability, for any $\epsilon > 0$, let $\delta = \bar{\beta}^{-1}(\epsilon)$, where $\bar{\beta}(\cdot) \stackrel{\text{def}}{=} \beta(\cdot, 0)$. Then, $|\phi(t; 0, x, \sigma)| \le \epsilon$ for any $x_0 \in \mathbf{B}_\delta$, $t \in \mathcal{T}_0$, and $\sigma \in \mathcal{S}$. Similarly, to achieve uniform attractivity, for any $\epsilon > 0$ and $\delta > 0$, let $T = \hat{\beta}^{-1}(\epsilon)$ where $\hat{\beta}(\cdot) \stackrel{\text{def}}{=} \beta(\delta, \cdot)$; then $|\phi(t; 0, x, \sigma)| \le \epsilon$ for any $x_0 \in \mathbf{B}_\delta$, $t \in \mathcal{T}_T$, and $\sigma \in \mathcal{S}$.

To summarize, we have the following proposition.

Proposition 2.6 *Switched system* (2.6) *is uniformly stable if it admits a common weak Lyapunov function, and it is uniformly asymptotically stable if it admits a common Lyapunov function.*

Example 2.7 For the planar continuous-time two-form switched system with

$$\begin{aligned} &\frac{d}{dt}x(t) = f_\sigma\big(x(t)\big), \\ &f_1(x) = \begin{pmatrix} -x_2 \\ x_1 - x_2^k \end{pmatrix}, \qquad f_2(x) = \begin{pmatrix} x_2 \\ -x_1 - x_2^k \end{pmatrix}, \end{aligned} \tag{2.12}$$

where k is any odd natural number, it can be verified that $V(x_1, x_2) = x_1^2 + x_2^2$ is a common weak Lyapunov function. By Proposition 2.6, the system is uniformly stable. However, the system does not admit any common Lyapunov function. Indeed, if the system admits a common Lyapunov function, then it follows that any convex combination system of the subsystems is asymptotically stable, that is, the dynamical system

$$\dot{x}(t) = \omega f_1\big(x(t)\big) + (1 - \omega) f_2\big(x(t)\big)$$

is asymptotically stable for any $\omega \in [0, 1]$. Let $\omega = \frac{1}{2}$. It can be verified that any state on the x_1-axis is an equilibrium. As a result, the convex combination system is not asymptotically stable. The contradiction exhibits that the switched system does not admit any common Lyapunov function.

2.2.2 Converse Lyapunov Theorem

While the existence of a common weak/strong Lyapunov function guarantees the uniform/asymptotic stability of the switched system, one may naturally ask whether the converse is also true. The answer is confirmative, as shown in the following proposition.

Proposition 2.8 *Any uniformly stable switched system admits a common weak Lyapunov function, and any uniformly asymptotically stable switched system admits a common Lyapunov function.*

Proof To prove the first half of the statement, suppose that the system is uniformly stable. Fix an $\epsilon > 0$, and let $\delta > 0$ be the number as in Definition 2.1. As in the standard Lyapunov method, define the function $V : \mathbf{B}_\delta \mapsto \mathbf{R}_+$ by

$$V(x) = \sup_{t\in\mathcal{T}_0,\sigma\in\mathcal{S}} \left|\phi(t; 0, x, \sigma)\right|. \tag{2.13}$$

It is clear that the function is well defined and positive definite in $\mathbf{B}_\delta$. Let $\Omega = \mathbf{B}_\delta^o$, where D^o denotes the interior of set D. We are to prove that V is a common weak Lyapunov function for the system.

To prove the lower semi-continuity of function V, fix an $x \in \mathbf{B}_\delta$. For any $\varepsilon > 0$, there exist a time $t_x \geq 0$ and $\sigma_x \in \mathcal{S}$ such that

$$\left|\phi(t_x; 0, x, \sigma_x)\right| \geq V(x) - \varepsilon.$$

Let $\eta = e^L$ in continuous time and $\eta = L$ in discrete time. It follows from the Lipschitz assumption (2.3) that

$$\begin{aligned} V(y) &\geq \left|\phi(t_x; 0, y, \sigma_x)\right| \geq \left|\phi(t_x; 0, x, \sigma_x)\right| - \left|\phi(t_x; 0, y, \sigma_x) - \phi(t_x; 0, x, \sigma_x)\right| \\ &\geq \left|\phi(t_x; 0, x, \sigma_x)\right| - \eta^{t_x}|x - y| \geq V(x) - 2\varepsilon \quad \forall y \in \mathbf{B}\left(x, \frac{\varepsilon}{\eta^{t_x}}\right) \cap \mathbf{B}_\delta. \end{aligned}$$

As a result, V is lower semi-continuous.

Next, for any $x \in \mathbf{B}_\delta$, $s \in \mathcal{T}_0$, and $i \in M$, we have

$$\begin{aligned} V\big(\phi(s; 0, x, \hat{i})\big) &= \sup_{t\in\mathcal{T}_0,\sigma\in\mathcal{S}} \left|\phi\big(t; 0, \phi(s; 0, x, \hat{i}), \sigma\big)\right| \leq \sup_{t\in\mathcal{T}_0,\sigma\in\mathcal{S}} \left|\phi(t + s; 0, x, \sigma)\right| \\ &\leq \sup_{(t+s)\in\mathcal{T}_0,\sigma\in\mathcal{S}} \left|\phi(t + s; 0, x, \sigma)\right| = V(x). \end{aligned}$$

It follows that $\mathcal{D}^+V(x)|_{f_i} \leq 0$ for any $i \in M$.

Finally, it is clear that $V(x) \geq |x|$ for any $x \in \mathbf{B}_\delta$, and hence it admits a class $\mathcal{K}$ lower bound. On the other hand, the uniform stability means the existence of a class $\mathcal{K}$ function α_2 such that $|\phi(t; 0, x, \sigma)| \leq \alpha_2(|x|)$ for any $t \in \mathcal{T}_0$, $x \in \mathbf{B}_\delta$, and $\sigma \in \mathcal{S}$. As a result, $V(x) \leq \alpha_2(|x|)$ for all $x \in \mathbf{B}_\delta$.

The above reasonale shows that the function V defined in (2.13) is indeed a common weak Lyapunov function for the switched system.

The proof of the second half of Proposition 2.8 is involved as we have to smooth the above function while preserving its properties such as decreasing along the state trajectories. We will not go into the details; instead we cite the following support lemma, which is a simper version of the well-known Lin–Sontag–Wang converse Lyapunov theorem [152].

Lemma 2.9 *For the uncertain system*

$$x^+(t) = f\big(x(t), d(t)\big), \quad x(t) \in \mathbf{R}^n, \; d(t) \in \mathcal{D} \subset \mathbf{R}^p, \tag{2.14}$$

suppose that f is locally Lipschitz continuous in x uniformly in d, $\mathcal{D}$ is a compact set, and $d \in PC(\mathcal{T}_0, \mathcal{D})$, the set of piecewise continuous functions mapping from $\mathcal{T}_0$ to $\mathcal{D}$. Assume that the system is forward complete and the compact subset $\mathcal{X}$ of $\mathbf{R}^n$ is a forward invariant set in the sense that $x(t) \in \mathcal{X}$ for all $x_0 \in \mathcal{X}$ and $d \in PC(\mathcal{T}_0, \mathcal{D})$. If there is a $\mathcal{KL}$ function β such that

$$\big|x(t)\big|_{\mathcal{X}} \leq \beta\big(|x_0|_{\mathcal{X}}, t\big) \quad \forall x_0 \in \mathbf{R}^n - \mathcal{X}, \; t \geq 0, \; d \in PC(\mathcal{T}_0, \mathcal{D}), \tag{2.15}$$

then, there is a Lyapunov function V for system (2.14), *i.e., V is continuous everywhere and infinitely differentiable on $\mathbf{R}^n - \mathcal{X}$, and there are $\mathcal{K}_\infty$ functions α_1, α_2 and a class $\mathcal{K}$ function α_3 such that*

$$\alpha_1\big(|x|_{\mathcal{X}}\big) \leq V(x) \leq \alpha_2\big(|x|_{\mathcal{X}}\big)$$

and

$$L_f V(x) \leq -\alpha_3\big(|x|_{\mathcal{X}}\big) \quad \forall x \in \mathbf{R}^n, \; d \in \mathcal{D}.$$

The lemma provides a general converse Lyapunov theorem, which is very important in the development of the stability and robust analysis and design. The proof of the lemma is quite technically involved, and the reader is referred to [124, 152] for details.

With the help of Lemma 2.9, the proof for the necessity of Proposition 2.8 is quite straightforward. Indeed, in Lemma 2.9, let $\mathcal{D} = M$ and $\mathcal{X} = \{0\}$. Then, Proposition 2.8 follows immediately from the lemma due to the fact that asymptotic stability implies (2.15). □

To summarize, in terms of the relationship between the uniform stability and the existence of a common Lyapunov function, the following conclusion can be drawn.

Theorem 2.10 *Switched system* (2.6) *is uniformly stable iff it admits a common weak Lyapunov function, and it is uniformly asymptotically stable iff it admits a common Lyapunov function.*

The theorem clearly brings the stability verification to the search of an appropriate common (weak) Lyapunov function and extends the conventional Lyapunov theory to the more general setting of switched systems. As in the conventional Lyapunov theory, there is generally no systematic way of finding the Lyapunov functions. Nevertheless, for some classes of systems with special structures or properties, the search of a Lyapunov function is tractable. We will come back to this topic in Sect. 2.3.5.

2.3 Switched Linear Systems

In this section, we focus on a special but very important class of switched systems where all the subsystems are linear time-invariant. These systems are termed as *switched linear systems* and are mathematically represented by

$$x^+(t) = A_{\sigma(t)}x(t), \qquad x(0) = x_0, \tag{2.16}$$

where $A_k \in \mathbf{R}^{n\times n}$, $k \in M$, are constant matrices.

Let $\mathbf{A} = \{A_1, \dots, A_m\}$. $\mathbf{A}$ can be seen as the system matrix set for the switched linear system. For briefness, we term the switched linear system as system $\mathbf{A}$.

Due to the linear nature of the subsystems, the state solution can be given in an analytic way. In fact, the solution is given by

$$\phi(t; t_0, x_0, \sigma) = \Phi(t; t_0, \sigma)x_0, \tag{2.17}$$

where $\Phi(t; t_0, \sigma)$ is known to be the *state transition matrix*. In discrete time, the state transition matrix is

$$\Phi(t; t_0, \sigma) = A_{\sigma(t-1)} \cdots A_{\sigma(t_0)},$$

while in continuous time, it is

$$\Phi(t; t_0, \sigma) = e^{A_{i_s}(t-t_s)}e^{A_{i_{s-1}}(t_s-t_{s-1})} \cdots e^{A_{i_1}(t_2-t_1)}e^{A_{i_0}(t_1-t_0)},$$

where $t_0, t_1, \dots, t_s$ and $i_0, i_1, \dots, i_s$ are the switching time/index sequences in $[t_0, t)$, respectively.

It follows from the above expressions that

$$\phi(t; t_0, \lambda x_0, \sigma) = \lambda\phi(t; t_0, x_0, \sigma) \quad \forall t, t_0, x_0, \sigma, \ \forall\lambda \in \mathbf{R}, \tag{2.18}$$

which we term as the *radial linearity property*, and

$$\phi(t; t_0, x_0, \sigma) = \phi(t - t_0; 0, x_0, \sigma') \quad \forall t, t_0, x_0, \sigma, \tag{2.19}$$

where the switching path σ' is the time transition of σ, that is, $\sigma'(t) = \sigma(t + t_0)$ for all t. The latter property is known as the *(time) transition invariance property*.

Due to the above two invariance properties, for switched linear systems, it is clear that local attractivity implies (and is equivalent to) global attractivity, and the initial time can always be taken as $t_0 = 0$ without loss of generality.

2.3.1 Relaxed System Frameworks

In the stability analysis of switched linear systems, the switching signals can be arbitrary. Taking the switching mechanism as the uncertainty, the guaranteed stability requires that the system is robust w.r.t. the uncertainty. As the switching signals are piecewise constant taking values from a finite discrete set, it is natural to "smooth" them in some sense so that the conventional perturbation analysis approaches apply. This leads to the relaxed or extended system frameworks as follows.

Let

$$\mathcal{W} = \left\{ w \in \mathbf{R}^m : w_i \geq 0,\ i = 1, \ldots, m,\ \sum_{i=1}^{m} w_i \leq 1 \right\}$$

$$A(w) = \sum_{i=1}^{m} w_i A_i, \quad w \in \mathcal{W}, \tag{2.20}$$

and

$$\mathcal{A}(x) = \{A(w)x : w \in \mathcal{W}\}, \quad x \in \mathbf{R}^n. \tag{2.21}$$

Let us consider the (*convex*) *differential/difference inclusion system*

$$x^+(t) \in \mathcal{A}(x(t)), \tag{2.22}$$

and the *polytopic linear uncertain system*

$$x^+(t) = A(w(t))x(t), \tag{2.23}$$

where $w(\cdot) \in \mathcal{W}$ is a piecewise continuous function. For convenience, system (2.22) is called (*relaxed*) *differential/difference inclusion*. Note that both the differential/difference inclusion and the polytopic linear uncertain system can be connected to the switched system in a one-to-one manner, and the switched system can be seen as the extreme system of the others.

A solution of (2.22) is a vector flow $x : [0, +\infty) \mapsto \mathbf{R}^n$ with absolutely continuous entries that satisfies (2.22) almost everywhere. Solutions of the polytopic system can be understood in the same way.

For comparison, let Γ_s denote the set of solutions of the switched linear system, Γ_p the set of solutions of the polytopic system, and Γ_d the set of solutions of the differential inclusion system. It is readily seen that

$$\Gamma_s \subset \Gamma_p \subset \Gamma_d$$

and the first subset relationship is strict. However, under mild assumptions, each solution of the relaxed differential inclusion system can be approximated by a trajectory of the switched linear system in the sense specified below.

Lemma 2.11 (See [117]) *Fix $\xi \in \mathbf{R}^n$ and let $z\colon [0,+\infty) \mapsto \mathbf{R}^n$ be a solution of*

$$\dot{z}(t) \in \mathcal{A}\big(z(t)\big), \qquad z(0) = \xi.$$

Let $r\colon [0,+\infty) \mapsto \mathbf{R}$ be a continuous function satisfying $r(t) > 0$ for all $t \geq 0$. Then, there exist η with $|\eta - \xi| \leq r(0)$ and a solution $x\colon [0,+\infty) \mapsto \mathbf{R}^n$ of

$$\dot{x}(t) \in \big\{A_1 x(t), \ldots, A_m x(t)\big\}, \qquad x(0) = \eta,$$

such that

$$\big|z(t) - x(t)\big| \leq r(t) \quad \forall t \in [0,+\infty).$$

This lemma sets up a connection between stability of a switched linear system and stability of its relaxed system. Indeed, suppose that each solution of the switched linear system is convergent; then, each solution of the differential inclusion (2.22) is also convergent. For discrete-time systems, the relationship also holds.

Corollary 2.12 *The following statements are equivalent*:

(1) *The switched linear system is attractive.*
(2) *The polytopic linear uncertain system is attractive.*
(3) *The differential inclusion system is attractive.*

For linear systems, it is well known that attractivity implies (and is equivalent to) exponential stability. For switched linear systems, it can be proven that the same property also holds as follows.

Proposition 2.13 *The following statements are equivalent*:

(1) *The switched linear system is attractive.*
(2) *The switched linear system is uniformly attractive.*
(3) *The switched linear system is asymptotically stable.*
(4) *The switched linear system is uniformly asymptotically stable.*
(5) *The switched linear system is exponentially stable.*
(6) *The switched linear system is uniformly exponentially stable.*

Proof First, it is clear that the uniform exponential stability implies any other stability in the proposition, and attractivity is implied by any other stability. As a result, we only need to prove that the attractivity implies the uniform exponential stability.

Second, for any state x on the unit sphere, it follows from the attractivity that there is a time t^x such that

$$\sup_{\sigma \in \mathcal{S}} \big|\phi\big(t^x; 0, x, \sigma\big)\big| < \frac{1}{2}. \tag{2.24}$$

For any fixed $x \in \mathbf{H}_1$, we are to prove that $\sup_{\sigma \in \mathcal{S}} |\phi(t^x; 0, y, \sigma)| \leq \frac{1}{2}$ if y is sufficiently close to x. Indeed, let $\eta = \max_{i \in M} |A_i|$. It follows from (2.24) and (2.17) that

$$\begin{aligned}
\left|\phi\left(t^x; 0, y, \sigma\right)\right| &= \left|\phi\left(t^x; 0, x, \sigma\right) + \phi\left(t^x; 0, y - x, \sigma\right)\right| \\
&\le \left|\phi\left(t^x; 0, x, \sigma\right)\right| + \left|\phi\left(t^x; 0, x - y, \sigma\right)\right| \\
&\le \left|\phi\left(t^x; 0, x, \sigma\right)\right| + e^{\eta t^x} |y - x| \\
&\le \frac{1}{2} \quad \forall \sigma \in \mathcal{S},\ |y - x| \le e^{-\eta t^x} \left(\frac{1}{2} - \sup_{\sigma \in \mathcal{S}} \left|\phi\left(t^x; 0, x, \sigma\right)\right| \right).
\end{aligned}$$

This implies that, for any $x \in \mathbf{H}_1$, there is a neighborhood N_x of x such that

$$\sup_{\sigma \in \mathcal{S}} \left|\phi\left(t^x; 0, y, \sigma\right)\right| \le \frac{1}{2} \quad \forall y \in N_x.$$

Third, letting x vary along the unit sphere, it is obvious that

$$\bigcup_{x \in \mathbf{H}_1} N_x \supseteq \mathbf{H}_1.$$

As the unit sphere is a compact set, by the Finite Covering Theorem, there exist a finite number l and a set of states $x_1, \ldots, x_l$ on the unit sphere such that

$$\bigcup_{i=1}^{l} N_{x_i} \supseteq \mathbf{H}_1.$$

Accordingly, we can partition the unit sphere into l regions $R_1, \ldots, R_l$ such that

(a) $\bigcup_{i=1}^{l} R_i = \mathbf{H}_1$, and $R_i \cap R_j = \emptyset$ for $i \neq j$; and
(b) for each i, $1 \le i \le l$, $x_i \in R_i$, and

$$\sup_{\sigma \in \mathcal{S}} \left|\phi\left(t^{x_i}; 0, y, \sigma\right)\right| \le \frac{1}{2}, \quad \forall y \in R_i.$$

Define the cones

$$\Omega_i = \{x \in \mathbf{R}^n : \exists \lambda \neq 0 \text{ and } y \in R_i \text{ such that } x = \lambda y\}, \quad i = 1, \ldots, l.$$

Let $\Omega_0 = \{0\}$. It can be seen that $\bigcup_{i=0}^{l} \Omega_i = \mathbf{R}^n$ and $\Omega_i \cap \Omega_j = \emptyset$ for $i \neq j$. In particular, Ω_0 is invariant under arbitrary switching and forms an invariant equilibrium of the system.

Fourth, for any $i = 1, \ldots, l$ and $x \in \Omega_i$, let $t_x = t^{x_i}$. It is clear that

$$\max_{x \neq 0} t_x = \max_{i=1}^{l} t^{x_i} \stackrel{\text{def}}{=} T_1 < +\infty. \tag{2.25}$$

According to properties (a) and (b), for any $x \in \Omega_i$, $i = 1, \ldots, l$, any switching signal σ will bring x into the ball $\mathbf{B}_{\frac{|x|}{2}}$ at time t_x.

Finally, for any initial state x_0 and switching path σ, define recursively a sequence of times and states

$$\begin{aligned} s_0 &= 0, \\ z_0 &= x_0, \\ s_k &= s_{k-1} + t_{z_{k-1}}, \\ z_k &= \phi(s_k; 0, x_0, \sigma), \quad k = 1, 2, \ldots. \end{aligned}$$

It is readily seen that

$$s_k \le kT_1, \qquad |z_{k+1}| \le \frac{|z_k|}{2}, \quad k = 0, 1, \ldots, \tag{2.26}$$

which implies that

$$\left|\phi(s_k; 0, x_0, \sigma)\right| \le \frac{|x_0|}{2^k} \le e^{-\gamma s_k}|x_0|, \quad k = 0, 1, 2, \ldots,$$

where $\gamma \stackrel{\text{def}}{=} \frac{\ln 2}{T_1}$. On the other hand, let $\eta = 2\exp(T_1 \max\{\|A_1\|, \ldots, \|A_m\|\})$. Then, we have

$$\left|\phi(t; 0, x_0, \sigma)\right| \le \eta e^{-\gamma t}|x_0| \quad \forall t \ge 0. \tag{2.27}$$

Note that the inequality holds for any x_0 and σ, and the parameters γ and η are independent of x_0 and σ. This clearly shows that the system is uniformly exponentially stable. The proof is completed. □

Remark 2.14 While the theorem establishes a nice property for switched linear systems, the proof itself does not provide a constructive approach for calculating or estimating the convergence rate. For special classes of switched linear systems with additional structure information, however, it is possible to compute the convergence rates explicitly. This issue will be addressed in Sect. 2.3.5.

Simple observation exhibits that the above analysis can be slightly adopted to prove the equivalence (between attractivity and exponential stability) for the polytopic system and the differential inclusion system, respectively. Together with Corollary 2.12 and Theorem 2.10, we reach the following conclusion.

Theorem 2.15 *The following statements are equivalent for the switched linear system, the polytopic system, and the differential/difference inclusion system, respectively*:

(1) *The system is attractive.*
(2) *The system is asymptotically stable.*
(3) *The system is exponentially stable.*
(4) *The switched system admits a common Lyapunov function.*

Remark 2.16 The theorem bridges the stability analysis for various classes of systems with different backgrounds. Indeed, the stability analysis for polytopic systems, for linear differential inclusions, and for switched linear systems is mostly independent of each other until quite recently. The theorem assures that the stability criteria developed for one class of systems are also applied to the others. This greatly enriches the stability theory for the systems.

Theorem 2.15 allows us to define the stabilities in a more refined manner, as in the linear time-invariant case. For this, define

$$\varrho(\mathbf{A}) = \limsup_{t\to+\infty,\sigma\in\mathcal{S},|x|=1} \frac{\ln|\phi(t;0,x,\sigma)|}{t}, \tag{2.28}$$

which is the largest Lyapunov exponent that specifies the highest possible rate of state divergence, and

$$R(\mathbf{A}) = \{\phi(t;0,x,\sigma): t\in\mathcal{T}_0,\ x\in\mathbf{H}_1,\ \sigma\in\mathcal{S}\}, \tag{2.29}$$

which is the attainability set of the system from the unit sphere.

Definition 2.17 A switched linear system $\mathbf{A}$ is said to be

(1) *(exponential) stable* if $\varrho(\mathbf{A}) < 0$
(2) *marginally stable* if $\varrho(\mathbf{A}) = 0$ and the set $R(\mathbf{A})$ is bounded
(3) *marginally unstable* if $\varrho(\mathbf{A}) = 0$ and the set $R(\mathbf{A})$ is unbounded
(4) *(exponentially) unstable* if $\varrho(\mathbf{A}) > 0$

It is clear that the notion of stability here is abused with the one defined for the nonlinear setting as it is referred to the situation of exponential convergence only. Similarly, the notion of instability is referred to exponential divergence only. In the sequel, unless otherwise stated, all the stability notions are referred in accordance with Definition 2.17.

2.3.2 *Universal Lyapunov Functions*

Through this subsection we assume, unless otherwise stated, that the switched linear system is stable. According to Theorem 2.10, the system admits a common Lyapunov function that is smooth ($\mathcal{C}^1$). As a result, the set of smooth positive definite functions is universal for switched linear systems. Here by universal Lyapunov functions we mean a set of functions such that each stable switched linear system admits a common Lyapunov function which belongs to the set. Due to the linear structure of the subsystems, it seems reasonable to expect a more restricted set of universal Lyapunov functions, e.g., the set of polynomials. In particular, as a stable linear time-invariant system always admits a quadratic Lyapunov function, it is natural to conjecture that the set of quadratic functions is also universal for switched linear systems. If so, then it is possible to develop a constructive approach for calculating a common quadratic Lyapunov function. Unfortunately, this conjecture finally was disproved through a counterexample, which exhibits that the stability analysis of switched linear systems is much more difficult than that of linear systems.

There are quite a few efforts in the literature to reveal the universal sets of common Lyapunov functions for switched linear systems or, equivalently, for polytopic systems or linear convex differential inclusions. Due to the nonsmooth nature of switched system, nonsmooth functions are also considered as Lyapunov function candidates.

The following theorem provides several sets of universal common Lyapunov functions for switched linear systems.

Theorem 2.18 *Each of the following function sets provides universal Lyapunov functions for stable switched linear systems.*

(1) *Convex and homogeneous functions of degree* 2.
(2) *Polynomials.*
(3) *Piecewise linear functions.*
(4) *Piecewise quadratic functions.*
(5) *Norms, that is, positive definite functions* $N : \mathbf{R}^n \mapsto \mathbf{R}_+$ *such that* $N(\lambda x) = |\lambda| N(x)$ *for any* $\lambda \in \mathbf{R}$ *and* $N(x+y) \le N(x) + N(y)$ *for any* $x, y \in \mathbf{R}^n$.

The key ideas of proving the existence of universal Lyapunov functions are outlined as follows. First, as in the conventional Lyapunov approach, for a stable switched linear system, we define the function $V : \mathbf{R}^n \mapsto \mathbf{R}_+$ by

$$V(x) = \sup_{\sigma \in \mathcal{S}} \int_0^{+\infty} \left|\phi(t; 0, x, \sigma)\right|^2 dt \tag{2.30}$$

in continuous time and

$$V(x) = \sup_{\sigma \in \mathcal{S}} \sum_{t=0}^{+\infty} \left|\phi(t; 0, x, \sigma)\right|^2 \tag{2.31}$$

in discrete time. It can be verified that the function is continuous, positive definite, strictly convex, and homogeneous of degree 2. In addition, for any nontrivial state trajectory $y(\cdot)$ of the switched system, we have

$$\begin{aligned} V\big(y(t_2)\big) &= \sup_{\sigma \in \mathcal{S}} \int_0^{+\infty} \left|\phi\big(t; 0, y(t_2), \sigma\big)\right|^2 dt \\ &= \sup_{\sigma \in \mathcal{S}} \int_0^{+\infty} \left|\phi\big(t + t_2 - t_1; 0, y(t_1), \sigma\big)\right|^2 dt \\ &= \sup_{\sigma \in \mathcal{S}} \int_{t_2 - t_1}^{+\infty} \left|\phi\big(t; 0, y(t_1), \sigma\big)\right|^2 dt \end{aligned}$$

$$< \sup_{\sigma\in\mathcal{S}} \int_0^{+\infty} \left|\phi\big(t;0,y(t_1),\sigma\big)\right|^2 dt$$
$$= V\big(y(t_1)\big) \quad \forall t_2 > t_1,$$

which clearly exhibits that the function V is strictly decreasing along the trajectory. Note that the function is continuous but may be nondifferentiable. The next step is to smooth the function by introducing the integral

$$\tilde{V}(x) = \int_{SO(n)} f(R)V(Rx)\,dR, \quad x \in \mathbf{R}^n,$$

where $SO(n)$ is the set of $n \times n$ orthogonal matrices with positive determinants, $f: SO(n) \mapsto \mathbf{R}_+$ is a smooth function with support in a small neighborhood of the identity matrix, and $\int_{SO(n)} f(R)\,dR = 1$. It can be shown that the function $\tilde{V}$ is continuously differentiable ($\mathcal{C}^1$) except possibly at the origin. A smooth function of $\mathcal{C}^k$ with any k can be obtained iteratively in the same manner. The newly defined function preserves the properties of convexity and homogeneity of degree 2. Moreover, the function strictly decreases along any nontrivial state trajectory of the switched system. By definition, the function is a common Lyapunov function of the system, and Item (1) of the theorem is established. The existence of other classes of universal Lyapunov functions can be guaranteed by the fact that a convex level set can be approximated to any degree by the level sets of polynomials, piecewise linear functions, and piecewise quadratic functions, respectively. We omit the technical details for briefness.

By the first statement of the theorem, there exists a common Lyapunov function of the form

$$V(x) = x^T P(x)x \quad \text{with } P(\lambda x) = P(x) > 0 \quad \forall \lambda \neq 0,\ x \neq 0.$$

Note that this function is in the quadratic form, and positive definite matrix $P(x)$ is homogeneous of degree zero; hence it is uniquely characterized by its image on the unit sphere. A special case is that $P(x)$ is independent of the state, which corresponds to a quadratic Lyapunov function. When this is the case, the search of an appropriate Lyapunov function can be reduced to solving the linear matrix inequalities (LMIs)

$$A_i^T P + PA_i < 0, \quad i = 1, \ldots, m, \tag{2.32}$$

which is computationally tractable.

A piecewise linear function is of the form

$$V(x) = \max\{l_i^T x : i = 1, \ldots, s\},$$

where l_i's are column vectors in $\mathbf{R}^n$. It is convex and positively homogeneous of degree 1. Its level set

$$\Gamma = \{x : l_i^T x \leq 1,\ i = 1, \ldots, s\}$$

is a convex and compact set containing the origin as an interior point. We call such a set *C-set*. The level set Γ is also a polyhedron, and it induces the gauge function, known as the *Minkowski function of* Γ, which is defined as

$$\mathrm{MF}_{\Gamma}(x) = \inf\{\mu \in \mathbf{R}_{+} : x \in \mu\Gamma\},$$

where $\mu\Gamma = \{\mu y : y \in \Gamma\}$. By definition, function MF_{Γ} is exactly the function V, that is, $\mathrm{MF}_{\Gamma}(x) = V(x)$ for all $x \in \mathbf{R}^n$. The Minkowski function is a norm iff the level set is 0-symmetric, i.e., $x \in \Gamma$ implies $-x \in \Gamma$. The universal of piecewise linear functions as Lyapunov candidates implies that any stable switched linear system admits a polyhedral C-set as its attractive level set, that is, $D^{+}V(x) \leq -\beta$ for any x on the boundary of the level set, where β is some positive real number. As a result, the stability verification reduces to the search of a proper attractive polyhedral C-set.

A particular and interesting situation is that the common Lyapunov function is a norm. In this case, it can be seen that the norm is exponentially contractive along any nontrivial state trajectory of the switched system.

A piecewise quadratic function is in the form

$$V_{\max}(x) = \max\{x^T P_i x : i = 1, \ldots, s\},$$

where P_i's are symmetric and positive definite matrices. It is clear that the functions are positive definite and homogeneous of degree 2. A level set of $V_{\max}$ is an intersection of a number of ellipsoids and is strict convex. As a result, the function is also strictly convex and thus can be classified into the first class of the theorem.

As a corollary of Theorem 2.18, we have the following useful lemma that establishes the connection between a stable continuous-time system and its discrete-time Euler approximating system.

Lemma 2.19 *Suppose that the continuous-time switched linear system is stable. Then, the Euler approximating system defined as*

$$x(t+1) = (I_n + \tau A_{\sigma})x(t) \tag{2.33}$$

is also stable for sufficiently small τ.

Proof The lemma can be proved based on the fact that each stable switched linear system admits a polyhedral common Lyapunov function, V. For the level set $\Gamma = \{x \colon V(x) \leq 1\}$, define $\beta = \min_{x \in \partial\Gamma} \alpha_3(|x|)$, where α_3 is the class $\mathcal{K}$ function as in (2.7), and $\partial\Gamma$ stands for the boundary of set Γ. It follows from (2.7) that $\mathcal{D}^{+}V(x) \leq -\beta$ for any $x \in \partial\Gamma$. This clearly indicates that the set Γ is also attractive for discrete-time system (2.33) when τ is sufficiently small, which in turn implies stability of the discrete-time system for a sufficiently small τ. □

Remark 2.20 An interesting question is whether or not the degrees of the common polynomial Lyapunov functions are uniformly bounded? In other word, does it exist a map $\chi : \mathbf{N}^{+} \times \mathbf{N}^{+} \mapsto \mathbf{N}^{+}$ such that each m-form attractive switched linear system of order n admits a common polynomial Lyapunov function of degree $\chi(n, m)$ or less? If the answer is confirmative, then, it is possible to numerically verify the stability by checking that, among all polynomials of degree $\chi(n, m)$ or less, whether there is a common Lyapunov function or not. Unfortunately, such a bound does not generally exist, even for planar switched linear systems. The reader is referred to [168] for detailed analysis.

Finally, we turn to marginal stability and propose a universal set of common weak Lyapunov functions.

Proposition 2.21 *A marginally stable switched linear system admits a norm as its common weak Lyapunov function.*

Proof We are to prove that the common weak Lyapunov function V defined in (2.13) is a norm. In fact, as

$$\begin{aligned} V(x) + V(y) &\geq \sup_{t \in \mathcal{T}_0, \sigma \in \mathcal{S}} \{ |\phi(t; x, \sigma)| + |\phi(t; y, \sigma)| \} \\ &\geq \sup_{t \in \mathcal{T}_0, \sigma \in \mathcal{S}} \{ |\phi(t; x, \sigma) + \phi(t; y, \sigma)| \} \geq V(x+y) \quad \forall x, y \in \mathbf{R}^n, \end{aligned}$$

V is convex. On the other hand, for any $\epsilon > 0$, there is $\delta > 0$ such that

$$|\phi(t; 0, x, \sigma)| \leq \epsilon \quad \forall t \in \mathcal{T}_0, \ x \in \mathbf{B}_\delta, \ \sigma \in \mathcal{S}.$$

It follows from the radial linearity property (2.18) that $V(x) \leq \frac{\epsilon}{\delta}|x|$ for all $x \in \mathbf{R}^n$. This, together with the convexity, implies that

$$|V(x) - V(y)| \leq V(x-y) \leq \frac{\epsilon}{\delta}|x-y| \quad \forall x, y \in \mathbf{R}^n.$$

As a result, V is globally Lipschitz continuous. It is obvious that V is 0-symmetric, and positively homogeneous of degree one. Thus, it is indeed a norm. □

By Theorem 2.18 and Proposition 2.21, a stable (marginally stable) switched system always admits a norm as its common (weak) Lyapunov function.

A switched linear system $\mathbf{A}$ is said to be *regular* if $\varrho(\mathbf{A})$ is finite, where $\varrho(\mathbf{A})$ is the largest divergence rate defined in (2.28). It is clear that any continuous-time switched linear system is regular and that a discrete-time system is regular if $\varrho(\mathbf{A}) \neq -\infty$. For a continuous-time switched linear system $\mathbf{A} = \{A_1, \ldots, A_m\}$, the *normalized switched system* is the switched system $\underline{\mathbf{A}} = \{A_1 - \varrho(\mathbf{A})I_n, \ldots, A_m - \varrho(\mathbf{A})I_n\}$. Similarly, for a regular discrete-time system $\mathbf{A} = \{A_1, \ldots, A_m\}$, its normalized system is defined to be the switched system $\underline{\mathbf{A}} = \{A_1/e^{\varrho(\mathbf{A})}, \ldots, A_m/e^{\varrho(\mathbf{A})}\}$. It is clear that any normalized system is either marginally stable or marginally unstable.

2.3.3 Algebraic Criteria

For a linear time-invariant system, it is well known that the stability is characterized by the location of the poles. However, such a concise characteristic is still missing for switched linear systems. Nevertheless, much effort has been paid to find algebraic criteria for stability of the switched systems. In this subsection, we present some of the criteria which provide necessary and sufficient algebraic conditions.

For discrete-time systems, the stability is closely related to the convergence of the transition matrices, for which numerous algebraic criteria were developed mainly by researchers in the mathematics community. In the following, we first address the stability and then move to other related properties.

Suppose that $\mathbf{A} = \{A_1, \ldots, A_m\}$ is a finite set of matrices in $\mathbf{R}^{n\times n}$. For $k \in \mathbf{N}^+$, denote by $\Pi_k(\mathbf{A})$ the set of length-k products of $\mathbf{A}$, that is,

$$\Pi_k(\mathbf{A}) = \{A_{i_1} \ldots A_{i_k} : i_1, \ldots, i_k \in M\},$$

and further

$$\Pi(\mathbf{A}) = \bigcup_{k\in\mathbf{N}^+} \Pi_k(\mathbf{A}),$$

which is the set of all products whose factors are elements of $\mathbf{A}$.

Let $|\cdot|$ be any induced norm, and $\rho(\cdot)$ be the spectral radius of a matrix. Define

$$\hat{\rho}_k(\mathbf{A}) = \max\bigl\{\|P\|: \ P \in \Pi_k(\mathbf{A})\bigr\}, \tag{2.34}$$

which represents the largest possible norm of all products of k matrices chosen in set $\mathbf{A}$. In particular, denote $\hat{\rho}_1(\mathbf{A})$ by $\|\mathbf{A}\|$. The *joint spectral radius* of $\mathbf{A}$ is then defined as

$$\hat{\rho}(\mathbf{A}) = \limsup_{k\to+\infty} \hat{\rho}_k(\mathbf{A})^{1/k}, \tag{2.35}$$

which is the maximal asymptotic norm of the products of matrices. It is clear that $\hat{\rho}(\mathbf{A})$ is norm-independent due to the equivalence of the norms. Analogously, define the number

$$\bar{\rho}_k(\mathbf{A}) = \max\bigl\{\rho(P): \ P \in \Pi_k(\mathbf{A})\bigr\},$$

which represents the largest possible spectral radius of all products of k matrices in a set $\mathbf{A}$. Furthermore, define the *generalized spectral radius* of $\mathbf{A}$ as

$$\bar{\rho}(\mathbf{A}) = \limsup_{k\to+\infty} \bar{\rho}_k(\mathbf{A})^{1/k}, \tag{2.36}$$

which is the maximal asymptotic spectral radius of the products of matrices.

Recall that, for a matrix A, we have

$$\lim_{k\to+\infty} \|A^k\|^{1/k} = \lim_{k\to+\infty} \rho\bigl(A^k\bigr)^{1/k} = \rho(A).$$

Therefore, both the joint spectral radius and generalized spectral radius degenerate into the standard spectral radius when $\mathbf{A}$ is a singleton.

We are to present some properties of the joint spectral radius and generalized spectral radius. To this end, let Υ be the set of vector norms in $\mathbf{R}^n$. For a set of matrices $\mathbf{A} = \{A_1, \dots, A_m\}$ and a norm $|\cdot| \in \Upsilon$, define the (*induced*) *norm* of $\mathbf{A}$ to be

$$\|\mathbf{A}\| = \max_{x \neq 0, i \in M} |A_i x|/|x|.$$

The *least norm* of $\mathbf{A}$ is defined to be

$$\mathrm{LN}_{\mathbf{A}} = \inf_{|\cdot| \in \Upsilon} \|\mathbf{A}\|.$$

A norm $\|\cdot\|^*$ is an *extreme norm* of $\mathbf{A}$ if

$$\|\mathbf{A}\|^* = \mathrm{LN}_{\mathbf{A}}.$$

For a real number μ, $\mu\mathbf{A}$ denotes the matrix set $\{\mu A_1, \dots, \mu A_m\}$.

Lemma 2.22 *Suppose that* $\mathbf{A}$ *is a set of real matrices. Then, the following statements hold.*

(1) $\bar{\rho}(\mathbf{A}) \le \hat{\rho}(\mathbf{A}) \le \mathrm{LN}_{\mathbf{A}}$.
(2) $\bar{\rho}(\mu\mathbf{A}) = |\mu|\bar{\rho}(\mathbf{A})$ *and* $\hat{\rho}(\mu\mathbf{A}) = |\mu|\hat{\rho}(\mathbf{A})$ $\forall \mu \in \mathbf{R}$.
(3) $\bar{\rho}(\mathbf{A}) < 1$ *implies the stability of* $\mathbf{A}$.

Proof First, note that $\rho(A) \le \|A\|$ for any norm $\|\cdot\|$. As a result, $\bar{\rho}_k(\mathbf{A}) \le \hat{\rho}_k(\mathbf{A})$, which implies that $\bar{\rho}(\mathbf{A}) \le \hat{\rho}(\mathbf{A})$. On the other hand, it follows from the norm sub-multiplicativity property that

$$\hat{\rho}(\mathbf{A}) \le \|\mathbf{A}\|,$$

which leads to the inequality $\hat{\rho}(\mathbf{A}) \le \mathrm{LN}_{\mathbf{A}}$. This proves the first statement.

The second statement trivially follows from the linearity of the spectral radius. To establish the third one, observe that $\bar{\rho}(\mathbf{A}) < 1$ implies that the state transition matrix $\Phi(t, 0, \sigma)$ approaches zero as t approaches infinity for any switching signal σ, which further implies that the switched linear system is attractive and hence stable. □

Based on the lemma and the converse Lyapunov theorem presented in Sect. 2.2.2, we are able to establish the equivalence among several fundamental indices.

Theorem 2.23 *For any switched linear system* $\mathbf{A}$, *we have*

$$\bar{\rho}(\mathbf{A}) = \hat{\rho}(\mathbf{A}) = \mathrm{LN}_{\mathbf{A}} = \exp\big(\varrho(\mathbf{A})\big). \tag{2.37}$$

Proof By Lemma 2.22, to establish $\bar{\rho}(\mathbf{A}) = \hat{\rho}(\mathbf{A}) = \mathrm{LN}_{\mathbf{A}}$, we only need to prove $\bar{\rho}(\mathbf{A}) \geq \mathrm{LN}_{\mathbf{A}}$. For this, suppose by contradiction that $\bar{\rho}(\mathbf{A}) < \mathrm{LN}_{\mathbf{A}}$. Fix a real number μ with $\bar{\rho}(\mathbf{A}) < \mu < \mathrm{LN}_{\mathbf{A}}$. Denote $B_i = A_i/\mu$ for $i = 1, \ldots, m$, and further $\mathbf{B} = \{B_1, \ldots, B_m\}$. It follows from Lemma 2.22 that

$$\bar{\rho}(\mathbf{B}) < 1 < \mathrm{LN}_{\mathbf{B}} . \tag{2.38}$$

Applying Lemma 2.22 once again, the switched linear system $\mathbf{B}$ is stable. By Theorem 2.18, there is a norm V that serves as a common Lyapunov function of the switched system, which is contractive. It is clear that

$$\mathrm{LN}_{\mathbf{B}} \leq \|\mathbf{B}\|_V \leq 1,$$

which contradicts inequality (2.38). This establishes

$$\bar{\rho}(\mathbf{A}) = \hat{\rho}(\mathbf{A}) = \mathrm{LN}_{\mathbf{A}} .$$

By the definition of the largest Lyapunov exponent as in (2.28), it is clear that

$$\exp\bigl(\varrho(\mathbf{A})\bigr) \leq \mathrm{LN}_{\mathbf{A}} .$$

Using a similar idea as in the former part of the proof, we arrive at the conclusion that the equality relation must hold. This completes the proof. □

The equality between the generalized spectral radius and the joint spectral radius allows us to term the quantity as the *spectral radius* of matrix set $\mathbf{A}$, denoted $\rho(\mathbf{A})$.

Corollary 2.24 *The discrete-time switched linear system is stable iff its spectral radius is less than one. It is unstable iff its spectral radius is greater than one.*

The corollary provides a new criterion for stability of a switched linear system in terms of the spectral radius, which extends the well-known spectral radius criterion for linear time-invariant systems. A semi-decidable verification procedure can be developed based on the criterion, which will be presented in Sect. 2.4.

Proposition 2.25 *A discrete-time switched linear system* $\mathbf{A}$ *is marginally stable iff it admits an extreme norm with* $\|\mathbf{A}\|^* = 1$.

Proof By Proposition 2.21, marginal stability implies the existence of a common weak Lyapunov norm V. From the definitions for the common (weak) Lyapunov function, we have $V(A_i x) \leq V(x)$ for all $i \in M$ and $x \in \mathbf{R}^n$, which implies that $\|\mathbf{A}\|_V \leq 1$. If $\mathrm{LN}_{\mathbf{A}} < 1$, then the switched system is asymptotically stable, which contradicts the assumption of marginal stability. As a result, we have $\mathrm{LN}_{\mathbf{A}} \geq 1$. As $\mathrm{LN}_{\mathbf{A}} \leq \|\mathbf{A}\|_V$, we have $\mathrm{LN}_{\mathbf{A}} = \|\mathbf{A}\|_V = 1$, which clearly shows that $\|\cdot\|_V$ is an extreme norm for the switched system.

Conversely, suppose that the switched system admits an extreme norm $\|\cdot\|$ with $\|\mathbf{A}\| = \mathrm{LN}_{\mathbf{A}} = 1$. It is clear that the system is either stable or marginally stable. If the system is stable, then, there is a common Lyapunov norm V_0 such that

$$V_0(A_i x) - V_0(x) \leq -\omega(x) \quad \forall x \in \mathbf{R}^n,\ i \in M,$$

where ω is a continuous positive definite function. Let $\beta = \min_{V_0(x)=1} \omega(x)$. It follows that

$$V_0(A_i x) - V_0(x) \leq -\beta V_0(x) \quad \forall x \in \mathbf{R}^n,\ i \in M,$$

which further implies that $\|\mathbf{A}\|_{V_0} \leq 1 - \beta$, a contradiction. Therefore, the switched system must be marginally stable. □

Next, we move to the case of continuous time. For any norm $|\cdot|$ in $\mathbf{R}^n$, the *induced matrix measure* on $\mathbf{R}^{n\times n}$ is defined as

$$\mu_{|\cdot|}(A) = \limsup_{\tau\to 0^+, |x|=1} \frac{|x + \tau A x| - |x|}{\tau}. \tag{2.39}$$

It is clear that the matrix measure possesses the following properties (see, e.g., [253]. The subscript $|\cdot|$ is dropped for briefness):

(1) *Well-definedness.* The matrix measure is well defined for any vector norm.
(2) *Positive homogeneousness.* $\mu(\alpha A) = \alpha\mu(A)$ for all $\alpha \geq 0$ and $A \in \mathbf{R}^{n\times n}$.
(3) *Convexity.* $\mu(\alpha A + (1-\alpha)B) \leq \alpha\mu(A) + (1-\alpha)\mu(B)$ for all $\alpha \in [0, 1]$ and $A, B \in \mathbf{R}^{n\times n}$.
(4) *Exponential estimation.* $|e^{At}x| \leq e^{\mu(A)t}|x|$ for all $t \geq 0$, $x \in \mathbf{R}^n$, and $A \in \mathbf{R}^{n\times n}$.

The definition of the matrix measure is extendable to a set of matrices $\mathbf{A} = \{A_1, \ldots, A_m\}$ as follows. For a set of matrices $\mathbf{A} = \{A_1, \ldots, A_m\}$ and a norm $|\cdot|$ in $\mathbf{R}^n$, the (*induced*) *measure* of $\mathbf{A}$ w.r.t. $|\cdot|$ is defined as

$$\mu_{|\cdot|}(\mathbf{A}) = \max\bigl\{\mu_{|\cdot|}(A_1), \ldots, \mu_{|\cdot|}(A_m)\bigr\}. \tag{2.40}$$

It can be verified that the measure possesses the positive homogeneity and convexity properties as the matrix measure. As for the exponential estimation property, it can be seen that

$$\bigl|\phi(t; 0, x, \sigma)\bigr| \leq e^{\mu_{|\cdot|}(\mathbf{A})t}|x| \quad \forall t \geq 0,\ x \in \mathbf{R}^n,\ \sigma \in \mathcal{S}. \tag{2.41}$$

Furthermore, for a set of matrices $\mathbf{A} = \{A_1, \ldots, A_m\}$, define the *least measure value* as

$$\nu(\mathbf{A}) = \inf_{|\cdot|\in\Upsilon} \mu_{|\cdot|}(\mathbf{A}). \tag{2.42}$$

Any matrix set measure μ with $\mu(\mathbf{A}) = \nu(\mathbf{A})$ is said to be an *extreme measure* for $\mathbf{A}$. It is well known that the switched linear system $\mathbf{A}$ is (exponentially) stable if there exists a norm $|\cdot|$ such that its matrix measure is negative. As an implication, when the least measure is negative, then the switched system is exponentially stable. In the following, we are to establish that the converse is also true.

Theorem 2.26 *For any continuous-time switched linear system* $\mathbf{A}$, *we have*

$$\nu(\mathbf{A}) = \varrho(\mathbf{A}). \tag{2.43}$$

In addition, the switched system admits (at least) one extreme measure iff its normalized system is marginally stable.

Proof Firstly, it can be verified that

$$\mu_{|\cdot|}(\mathbf{A} + \lambda I_n) = \lambda + \mu_{|\cdot|}(\mathbf{A}) \quad \forall |\cdot| \in \Upsilon,\ \lambda \in \mathbf{R},$$

where $\mathbf{A} + \lambda I_n$ denotes the switched linear system $(A_1 + \lambda I_n, \ldots, A_m + \lambda I_n)$. As a result, we have

$$\nu(\mathbf{A} + \lambda I_n) = \lambda + \nu(\mathbf{A}) \quad \forall \lambda \in \mathbf{R}.$$

This, together with the fact that

$$\varrho(\mathbf{A} + \lambda I_n) = \lambda + \varrho(\mathbf{A}) \quad \forall \lambda \in \mathbf{R},$$

indicates that (2.43) holds for general case if it holds when $\varrho(\mathbf{A}) = 0$, that is, the switched system is either marginally stable or marginally unstable.

Secondly, suppose that the system is either stable or marginally stable. Fix a vector norm $|\cdot|$ in $\mathbf{R}^n$ and define the function $V : \mathbf{R}^n \mapsto \mathbf{R}_+$ by

$$V(x) = \sup_{t \in \mathbf{R}_+, \sigma \in \mathcal{S}} \left|\phi(t; 0, x, \sigma)\right|. \tag{2.44}$$

It can be seen that the function is well defined, positive definite, convex, 0-symmetric, and positively homogeneous of degree one. In addition, there is a positive real number L such that

$$|x| \le V(x) \le L|x| \quad \forall x \in \mathbf{R}^n.$$

This, together with the radial linearity property (2.18), implies that V is globally Lipschitz continuous. As a result, the function V in fact forms a vector norm of $\mathbf{R}^n$.

Thirdly, for any $x \in \mathbf{R}^n$, $s \in \mathbf{R}_+$, and $i \in M$, we have

$$\begin{aligned} V\big(\phi(s; 0, x, \hat{i})\big) &= \sup_{t \in \mathbf{R}_+, \sigma \in \mathcal{S}} \left|\phi\big(t; 0, \phi(s; 0, x, \hat{i}), \sigma\big)\right| \\ &\le \sup_{t \in \mathbf{R}_+, \sigma \in \mathcal{S}} \left|\phi(t + s; 0, x, \sigma)\right| \\ &\le \sup_{(t+s) \in \mathbf{R}_+, \sigma \in \mathcal{S}} \left|\phi(t + s; 0, x, \sigma)\right| \\ &= V(x), \end{aligned}$$

where $\hat{i}$ stands for the switching signal $\hat{i}(t) \equiv i$. As V is Lipschitz continuous, we further have

$$\begin{aligned}\mu_V(A_i) &= \limsup_{\tau \to 0^+, x \neq 0} \frac{V(x + \tau A_i x) - V(x)}{\tau V(x)} \\ &= \limsup_{\tau \to 0^+, x \neq 0} \frac{V(\phi(\tau; 0, x, \hat{i})) - V(x)}{\tau V(x)} \\ &\leq 0 \quad \forall i \in M.\end{aligned}$$

That is, the norm V induces a matrix set measure that satisfies

$$\mu_V(\mathbf{A}) \leq 0, \tag{2.45}$$

which further implies that $\nu(\mathbf{A}) \leq 0$.

Fourthly, we focus on the situation that the switched system is marginally stable. We claim that the least measure value is exactly zero. Indeed, if it is negative, then, it follows from relationship (2.41) that the system is (exponentially) stable, which yields a contradiction. This means that the matrix set measure in (2.45) is an extreme measure.

Finally, consider the case that the switched system is marginally unstable. It is clear that the least measure value is nonnegative. Assume that it is positive. Then, there is a positive real number ϵ with $\epsilon < \nu(\mathbf{A})$ such that

$$\nu_{|\cdot|}(\mathbf{A} - \epsilon I_n) > 0,$$

which further means that the switched system $\mathbf{A} - \epsilon I_n$ is either unstable or marginally unstable. This is a contradiction since $\varrho(\mathbf{A} - \epsilon I_n) = -\epsilon < 0$. The contradiction means that the least measure value is zero. On the other hand, when the least measure value is zero, the existence of an extreme measure implies the boundedness of the attainability set $R(\mathbf{A})$, and hence the switched system is either stable or marginally stable. As a result, a marginally unstable switched system does not admit any extreme measure.

To summarize, marginal stability implies the existence of an extreme measure of zero value, and marginal instability implies zero least measure value but does not admit any extreme measure. As the normalization does not alter the existence of an extreme measure, the second statement of the theorem follows. □

Remark 2.27 For a real matrix, its largest divergence rate is the largest real part of its eigenvalues, which is exactly the least measure value. This fact was pointed out in [274]. Theorem 2.26 extends the fact to the case of switched linear systems, though the concept of system spectrum is missing for switched linear systems.

Remark 2.28 The function V in the proof is in fact a (weak) common Lyapunov function for the switched linear system (cf. Proposition 2.21). The observation that it serves as a vector norm is crucial in the development.

With the help of Theorem 2.26, we can fully characterize the stabilities in terms of matrix set measure.

Corollary 2.29 *For a continuous-time switched linear system, we have the following statements*:

(1) *The system is stable iff its least measure value is negative.*
(2) *The system is marginally stable iff its least measure value is zero and it admits an extreme measure.*
(3) *The system is marginally unstable if and only if its least measure value is zero and it does not admit any extreme measure.*
(4) *The systems is unstable iff its least measure value is positive.*

2.3.4 Extended Coordinate Transformation and Set Invariance

In linear systems theory, coordinate transformation and system equivalence are powerful tools in stability analysis. For switched linear systems, coordinate transformation also plays an important role in converting a switched system into a new one with clearer and/or simpler structural information, which enables us to analyze the stability properties in a more convenient manner. However, the standard notion of equivalence coordinate change does not directly work, and we need to extend the notion in a way that it is capable of rigorous stability analysis.

Definition 2.30 Suppose that $T \in R^{n\times r}$ is a constant matrix. The linear coordinate change $x = Ty$ is said to be an *extended coordinate transformation* for the switched linear system if the matrix T is of full row rank, that is, rank $T = n$.

It is clear that the transforming matrix could be nonsquare in that the number of columns might be larger than that of rows. Under the extended coordinate change, the switched system is converted into

$$y^+ = T^+ A_\sigma T y, \tag{2.46}$$

where T^+ denotes the Moore–Penrose pseudo-inverse of a matrix T. Note that the transformed system is r-dimensional. The system is said to be the *extended transformed system* of the original switched linear system. Note that the process of extended coordinate transformation is nonreversible, that is, the original system is not necessarily an extended transformed system of (2.46).

Recall that a matrix $A = (a_{i,j})$ in $\mathbf{R}^{n\times n}$ is said to be *strictly (column) negatively diagonal dominant* if

$$a_{ii} + \sum_{j\neq i} |a_{ji}| < 0, \quad i = 1, \ldots, n.$$

Note that this can be equivalently characterized by $\mu_1(A) < 0$, where μ_1 is the matrix measure corresponding to the ℓ_1-norm.

Theorem 2.31 *The switched linear system is stable iff there exists an extended coordinate transformation such that the extended transformed system admits a contractive* 1*-norm in discrete time or a negative* 1*-measure in discrete time.*

Proof Note that, if the extended transformed system admits a contractive 1-norm in discrete time or a negative 1-measure in discrete time, then the extended transformed system is stable. It follows from $x = Ty$ that the original switched system is stable. To establish the converse relationship, we first consider the discrete-time case. As the system is stable, it follows that it is also exponentially convergent. Therefore, there is a piecewise linear function that serves as a common Lyapunov function, $V(x) = |Fx|_\infty$. The Lyapunov function can be rewritten in the dual representation

$$V(x) = \min\{|h|_1 : x = Xh,\ h \in \mathbf{R}^r\},$$

where X is a matrix of full row rank. The duality relationship comes from the fact that the set of column vectors of $[X, -X]$ is the set of vertices of the level set $\Gamma = \{x \in \mathbf{R}^n\colon V(x) \le 1\}$. Let $\{x_j\colon j = 1, \dots, s\}$ be the set of vertices of Γ. By the symmetry of Γ, we have that $s = 2r$, and we can reindex the vertices such that $x_{k+r} = -x_k$ for $k = 1, \dots, r$. As the level set is attractive w.r.t. the switched system, there is $\lambda \in [0, 1)$ such that

$$A_i x_j \in \lambda\Gamma \quad \forall i \in M,\ j = 1, \dots, s. \tag{2.47}$$

This is equivalent to the existence of vectors p_j^i with $|p_j^i|_1 \le \lambda$ such that

$$A_i x_j = X p_j^i \quad \forall i \in M,\ j = 1, \dots, r. \tag{2.48}$$

Define

$$P_i = [p_1^i, \dots, p_r^i], \quad i = 1, \dots, m.$$

It is clear that $\| P_i \|_1 \le \lambda$. Equations (2.48) can be rewritten as

$$A_i X = X P_i, \quad i \in M,$$

which leads to the first statement of the theorem.

The case of continuous time can be treated based on the above reasonale for discrete time. In fact, by Lemma 2.19, there always exists a sufficiently small positive real number τ such that the discrete-time Euler approximating system

$$x(t+1) = (I_n + \tau A_\sigma)x(t)$$

is also stable. This, together with the fact that the extended transformed system admits a contractive 1-norm, yields

$$(I_n + \tau A_i)X = X P_i, \quad i = 1, \dots, m \tag{2.49}$$

for some full row rank matrix $X \in \mathbf{R}^{n\times r}$ and matrices P_i with $\|P_i\|_1 < 1$. It is clear that (2.49) can be equivalently expressed by

$$A_i X = X(P_i - I_n)/\tau, \quad i = 1, \dots, m,$$

which directly leads to the conclusion with $H_i = (P_i - I_n)/\tau$. □

Remark 2.32 It is interesting to note that, for a linear time-invariant system $x^+ = Ax$, the criterion degenerates into the matrix relation $AX = XP$ in discrete time and $AX = XH$ in continuous time, where X is a matrix of full row rank, P is a square matrix with $\|P\|_1 < 1$, and H is a strictly negatively diagonal dominant matrix. This criterion for stability contains interesting information as discussed below. The equality $AX = XP$ can be reasonably seen as the generalized similarity between A and P. In this sense, a matrix is Hurwitz iff it is generalized similar to a strictly negatively diagonal dominant matrix. For the switched linear system, it is stable iff the subsystem matrices are simultaneously generalized similar to strictly negatively diagonal dominant matrices. In discrete time, it is interesting to note that any stable switched system is generalized similar to a system which admits the norm $|\cdot|_1$ as its common Lyapunov function.

To further identify the subtle properties of marginal stability and marginal instability, we take a view of invariant sets that start from an origin-symmetric polyhedron Λ_0 which contains the origin as an interior point. An example of Λ_0 is the polyhedron with extreme points whose entries are either 1 or -1. Define the set

$$\Lambda_\infty = \{x \in \mathbf{R}^n : \phi(t; 0, x, \sigma) \in \Lambda_0 \ \forall t \in \mathcal{T}_0, \ \sigma \in \mathcal{S}\}. \tag{2.50}$$

It can be seen that Λ_∞ is the largest (positively) invariant set contained in Λ_0 for the switched system.

Proposition 2.33 *The following statements hold*:

(1) *If switched linear system* $\mathbf{A}$ *is stable or marginally stable, then,* Λ_∞ *contains the origin as an interior point, and* $\Lambda_\infty \cap \partial\Lambda_0 \neq \emptyset$, *where* $\partial\Lambda_0$ *is the boundary of* Λ_0. *Conversely, if* Λ_∞ *contains the origin as an interior point, then, the system is stable or marginally stable, and* $\Lambda_\infty \cap \partial\Lambda_0 \neq \emptyset$.
(2) *If the system is marginally unstable, then,* $\Lambda_\infty \cap \partial\Lambda_0 \neq \emptyset$, *and* $\Lambda_\infty \subset \Lambda_0 \cap H$, *where* H *is a nontrivial subspace of* $\mathbf{R}^n$.
(3) *The switched system is unstable if* $\Lambda_\infty = \{0\}$.

Proof It is clear that Λ_∞ is origin-symmetric. Furthermore, it can be seen that it is convex and closed. Another observation is that, as Λ_∞ is an invariant set for the system, $\lambda\Lambda_\infty$ is also an invariant set for any $\lambda \geq 0$. Therefore, $\Lambda_\infty \cap \partial\Lambda_0 \neq \emptyset$ iff $\Lambda_\infty \neq \{0\}$.

For the first statement, note that the (marginally) stable system admits a common weak Lyapunov function, V. Define

$$v = \max\{r \in \mathbf{R}: V(x) \leq r \Rightarrow x \in \Lambda_0\},$$

$$\Lambda_V = \{x \in \mathbf{R}^n : V(x) \le v\}.$$

As Λ_V is an invariant set contained in Λ_0, we have that $\Lambda_\infty \supset \Lambda_V$, and thus it contains the origin as an interior point. Conversely, suppose that Λ_∞ contains the origin as an interior point. As Λ_∞ is invariant w.r.t. the system, the system is stable or marginally stable.

For the third statement, suppose that $\Lambda_\infty = \{0\}$. Define

$$\Lambda_t = \{x \in \mathbf{R}^n : \phi(s; 0, x, \sigma) \in \Lambda_0 \ \forall s \in [0, t],\ \sigma \in \mathcal{S}\}, \quad t \in \mathcal{T}_0,$$

and further

$$\eta(t) = \sup\{|x| : x \in \Lambda_t\}, \quad t \in \mathcal{T}_0.$$

It can be seen that the function η is decreasing and approaching to zero. As the function is continuous, there is a time $\tau \in \mathcal{T}_0$ such that

$$\eta(\tau) \le \frac{1}{2} \inf\{x : x \in \partial\Lambda_0\}.$$

That is, for any $\lambda \le 2$, we have $\lambda\Lambda_\tau \subset \Lambda_0$. Define

$$\Lambda_\tau^\lambda = \{x \in \mathbf{R}^n : \phi^\lambda(s; 0, x, \sigma) \in \Lambda_0 \ \forall s \in [0, \tau],\ \sigma \in \mathcal{S}\},$$

where ϕ^λ denotes the state for switched system $\{(1-\lambda)A_1, \ldots, (1-\lambda)A_m\}$ in discrete time and for $\{A_1 - \lambda I_n, \ldots, A_m - \lambda I_n\}$ in continuous time, where I_n is the nth-order identity matrix. It can be seen that $\Lambda_\tau^\lambda \subset (\Lambda_0)^o$ when λ is a sufficiently small positive real number. This means that the above matrix set is neither stable nor marginal stable, which further implies the instability of the original system.

Finally, we prove the second statement. As $\Lambda_\infty = \{0\}$ implies instability as proved previously, $\Lambda_\infty \ne \{0\}$ for marginal instability. As Λ_∞ is convex, 0-symmetric, and containing the origin as a boundary point (w.r.t. $\mathbf{R}^n$), it induces the nontrivial subspace H of $\mathbf{R}^n$ given by

$$H \overset{\text{def}}{=} \bigcup_{\lambda \ge 0} \lambda\Lambda_\infty. \tag{2.51}$$

It is clear that $\Lambda_\infty \subset H$, and this completes the proof. □

A byproduct of Proposition 2.33 is the following system decomposition lemma.

Lemma 2.34 *A marginally unstable switched linear system is simultaneously block-triangularizable. That is, there exist a nonsingular matrix T and two positive integers n_1, n_2 with $n_1 + n_2 = n$ such that*

$$\bar{A}_i \overset{\text{def}}{=} T^{-1} A_i T = \begin{bmatrix} \bar{A}_i^1 & \bar{A}_i^3 \\ 0 & \bar{A}_i^2 \end{bmatrix}, \quad i \in M, \tag{2.52}$$

where $\bar{A}_i^1$ *and* $\bar{A}_i^2$ *are* $n_1 \times n_1$ *and* $n_2 \times n_2$, *respectively. In addition, both* $\bar{\mathbf{A}}^1 = \{\bar{A}_1^1, \ldots, \bar{A}_m^1\}$ *and* $\bar{\mathbf{A}}^2 = \{\bar{A}_1^2, \ldots, \bar{A}_m^2\}$ *are marginally stable as switched linear systems of dimensions* n_1 *and* n_2, *respectively.*

Proof Let H be the subspace defined in (2.51), and let further $n_1 = \dim H$ and $n_2 = n - n_1$. The nontriviality of H implies that $1 \leq n_1 < n$. Let $H^\perp$ be the subspace orthogonal to subspace H with $H \oplus H^\perp = \mathbf{R}^n$. Let T be a nonsingular matrix whose first n_1 columns belong to H and the others belong to $H^\perp$. The block-triangular structure of (2.52) comes from the fact that H is A_i-invariant for all $i \in M$. Indeed, for any $x \in H$, there exist $y \in \Lambda_\infty$ and $\lambda \in \mathbf{R}_+$ such that $x = \lambda y$. As Λ_∞ is A_i-invariant, we have

$$A_i x = \lambda A_i y \in \lambda \Lambda_\infty \subset H.$$

It is clear that the switched system $\bar{\mathbf{A}}^1 = \{\bar{A}_1^1, \ldots, \bar{A}_m^1\}$ is either stable or marginally stable, and the marginal stability of the switched system $\bar{\mathbf{A}}^2 = \{\bar{A}_1^2, \ldots, \bar{A}_m^2\}$ can be derived as follows. If the switched system $\bar{\mathbf{A}}^2$ is marginally unstable, then it also admits a nontrivial invariant subspace. In this case, it can be seen that, besides the subspace $T^{-1}H$, the switched block-triangular system $\bar{\mathbf{A}} = \{\bar{A}_1, \ldots, \bar{A}_m\}$ admits another nontrivial invariant subspace, denoted H'. We can prove that $T^{-1}H + H'$ is $\bar{A}_i$-invariant for all $i \in M$, which further implies that $H + TH'$ is A_i-invariant for all $i \in M$. This contradicts the definition of H as it is the largest invariant subspace under A_i for $i \in M$. The contradiction exhibits that the switched system $\bar{\mathbf{A}}^2$ is either stable or marginally stable. However, the (exponential) stability of the switched system $\bar{\mathbf{A}}^2$ would imply marginal stability of the switched system $\bar{\mathbf{A}}$. Thus the switched system $\bar{\mathbf{A}}^2$ must be marginally stable. Finally, note that the original system is marginally stable if the switched system $\bar{\mathbf{A}}^1$ is stable, and thus the switched system $\bar{\mathbf{A}}^1$ must be marginally stable. □

Finally, by means of the triangular structure in (2.52), it is possible to estimate the rate of growth for state norm.

Proposition 2.35 *Suppose that the switched linear system* $\mathbf{A}$ *is marginally unstable. Then, there is a polynomial* P *with degree less than* n *such that*

$$\left|\phi(t; 0, x, \sigma)\right| \leq P(t)|x| \quad \forall x \in \mathbf{R}^n,\ t \in \mathcal{T}_0,\ \sigma \in \mathcal{S}. \tag{2.53}$$

Proof For continuous time, write $e^{\bar{A}_i t} = \begin{bmatrix} e^{\bar{A}_i^1 t} & G_i(t) \\ 0 & e^{\bar{A}_i^2 t} \end{bmatrix}$. As both $\bar{\mathbf{A}}^1 = \{\bar{A}_1^1, \ldots, \bar{A}_m^1\}$ and $\bar{\mathbf{A}}^2 = \{\bar{A}_1^2, \ldots, \bar{A}_m^2\}$ are marginally stable as switched linear systems, each $\bar{A}_i^j$ is either stable or marginally stable for $i \in M$ and $j = 1, 2$. As a result, there exists a polynomial vanishing at zero and with degree less than n, denoted $P_1(t) = p_1 t + \cdots + p_{n-1} t^{n-1}$, such that the absolute value of each entry of $G_i(t)$ is upper bounded by $P_1(t)$ for all $i \in M$ and $t \in \mathcal{T}_0$. Without loss of generality, we assume that all the coefficients of P_1 are nonnegative. Fix an arbitrarily given switching signal $\sigma \in \mathcal{S}$ and a time $t \in \mathcal{T}_0$, and let $t_0 = 0, \ldots, t_s$ be the switching time sequence in $[t_0, t)$ and

$i_0, \ldots, i_s$ be the corresponding switching index sequence. It is clear that the state transition matrix $\bar{\Phi}(t; t_0, \sigma) = e^{\bar{A}_{i_s}(t-t_s)} \ldots e^{\bar{A}_{i_0}(t_1-t_0)}$ is of the form

$$\bar{\Phi}(t; t_0, \sigma) = \begin{bmatrix} e^{\bar{A}^1_{i_s}(t-t_s)} \ldots e^{\bar{A}^1_{i_0}(t_1-t_0)} & Q \\ 0 & e^{\bar{A}^2_{i_s}(t-t_s)} \ldots e^{\bar{A}^2_{i_0}(t_1-t_0)} \end{bmatrix},$$

where

$$\begin{aligned} Q &= G_{i_s}(t-t_s)e^{\bar{A}^2_{i_{s-1}}(t_s-t_{s-1})} \ldots e^{\bar{A}^2_{i_0}(t_1-t_0)} \\ &\quad + e^{\bar{A}^1_{i_s}(t-t_s)} G_{i_{s-1}}(t_s-t_{s-1})e^{\bar{A}^2_{i_{s-2}}(t_{s-1}-t_{s-2})} \ldots e^{\bar{A}^2_{i_0}(t_1-t_0)} \\ &\quad + \cdots + e^{\bar{A}^1_{i_s}(t-t_s)} \ldots e^{\bar{A}^1_{i_2}(t_3-t_2)} G_{i_1}(t_2-t_1)e^{\bar{A}^2_{i_0}(t_1-t_0)} \\ &\quad + e^{\bar{A}^1_{i_s}(t-t_s)} \ldots e^{\bar{A}^1_{i_1}(t_2-t_1)} G_{i_0}(t_1-t_0). \end{aligned}$$

It can be seen that the marginal stability of $\bar{\mathbf{A}}^1$ and $\bar{\mathbf{A}}^2$ implies the existence of an upper bound, denoted κ, for the entries of the corresponding state transition matrices. It follows that each entry of Q satisfies

$$\begin{aligned} \big|Q(j,l)\big| &\le n\kappa^2\big(P_1(t-t_s) + P_1(t_s-t_{s-1}) + \cdots + P_1(t_1-t_0)\big) \\ &\le n\kappa^2 P_1(t-t_0), \quad j = 1, \ldots, n_1,\ l = 1, \ldots, n_2, \end{aligned}$$

where the latter inequality comes from the fact that P_1 vanishes at the origin and its coefficients are nonnegative. It follows that the norm of $\bar{\Phi}(t; t_0, \sigma)$ is upper bounded by $n^2\kappa^2(1 + P_1(t-t_0))$. As a result, the norm of the state transition matrix $\Phi(t; t_0, \sigma)$ is upper bounded by $P(t-t_0) \stackrel{\text{def}}{=} \|T\|\|T^{-1}\|n^2\kappa^2(1 + P_1(t-t_0))$, which is a polynomial with degree less than n. This clearly leads to the conclusion.

The discrete-time case can be proceeded in a similar manner, and the details are left to the reader. □

The proposition reveals the fact that there is a gap between exponential divergence of instability and divergence rate of marginal instability which is bounded by a polynomial with degree less than the system dimension.

Corollary 2.36 *For any n-dimensional regular switched linear system $\mathbf{A}$, there is a polynomial P with degree less than n such that*

$$\big|\phi(t; 0, x, \sigma)\big| \le P(t)e^{\varrho(\mathbf{A})t}|x| \quad \forall x \in \mathbf{R}^n,\ t \in \mathcal{T}_0,\ \sigma \in \mathcal{S}. \tag{2.54}$$

Moreover, P can be chosen to be of degree zero iff switched system $\underline{\mathbf{A}}$ is marginally stable.

Example 2.37 Let us examine the continuous-time two-form switched linear system with subsystem matrices

$$A_1 = \begin{bmatrix} -1 & 0 & 0 & \alpha \\ 2 & -1 & 0 & 0 \\ 0 & 0 & -2 & 3 \\ 0 & 0 & 0 & -1 \end{bmatrix}, \qquad A_2 = \begin{bmatrix} -1 & 2 & 0 & 0 \\ 0 & -1 & 0 & 0 \\ 0 & 0 & -1 & 0 \\ 0 & 0 & 3 & -2 \end{bmatrix},$$

where α is a real number. It is clear that both subsystems are stable, and the switched linear system admits a block triangular structure,

$$A_i = \begin{bmatrix} A_{i,1} & A_{i,3} \\ 0 & A_{i,2} \end{bmatrix}, \qquad i = 1, 2.$$

When $\alpha = 0$, it can be verified that the quadratic function

$$V(x) = x^T P x, \qquad P = \begin{bmatrix} 1 & 0 & 0 & 0 \\ 0 & 1 & 0 & 0 \\ 0 & 0 & 3 & -1 \\ 0 & 0 & -1 & 3 \end{bmatrix}$$

is a common weak Lyapunov function. By Proposition 2.6, the switched system is stable or marginally stable. On the other hand, it can be seen that the convex combination $\frac{1}{2}(A_1 + A_2)$ is marginally stable, and hence the switched system is not (asymptotically) stable [80]. As a result, the system is marginally stable.

When $\alpha \neq 0$, the switched system is marginally unstable due to the coupling between the two marginally stable modes. It is readily seen that the system is already in the form specified by Lemma 2.34. Let $\Lambda_0 = \{x \in \mathbf{R}^4 \colon \sum_{k=1}^4 |x_i| \leq 1\}$. The largest invariant set Λ_∞ contained in Λ_0 satisfies

$$\Lambda_{V_2} \stackrel{\text{def}}{=} \left\{ \begin{bmatrix} 0 \\ 0 \\ x_3 \\ x_4 \end{bmatrix} : V_2(x) \leq 1 \right\} \subset \Lambda_\infty \subset \left\{ \begin{bmatrix} 0 \\ 0 \\ x_3 \\ x_4 \end{bmatrix} : |x_3| + |x_4| \leq 1 \right\},$$

where $V_2(x) = 3x_3^2 + 3x_4^2 - 2x_3x_4$. Figure 2.1 depicts the sets in the $x_3 - x_4$ plane.

Finally, due to the block-triangular structure of the switched system, the system solution satisfies

$$\begin{aligned} &\left|x^1(t)\right| \leq \kappa_1 \left|x^1(0)\right| + \int_0^t \kappa_1 \left|\alpha x_4(\tau)\right| d\tau \leq \kappa_1 \left|x^1(0)\right| + \alpha \kappa_1 \kappa_2 t \left|x^2(0)\right|, \\ &\left|x^2(t)\right| \leq \kappa_2 \left|x^2(0)\right|, \end{aligned}$$

where $x^1 = [x_1, x_2]^T$, $x^2 = [x_3, x_4]^T$, and κ_1 and κ_2 are the largest possible norms of the state transition matrices w.r.t. subsystems $\{A_{i,1}\}$ and $\{A_{i,2}\}$, respectively. It is clear that the state norm is bounded by a polynomial of degree one, which is consistent with Proposition 2.35. Figure 2.2 presents a sample state trajectory with $\alpha = 1$. It is clear that the state norm grows linearly.

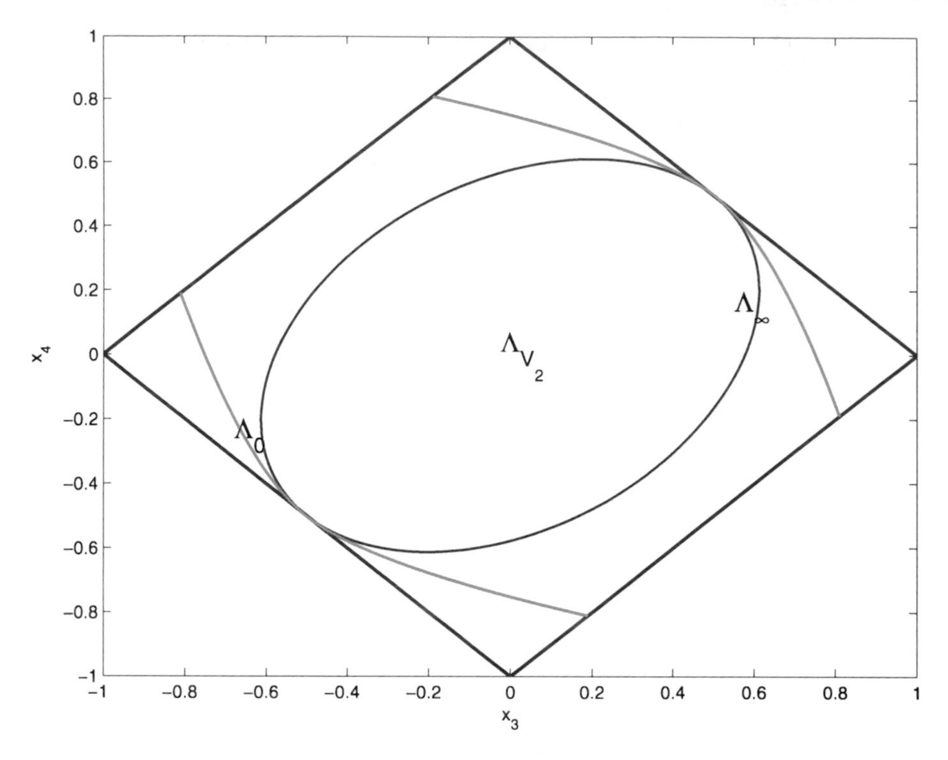

Fig. 2.1 Sets Λ_0, Λ_∞, and Λ_{V_2}

2.3.5 Triangularizable Systems

In this subsection, we focus on a special class of switched linear systems which either possess an upper (lower) triangular structure or are simultaneously equivalent to triangular systems. Triangular systems are interesting because they have simple structures, and many nontriangular systems can be made to be triangular by means of equivalence transformations (simultaneous triangularization).

Definition 2.38 The switched system is said to be *simultaneously* (*upper*) *triangularizable* if the matrix set $\mathbf{A} = \{A_1, \ldots, A_m\}$ is simultaneously triangularizable, that is, there exists a complex nonsingular matrix $T \in \mathbf{C}^{n\times n}$ such that $B_k \stackrel{\text{def}}{=} T^{-1}A_kT$, $k \in M$, are of the upper triangular form,

$$B_k = \begin{bmatrix} b_k(1,1) & \ldots & b_k(1,n) \\ & \ddots & \\ 0 & \ldots & b_k(n,n) \end{bmatrix} \in \mathbf{C}^{n\times n}, \quad k \in M. \tag{2.55}$$

Note that we allow complex matrices as the equivalence transformation. For a simultaneously triangularizable matrix set, we can transform it into the following real normal form.

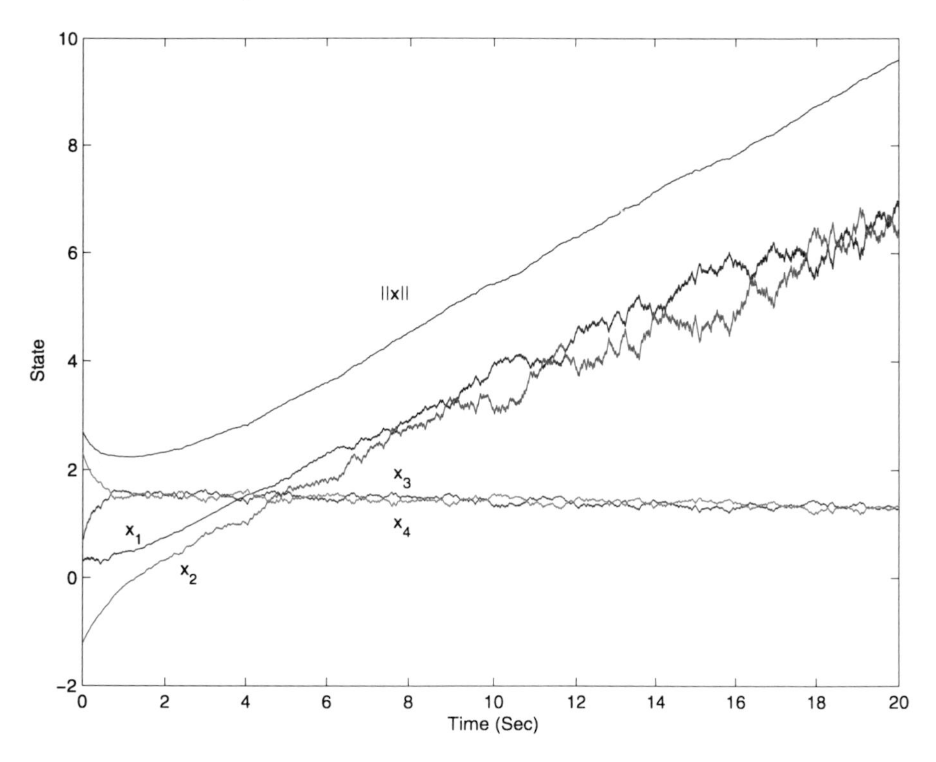

Fig. 2.2 Sample state trajectory

Lemma 2.39 *Suppose that a system* $\mathbf{A} = \{A_1, \dots, A_m\}$ *is simultaneously triangularizable. Then, there exists a real nonsingular matrix* G, *such that* $G^{-1}A_kG$ *is of the normal form*

$$\bar{A}_k \stackrel{\text{def}}{=} G^{-1}A_kG = \begin{bmatrix} A_{1k} & * & \dots & * \\ 0 & A_{2k} & \dots & * \\ \vdots & \vdots & \ddots & \vdots \\ 0 & 0 & \dots & A_{lk} \end{bmatrix}, \tag{2.56}$$

where $l \le n$, A_{jk} *is either a* 1×1 *or* 2×2 *block, and the size of the* j*th block* A_{jk} *is the same for all* $k \in M$. *In addition, if* A_{jk} *is of* 2×2, *then it is of the form*

$$A_{jk} = \begin{bmatrix} \mu_{jk} & \omega_{jk} \\ -\omega_{jk} & \mu_{jk} \end{bmatrix}. \tag{2.57}$$

Proof As the matrix set $\{A_1, \dots, A_m\}$ is simultaneously triangularizable, for any polynomial $p(y_1, \dots, y_m)$ over $\mathbf{R}$, the eigenvalues of $p(A_1, \dots, A_m)$ are $p(b_1(i,i), \dots, b_m(i,i))$, $i = 1, \dots, n$. This exhibits that the matrix set possesses Property III in [85, p. 442]. By Theorems 1 and 9 in [85], there is an orthogonal matrix $H \in \mathbf{R}^{n\times n}$ such that

$$\bar{B}_k \stackrel{\text{def}}{=} H^{-1} A_k H = \begin{bmatrix} B_{1k} & * & \dots & * \\ 0 & B_{2k} & \dots & * \\ \vdots & \vdots & \ddots & \vdots \\ 0 & 0 & \dots & B_{lk} \end{bmatrix}, \tag{2.58}$$

where $l \le n$, and for any fixed $j \le l$, we have

(i) $B_{j1}, \dots, B_{jm}$ are 1×1; or
(ii) $B_{j1}, \dots, B_{jm}$ are 2×2, with one of these matrices, say, B_{jq}, of the form

$$B_{jq} = \begin{bmatrix} r_{jq} & u_{jq} \\ -v_{jq} & r_{jq} \end{bmatrix}, \quad u_{jq} > 0, \ v_{jq} > 0,$$

and each of B_{jk}, $k \in M$, is a real linear polynomial g_{jk} in B_{jq}, that is, $B_{jk} = g_{jk}(B_{jq})$.

As B_{jq} in (ii) possesses a pair of (conjugated) complex eigenvalues, it follows from the standard matrix theory that there exists a real nonsingular matrix $T_j \in \mathbf{R}^{2\times 2}$ such that

$$T_j^{-1} B_{jq} T_j = \begin{bmatrix} \mu_{jq} & \omega_{jq} \\ -\omega_{jq} & \mu_{jq} \end{bmatrix},$$

where $\mu_{jq}, \omega_{jq} \in \mathbf{R}$. Furthermore, as any polynomial of the matrix $\left[\begin{smallmatrix} \mu_{jq} & \omega_{jq} \\ -\omega_{jq} & \mu_{jq} \end{smallmatrix}\right]$ is still of the same form, we have

$$T_j^{-1} B_{jk} T_j = g_{jk}\big(T_j^{-1} B_{jq} T_j\big) = \begin{bmatrix} \mu_{jk} & \omega_{jk} \\ -\omega_{jk} & \mu_{jk} \end{bmatrix}, \quad \mu_{jk}, \omega_{jk} \in \mathbf{R}, \ k \in M.$$

Define $K = \text{diag}[K_1, \dots, K_l]$, where $K_j = 1$ if the corresponding block in (2.58) is 1×1, and $K_j = T_j$ if the block is 2×2. Letting $G = HK$, the theorem follows. □

Remark 2.40 Simultaneous triangularization of matrix sets has been investigated extensively; see, for example, [5, 176, 195, 204] and the references therein. In particular, the following classes of matrix sets (and the corresponding switched systems) have been proven to be simultaneously triangularizable:

(a) The system matrices are commutative pairwise, that is, $A_i A_j = A_j A_i$, $i, j \in M$ [182].
(b) The Lie algebra generated by the system matrices is solvable [1, 147].
(c) $\mathcal{A} = \{A_1, A_2\}$ and $\text{rank}(A_1 A_2 - A_2 A_1) = 1$ [137].

For switched linear systems that are simultaneously triangularizable, it is possible to judge the stability directly upon the eigenvalues of the subsystems. In fact, we can go further to characterize the largest rate of convergence (cf. (2.28)) explicitly.

Theorem 2.41 *For any simultaneously triangularizable switched linear system* $\mathbf{A}$, *the largest divergence rate is*

$$\varrho(\mathbf{A}) = \max_{k\in M} \max_{i=1}^{n} \Re\lambda_i(A_k) \tag{2.59}$$

in continuous time and

$$\varrho(\mathbf{A}) = \max_{k\in M} \max_{i=1}^{n} \left|\lambda_i(A_k)\right| \tag{2.60}$$

in discrete time.

Remark 2.42 Note that (2.59) and (2.60) only involve eigenvalues of the system which are easily calculated. To apply the theorem, we do not necessarily need to find a linear transformation that converts the system into the triangular form. Instead, we only need to confirm that the system is simultaneously triangularizable.

To prove this theorem, we need the following technical lemma.

Lemma 2.43 *A switched system* $\bar{\mathbf{A}} = \{\bar{A}_1, \ldots, \bar{A}_m\}$, *where* $\bar{A}_k$ *is in normal form* (2.56), *is stable iff each* $\bar{A}_k$ *is stable.*

Proof Suppose that the switched system is stable. Then it is clear that each subsystem is also stable.

On the other hand, if each $\bar{A}_k$ is Hurwitz in continuous time, then we can construct a common Lyapunov function of the form $v(x) = x^T Px$ for the switched system. Indeed, for the $\bar{B}_k$ defined in (2.58), let $j_1, \ldots, j_l$ denote the size of $\bar{B}_k(1,1), \ldots, \bar{B}_k(l,l)$, respectively. Define

$$P = \mathrm{diag}[I_{j_1}, q_2 I_{j_2}, \ldots, q_l I_{j_l}],$$

where positive numbers $q_2, \ldots, q_l$ are chosen so that the minors of the matrix $-(\bar{A}_k^T P + P\bar{A}_k)$ of order $1, \ldots, n$ are positive for all $k \in M$. In this way, the switched system possesses a common Lyapunov function and thus is stable.

The discrete-time counterpart can be established in a similar manner, and the details are omitted. □

Proof of Theorem 2.41 We proceed with the continuous-time case, and the discrete-time case could be proven in a similar manner. Suppose that there exists a real nonsingular matrix G such that for each $k \in M$, the matrix $\bar{A}_k = G^{-1}A_k G$ is of the triangular form (2.56).

Consider the switched system $\mathbf{A} - \gamma\mathbf{I} = \{A_1 - \gamma I_n, \ldots, A_m - \gamma I_n\}$, where γ is any given real number. Let $\lambda_k^{\max}$ denote the largest real part of the matrix A_k. Suppose that $\gamma > \max_{k\in M} \lambda_k^{\max}$, then it can be seen that each $\bar{A}_k - \gamma I_n$ is Hurwitz. It follows from Lemma 2.43 that the switched system $\bar{\mathbf{A}} - \gamma\mathbf{I_n} = \{\bar{A}_1 - \gamma I_n, \ldots, \bar{A}_m - \gamma I_n\}$ is stable. This in turn implies that

$$\varrho(\bar{\mathbf{A}} - \gamma\mathbf{I_n}) < 0$$

and

$$\varrho(\mathbf{A}) = \varrho(\bar{\mathbf{A}}) < \gamma.$$

Accordingly, we have

$$\varrho(\mathbf{A}) \leq \max_{k \in M} \lambda_k^{\max}.$$

On the other hand, it can be seen that

$$\varrho(\mathbf{A}) \geq \max_{k \in M} \varrho(A_k) = \max_{k \in M} \lambda_k^{\max}.$$

As a result, we have

$$\varrho(\mathbf{A}) = \max_{k \in M} \lambda_k^{\max}.$$

□

Corollary 2.44 *A simultaneously triangularizable switched linear system is asymptotically stable iff each subsystem is stable.*

Example 2.45 For the continuous-time switched system $\mathbf{A} = \{A_1, A_2\}$ with

$$A_1 = \begin{bmatrix} -2 & -1 & -1 \\ 0 & -1 & 0 \\ -1 & -1 & -2 \end{bmatrix}, \qquad A_2 = \begin{bmatrix} 0 & 2 & 0 \\ -2 & -1 & 0 \\ 1 & -2 & -1 \end{bmatrix},$$

it can be verified that the eigenvalues are $\{-1, -1, -3\}$ for A_1 and $\{-1, -0.5 + 1.9365\sqrt{-1}, -0.5 - 1.9365\sqrt{-1}\}$ for A_2, and $A_1A_2 - A_2A_1$ is of rank one. It follows that the switched linear system is simultaneously triangularizable (cf. Remark 2.40), which further implies that the switched system is exponentially convergent with the rate of -0.5.

2.4 Computational Issues

In the previous section, we presented quite a few stability criteria, mostly for stability and some for marginal stability. In this section, we briefly discuss the possibility of verifying the conditions of the criteria in terms of appropriate computational procedures.

2.4.1 Approximating the Spectral Radius

For discrete-time switched linear systems, it follows from Corollary 2.24 that the verification of asymptotic stability can be reduced to the calculation of the spectral radius. Indeed, if the spectral radius can be exactly calculated in a finite time, then the verification problem is decidable, that is, there exists a computational procedure that produces either "yes" or "no" answer in a finite time. In the literature, it was conjectured that, for any finite matrix set $\mathbf{A}$, there exists a finite k such that $\bar{\rho}(\mathbf{A}) = \bar{\rho}_k(\mathbf{A})^{1/k}$. This conjecture is well known as the *Finiteness Conjecture*, which was finally disproved by counterexamples. As a result, it remains an open problem for the verification of asymptotic stability. Nevertheless, by Corollary 2.24 and the fact that $\rho(\mathbf{A}) = \inf_{k\in\mathbf{N}^+} \hat{\rho}_k(\mathbf{A})^{1/k}$, the verification problem is semi-decidable by verifying the relationship $\hat{\rho}_k(\mathbf{A}) < 1$ for $k = 1, 2, \ldots$ and terminating at the first confirmative instant. On the other hand, the problem of instability verification is also semi-decidable by verifying the relationship $\bar{\rho}_k(\mathbf{A}) > 1$ for $k = 1, 2, \ldots$ and terminating at the first confirmative instant. The problem of verifying the marginal stability, however, was shown to be undecidable.

While it is hard to exactly calculate the spectral radius of a matrix set, it is possible to compute an approximation as follows.

Proposition 2.46 *For any norm and natural number k, we have*

$$\max_{B\in\Pi_k(\mathbf{A})} \rho(B)^{1/k} \le \rho(\mathbf{A}) \le \max_{B\in\Pi_k(\mathbf{A})} \|B\|^{1/k}, \quad k \in \mathbf{N}_+. \tag{2.61}$$

Proof First, observe that, for any natural number j, we have

$$\max_{B\in\Pi_k(\mathbf{A})} \rho(B)^j \le \max_{B\in\Pi_{kj}(\mathbf{A})} \rho(B). \tag{2.62}$$

Indeed, the set of matrix products that can be expressed as the jth power of an element in $\Pi_k(\mathbf{A})$ is a subset of $\Pi_{kj}(\mathbf{A})$. Inequality (2.62) follows from the equality $\rho(A^j) = \rho(A)^j$. Rewrite (2.62) to be

$$\max_{B\in\Pi_k(\mathbf{A})} \rho(B)^{1/k} \le \max_{B\in\Pi_{kj}(\mathbf{A})} \rho(B)^{1/kj}, \tag{2.63}$$

which implies that

$$\max_{B\in\Pi_k(\mathbf{A})} \rho(B)^{1/k} \le \limsup_{j\to+\infty} \max_{B\in\Pi_{kj}(\mathbf{A})} \rho(B)^{1/kj}.$$

On the other hand, it follows from $\Pi_{kj}(\mathbf{A}) \subset \bigcup_{i=1}^{+\infty} \Pi_i(\mathbf{A})$ that

$$\limsup_{j\to+\infty} \max_{B\in\Pi_{kj}(\mathbf{A})} \rho(B)^{1/kj} \le \limsup_{i\to+\infty} \max_{B\in\Pi_i(\mathbf{A})} \rho(B)^{1/i},$$

which, together with inequality (2.63), leads to the first inequality of (2.61).

To establish the second inequality of (2.61), define

$$\zeta = \max\{\|A_1\|, \ldots, \|A_m\|\}.$$

For any nature number l, there are nonnegative integers v and j with $j < k$ such that $l = kv + j$. For any index sequence $i_1, \ldots, i_l$ in M, we have

$$\|A_{i_1}\cdots A_{i_l}\| \le \prod_{\iota=0}^{\nu-1}\big(\|A_{k\iota+1}A_{k\iota+2}\cdots A_{k\iota+k}\|\big)\prod_{\iota=1}^{j}\|A_{k\nu+\iota}\|$$
$$\le \Big(\max_{B\in\Pi_k(\mathbf{A})}\|B\|\Big)^{\nu}\zeta^{j}.$$

It follows that

$$\Big(\max_{B\in\Pi_l(\mathbf{A})}\|B\|\Big)^{1/l} \le \Big(\max_{B\in\Pi_k(\mathbf{A})}\|B\|\Big)^{1/k-j/kl}\zeta^{j/l},$$

which, by taking $l\to+\infty$, yields

$$\rho(\mathbf{A}) \le \max_{B\in\Pi_k(\mathbf{A})}\|B\|^{1/k}. \qquad \square$$

Remark 2.47 The proposition provides two-side bounds for the spectral radius. Based on this relationship, a computational procedure is readily developed to approximate the spectral radius. With a sufficiently large k, the approximation can be made arbitrarily accurate. For stability verification, it suffices to terminate the procedure when either $\max_{B\in\Pi_k(\mathbf{A})}\rho(B)>1$, which implies instability, or $\max_{B\in\Pi_k(\mathbf{A})}\|B\|<1$ that implies stability. While a merit of this approximation is that at each step the approximating accuracy can be estimated, the procedure is not computationally efficient as the cardinality of $\Pi_k(\mathbf{A})$ grows exponentially with k. Another drawback of the estimate is that both sequences $\max_{B\in\Pi_k(\mathbf{A})}\rho(B)^{1/k}$ and $\max_{B\in\Pi_k(\mathbf{A})}\|B\|^{1/k}$ are not necessarily monotone w.r.t. k, which means that the approximate accuracy of estimate (2.61) is not necessarily increasing as k increases. The following example clearly illustrates this.

Example 2.48 Let $M=\{1,2\}$, and

$$A_1=\begin{bmatrix}-1 & -\sqrt{3}\\ -0.9\sqrt{3} & -0.9\end{bmatrix},\qquad A_2=\begin{bmatrix}-0.9 & -0.9\sqrt{3}\\ \sqrt{3} & -1\end{bmatrix}.$$

Table 2.1 shows the calculated joint/generalized spectral radii and the differences that represent the accuracy errors, where $\bar{\rho}_k=\max_{B\in\Pi_k(\mathbf{A})}\rho(B)^{1/k}$, $\hat{\rho}_k=\max_{B\in\Pi_k(\mathbf{A})}\|B\|^{1/k}$, and $e_k=\hat{\rho}_k-\bar{\rho}_k$. It is clear that all the sequences are oscillating.

To obtain a guaranteed precision of a spectral radius estimate, we take a polynomial common Lyapunov approach, which could provide guaranteed accuracy by

Table 2.1 Estimated spectral radii and error bounds

k	1	2	3	4	5	6	7
$\bar{\rho}_k$	1.8974	1.8974	1.9652	1.8974	1.8974	1.9652	1.8974
$\hat{\rho}_k$	2.0000	1.9860	1.9654	1.9739	1.9709	1.9654	1.9702
e_k	0.1026	0.0886	0.0002	0.0765	0.0735	0.0002	0.0728

searching a proper common Lyapunov function in a preassigned set of polynomials. For this, we first restrict ourselves to the case of quadratic common Lyapunov functions and then extend the searching to sum-of-squares polynomials.

If the switched linear system admits a quadratic Lyapunov function, then, by solving the linear matrix inequalities

$$A_i^T P + P A_i < 0, \quad i = 1, \ldots, m$$

for symmetric matrix P, the asymptotic stability is decidable via effective algorithms. Based on this idea, we define the *ellipsoid norm* for matrix set $\mathbf{A}$ by

$$\rho_L(\mathbf{A}) = \inf\{\mu \in \mathbf{R}^+ : \exists\, P > 0 \text{ s.t. } A_i^T P A_i \le \mu^2 P,\ i = 1, \ldots, m\}.$$

It can be seen that the ellipsoid norm could be equivalently defined to be

$$\rho_L(\mathbf{A}) = \inf_{P>0} \max_{i=1}^{m} \|A_i\|_P, \tag{2.64}$$

where $\|A_i\|_P = \max_{x \neq 0} \sqrt{x^T A_i^T P A_i x} / \sqrt{x^T P x}$ is the P-norm.

Proposition 2.49 $\frac{\rho_L(\mathbf{A})}{\sqrt{n}} \le \rho(\mathbf{A}) \le \rho_L(\mathbf{A})$.

To proceed with the proof, we need the following supporting lemma that is part of the well-known *John's theorem* [128].

Lemma 2.50 *Suppose that $\Omega \subset \mathbf{R}^n$ is an origin-symmetric compact convex set with nonempty interior. Then, there is an origin-centered ellipsoid E such that*

$$E \subset \Omega \subset \sqrt{n} E. \tag{2.65}$$

Proof of Proposition 2.49 It follows from the definition of $\rho_L(\mathbf{A})$ that $\rho(\mathbf{A}) \le \rho_L(\mathbf{A})$. Therefore, we need only to establish that $\rho(\mathbf{A}) \ge \frac{\rho_L(\mathbf{A})}{\sqrt{n}}$.

By Theorem 2.23, for any positive real number ϵ, there is a norm $|\cdot|$ such that

$$\|\mathbf{A}\| \le \rho(\mathbf{A}) + \epsilon. \tag{2.66}$$

It is clear that the unit ball $\Omega = \{x \in \mathbf{R}^n : |x| \le 1\}$ is origin-symmetric, compact, and convex. It follows from Lemma 2.50 that there exists an ellipsoid $E = \{x \in \mathbf{R}^n : x^T P x \le \frac{1}{n}\}$ such that relation (2.65) holds. This implies that

$$\sqrt{x^T P x} \le |x| \le \sqrt{n}\sqrt{x^T P x},$$

which, together with inequality (2.66), further implies that

$$A_i^T P A_i \le \big(\rho(\mathbf{A}) + \epsilon\big) n P.$$

This leads to

$$\max_{i\in M} \|A_i\|_P \leq \sqrt{n}\big(\rho(\mathbf{A}) + \epsilon\big).$$

By the arbitrariness of ϵ, we have

$$\rho_L(\mathbf{A}) = \inf_{P>0} \max_{i\in M} \|A_i\|_P \leq \sqrt{n}\rho(\mathbf{A}).$$

This completes the proof. □

Remark 2.51 Proposition 2.49 establishes that the ellipsoid norm approximation admits a guaranteed precision that relies on the system dimension. In particular, the approximation is exact for scalar systems. For higher-order systems, the approximation is tight in that Proposition 2.49 does not generally hold when $\sqrt{n}$ is substituted by a smaller number. Therefore, finding a common Lyapunov function becomes more and more difficult as the system dimension increases.

To improve the approximation precision, a natural idea is to use higher-order polynomials as common Lyapunov functions. A suitable class of polynomial Lyapunov functions is the set of homogeneous polynomials that can be expressed as sums of squares.

A polynomial p is said to admit a *sum-of-squares representation* if there are polynomials $p_1, \ldots, p_k$ such that

$$p(x) = \sum_{i=1}^{k} \big(p_i(x)\big)^2 \quad \forall x \in \mathbf{R}^n.$$

A polynomial is said to be a sum-of-squares if it admits a sum-of-squares representation.

It is clear that a sum-of-squares is always positive semi-definite, and it is positive definite if polynomials p_i, $i = 1, \ldots, k$, do not admit a common root. Moreover, a homogeneous sum-of-squares admits a quadratic representation as stated below.

Lemma 2.52 *A homogeneous polynomial $p(x)$ of degree $2d$ is a sum-of-squares if and only if*

$$p(x) = \big(x^{[d]}\big)^T P x^{[d]}, \tag{2.67}$$

where $x^{[d]}$ is a vector whose entries are monomials of degree d in x, and $P \geq 0$.

For a proof of the lemma, the reader is referred to [184, 203].

Example 2.53 Suppose that we try to find a sum-of-square representation for the polynomial

$$p(x_1, x_2) = 2x_1^4 + 2x_1^3 x_2 - x_1^2 x_2^2 + 5x_2^4.$$

For this, let $x^{[d]} = [x_1^2, x_1x_2, x_2^2]^T$. Then, try the representation as in (2.67) that reads

$$\begin{bmatrix} x_1^2 & x_1x_2 & x_2^2 \end{bmatrix} \begin{bmatrix} p_{11} & p_{12} & p_{13} \\ p_{12} & p_{22} & p_{23} \\ p_{13} & p_{23} & p_{33} \end{bmatrix} \begin{bmatrix} x_1^2 \\ x_1x_2 \\ x_2^2 \end{bmatrix}$$
$$= p_{11}x_1^4 + 2p_{12}x_1^3x_2 + 2p_{23}x_1x_2^3 + (p_{22} + 2p_{13})x_1^2x_2^2 + p_{33}x_2^4.$$

Solving $p(x) = (x^{[d]})^T P x^{[d]}$ gives

$$p_{11} = 2, \qquad p_{12} = 1, \qquad 2p_{13} + p_{22} = -1, \qquad p_{23} = 0, \qquad p_{33} = 5. \tag{2.68}$$

To obtain a positive semi-definite P satisfying the equalities, we can use the semi-definite programming technique, which corresponds to the optimization of a linear function over the intersection of an affine subspace and the cone of positive semi-definite matrices. Specifically, we take the following semi-definite programming:

$$\begin{aligned} &\text{minimize} && 0 \\ &\text{subject to} && \operatorname{tr}(B_i P) = b_i, \quad i = 1, \ldots, 5, \\ &&& P \geq 0, \end{aligned}$$

where B_i and b_i are chosen to represent the ith equation, $i = 1, \ldots, 5$. For instance, to represent the third equation, $2p_{13} + p_{22} = -1$, we should choose

$$B_3 = \begin{bmatrix} 0 & 0 & 1 \\ 0 & 1 & 0 \\ 1 & 0 & 0 \end{bmatrix}, \qquad b_3 = -1.$$

It turns out that a feasible solution of the semi-definite programming is

$$P = \begin{bmatrix} 2 & 1 & -3 \\ 1 & 5 & 0 \\ -3 & 0 & 5 \end{bmatrix}.$$

In this way, polynomial p is represented by a sum-of-squares as

$$p(x) = \frac{1}{2}\left[\left(2x_1^2 - 3x_2^2 + x_1x_2\right)^2 + \left(x_2^2 + 3x_1x_2\right)^2\right].$$

To utilize the sums of squares to approximate the spectral radius, we need the following lemma.

Lemma 2.54 *Suppose that p is a positive definite homogeneous polynomial of degree $2d$ that satisfies*

$$p(A_i x) \leq \gamma p(x) \quad \forall x \in \mathbf{R}^n, \ i \in M, \tag{2.69}$$

for some $\gamma > 0$. Then, we have $\rho(\mathbf{A}) \leq \gamma^{\frac{1}{2d}}$.

Proof By the positive definiteness and continuity of p, we have

$$0 < \min_{|x|=1} p(x) \stackrel{\text{def}}{=} \alpha_1 \le \max_{|x|=1} p(x) \stackrel{\text{def}}{=} \alpha_2 < +\infty,$$

which, together with the homogeneity, implies that

$$\alpha_1 |x|^{2d} \le p(x) \le \alpha_2 |x|^{2d}.$$

For any natural number k and $B \in \Pi_k(\mathbf{A})$, there exist an index sequence $i_1, \dots, i_k$ with elements in M such that $B = A_{i_k} \cdots A_{i_1}$. It is clear that

$$\begin{aligned} \|A_{i_k} \cdots A_{i_1}\| &= \max_{x \neq 0} \frac{|A_{i_k} \cdots A_{i_1} x|}{|x|} \\ &\le \left(\frac{\alpha_2}{\alpha_1}\right)^{\frac{1}{2d}} \max_{x \neq 0} \frac{p(A_{i_k} \cdots A_{i_1} x)}{p(x)} \\ &\le \left(\frac{\alpha_2}{\alpha_1}\right)^{\frac{1}{2d}} \gamma^{k/2d}. \end{aligned}$$

As a result, we have

$$\rho(\mathbf{A}) = \lim_{k \to +\infty} \sup \max_{B \in \Pi_k(\mathbf{A})} \|B\|^{1/k} \le \gamma^{\frac{1}{2d}}.$$

□

It is clear that, for any homogeneous polynomial p, there always exists a positive real number γ that satisfies inequality (2.69).

Example 2.55 For the planar discrete-time two-form switched linear system $\mathbf{A}$ with

$$A_1 = \begin{bmatrix} a & 0 \\ a & 0 \end{bmatrix}, \qquad A_2 = \begin{bmatrix} 0 & a \\ 0 & -a \end{bmatrix},$$

where a is a positive real number, simple calculation gives

$$\max_{B \in \Pi_k(\mathbf{A})} \rho(B)^{1/k} = a, \quad k = 1, 2, \dots.$$

As a result, the spectral radius of the switched system is a. Using a common quadratic Lyapunov function, the upper bound on the spectral radius is equal to $\sqrt{2}a$. As a result, only when $a \le \frac{\sqrt{2}}{2}$, we can find a common quadratic Lyapunov function for the switched system, which implies the (marginal) stability of system. On the other hand, take the sum-of-squares homogeneous polynomial of degree four

$$V(x) = \left(x_1^2 - x_2^2\right)^2 + \epsilon\left(x_1^2 + x_2^2\right)^2,$$

where ϵ is a positive real number. We could verify that, for any $b \le \frac{4}{3}$, there is a sufficiently small ϵ such that $ba^4 V(x) - V(A_i x)$ is a sum-of-squares. It follows

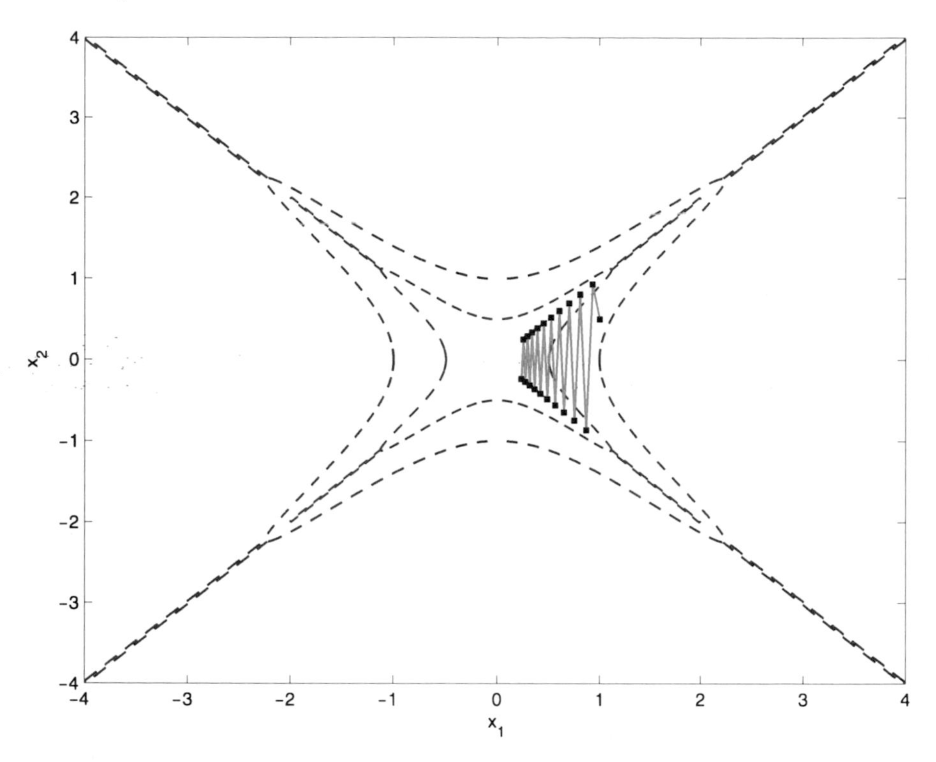

Fig. 2.3 Sample phase portrait and level sets

from Lemma 2.54 that

$$\rho(\mathbf{A}) \leq \left(\frac{4}{3}\right)^{1/4} a.$$

In particular, when $a < (\frac{3}{4})^{1/4} \approx 0.9306$, we conclude that the switched system is stable. It is interesting to note that the level sets of $V(x)$ are nonconvex, as shown in Fig. 2.3, where $a = 0.93$ and $\epsilon = 0.01$.

The following lemma, presented in [21], characterizes the ability of sums of squares to approximate a norm.

Lemma 2.56 *For any norm* $|\cdot|$ *and natural number* d*, there exists a homogeneous polynomial* $p(x)$ *of degree* $2d$ *such that*

(1) p *is a sum-of-squares and*
(2) *for all* $x \in \mathbf{R}^n$*, we have*

$$p(x)^{\frac{1}{2d}} \leq |x| \leq \kappa_n^d p(x)^{\frac{1}{2d}}, \tag{2.70}$$

where $\kappa_n^d = \binom{n+d-1}{d}^{\frac{1}{2d}}$.

Recall that $\binom{k}{j}$ is the number of combinations, $\binom{k}{j} = \frac{k!}{j!(k-j)!}$, where $k!$ is a factorial. When n is fixed and d is sufficiently large, we have

$$\kappa_n^d \approx 1 + \frac{2d}{(n-1)\ln d}.$$

This means that, for any positive real number ϵ, there is a sufficient large integer d such that $\kappa_n^d \leq 1 + \epsilon$. As an implication, by choosing sufficiently large d, the estimate in (2.70) can achieve any preassigned accuracy.

Let HP_k^n be the set of homogeneous polynomials of degree k defined on $\mathbf{R}^n$, and SOS_k^n be the subset of HP_k^n that are sums-of-squares. Define the quantity

$$\begin{aligned} \rho_S^{2d}(\mathbf{A}) = & \inf_{p \in \mathrm{HP}_k^n} \gamma^{\frac{1}{2d}} \\ \text{subject to} \quad & p \in \mathrm{SOS}_k^n, \\ & \gamma p(x) - p(A_i x) \in \mathrm{SOS}_k^n. \end{aligned}$$

With the help of the above notation, we are ready to state the main result of the subsection.

Theorem 2.57 *Suppose that d is a natural number. Then, we have*

$$\frac{\rho_S^{2d}(\mathbf{A})}{\kappa_n^d} \leq \rho(\mathbf{A}) \leq \rho_S^{2d}(\mathbf{A}). \tag{2.71}$$

To proceed with a proof of the theorem, we need some further auxiliary material. First, it follows from Lemma 2.54 that

$$\rho(\mathbf{A}) \leq \rho_S^{2d}(\mathbf{A}). \tag{2.72}$$

Another observation is that, when $d = 1$, HP_2^n is exactly the set of quadratic polynomials, and SOS_2^n is exactly the set of positive (semi-)definite quadratic polynomials. For any positive real number ϵ, there is a norm $|\cdot|$ such that

$$\rho(\mathbf{A}) \geq \|\mathbf{A}\| - \epsilon.$$

By Lemma 2.56, we have

$$\rho(\mathbf{A}) \geq \frac{1}{\sqrt{n}} \rho_S^2(\mathbf{A}) - \epsilon,$$

where the equality $\binom{n}{1}^{1/2} = \sqrt{n}$ is used. By the arbitrariness of ϵ, we obtain

$$\rho(\mathbf{A}) \geq \frac{1}{\sqrt{n}} \rho_S^2(\mathbf{A}). \tag{2.73}$$

Note that the approximation is the same as in Proposition 2.49.

Next, for a vector $x \in \mathbf{R}^n$ and a natural number k, define the k-lift of x, denoted $x^{[k]}$, to be the vector with components $\{\sqrt{\alpha!}x^\alpha\}_\alpha$, where $\alpha = (\alpha_1, \dots, \alpha_n) \in \mathbf{N}_+^n$, $\sum_{i=1}^n \alpha_i = k$, and $\alpha!$ denotes the multinomial coefficient $\alpha! = \frac{k!}{\alpha_1! \cdots \alpha_n!}$. For example, when $n = 2$, we have

$$x^{[1]} = \begin{bmatrix} x_1 \\ x_2 \end{bmatrix}, \qquad x^{[2]} = \begin{bmatrix} x_1^2 \\ \sqrt{2}x_1x_2 \\ x_2^2 \end{bmatrix}, \qquad x^{[3]} = \begin{bmatrix} x_1^3 \\ \sqrt{3}x_1^2x_2 \\ \sqrt{3}x_1x_2^2 \\ x_2^3 \end{bmatrix}.$$

It is clear that $x^{[k]}$ is of dimension $N_n^k = \binom{n+k-1}{k}$. For standard Euclidean norm, it can be verified that

$$\left|x^{[k]}\right| = |x|^k, \quad k = 1, 2, \dots. \tag{2.74}$$

On the other hand, for any matrix $A \in \mathbf{R}^{n \times n}$, there is an induced matrix $A^{[k]} \in \mathbf{R}^{N_n^k \times N_n^k}$ satisfying

$$A^{[k]}x^{[k]} = (Ax)^{[k]} \quad \forall x \in \mathbf{R}^n. \tag{2.75}$$

It can be shown that the operation defines an algebra homomorphism that preserves the structure of matrix multiplication. In particular, for any $n \times n$ matrices A and B, we have

$$(AB)^{[k]} = A^{[k]}B^{[k]}, \quad k = 1, 2, \dots. \tag{2.76}$$

Integrating properties (2.74), (2.75) and the definition of spectral radius, we can obtain the following lemma.

Lemma 2.58 *For a matrix set $\{A_1, \dots, A_m\}$ and a natural number k, we have*

$$\rho\big(A_1^{[k]}, \dots, A_m^{[k]}\big) = \rho(A_1, \dots, A_m)^k. \tag{2.77}$$

Finally, with the help of the above preparations, we are ready to prove Theorem 2.57.

Proof of Theorem 2.57 Note that

$$\begin{aligned} \rho_S^2\big(A_1^{[d]}, \dots, A_1^{[d]}\big) &= \inf\big\{\gamma : P > 0, \gamma^{2d}P - \big(A_i^{[d]}\big)^T PA_i^{[d]} \geq 0,\ i = 1, \dots, m\big\} \\ &= \inf\big\{\gamma : p(x) > 0,\ \gamma^{2d}p(x) - p\big(A_i^{[d]}\big) \geq 0,\ i = 1, \dots, m\big\} \\ &\geq \big(\rho_S^{2d}(\mathbf{A})\big)^d, \end{aligned} \tag{2.78}$$

where the notation $p(x) = (A_i^{[d]})^T PA_i^{[d]}$ and the fact that $\gamma^{2d}p(x) - p(A_i^{[d]}) \in \mathrm{SOS}_2^{N_n^d}$ have been used. Combining Lemma 2.58 with inequalities (2.73) and (2.78) yields

$$\begin{aligned}\rho(\mathbf{A})^d &= \rho\big(A_1^{[d]}, \ldots, A_m^{[d]}\big)\\ &\geq \frac{1}{\sqrt{N_n^d}} \rho_S^2\big(A_1^{[d]}, \ldots, A_m^{[d]}\big)\\ &\geq \frac{1}{\sqrt{N_n^d}} \big(\rho_S^{2d}(\mathbf{A})\big)^d.\end{aligned}$$

As a result, we have

$$\rho(\mathbf{A}) \geq \frac{\rho_S^{2d}(\mathbf{A})}{\kappa_n^d},$$

which, together with inequality (2.72), leads to inequalities (2.71). The proof of Theorem 2.57 is completed. □

Example 2.59 For the four-dimensional three-form switched system with

$$A_1 = \begin{bmatrix} 0 & 1 & 7 & 4 \\ 1 & 6 & -2 & -3 \\ -1 & -1 & -2 & -6 \\ 3 & 0 & 9 & 1 \end{bmatrix}, \qquad A_2 = \begin{bmatrix} -3 & 3 & 0 & -2 \\ -2 & 1 & 4 & 9 \\ 4 & -3 & 1 & 1 \\ 1 & -5 & -1 & -2 \end{bmatrix},$$

$$A_3 = \begin{bmatrix} 1 & 4 & 5 & 10 \\ 0 & 5 & 1 & -4 \\ 0 & -1 & 4 & 6 \\ -1 & 5 & 0 & 1 \end{bmatrix},$$

it can be calculated that $\rho_S^2 = 9.761$ and $\rho_S^4 = \rho_S^6 = 8.92$. As $\rho(A_1 A_3)^{\frac{1}{2}} = 8.915$, we conclude that the spectral radius is between 8.915 and 8.92. It is clear that ρ_S^4 provides a much closer upper bound for the spectral radius than ρ_S^2.

2.4.2 An Invariant Set Approach

An approach for verifying the stability of the switched system is to find an appropriate piecewise linear common Lyapunov function.

Let us start from an initial polyhedron Λ_0 which is origin-symmetric. An example is the polyhedron with extreme points whose entries are either 1 or -1. Recall that for any 0-symmetric polyhedral set Λ, there is a full column rank matrix F_Λ such that $\Lambda = \{x : |F_\Lambda x|_\infty \leq 1\}$. Another representation is $\Lambda = \{X_\Lambda \alpha : |\alpha|_1 \leq 1\}$, where $X_\Lambda = [x_1, \ldots, x_r]$ is full row rank and $\{\pm x_1, \ldots, \pm x_r\}$ are vertices of Λ. For a 0-symmetric polyhedral set Λ, define the set

$$C(\Lambda) = \{x : A_i x \in \Lambda,\ i = 1, \ldots, m\}.$$

It is clear that

$$C(\Lambda) = \{x : |F_\Lambda A_i x|_\infty \le 1,\ i = 1, \dots, m\},$$

which is also a 0-symmetric polyhedron. Define recursively a set of polyhedra by

$$\Lambda_k = C(\Lambda_{k-1}) \cap \Lambda_0, \quad k = 1, 2, \dots. \tag{2.79}$$

The largest invariant set contained in Λ_0, as defined in (2.50), is the set

$$\Lambda_\infty = \bigcap_{k \in \mathbf{N}_+} \Lambda_k.$$

Note that, if $\Lambda_k = \Lambda_{k+1}$ for some $k \in \mathbf{N}_+$, then $\Lambda_\infty = \Lambda_k$, and the set is tractably computed.

Proposition 2.60 *For any initial polyhedron Λ_0, we have the following statements*:

(1) *If the discrete-time switched linear system* $\mathbf{A}$ *is stable, then, there is a finite number k such that $\Lambda_k = \Lambda_{k-1}$, which means that $\Lambda_k = \Lambda_\infty$. Conversely, if $\Lambda_k = \Lambda_{k-1}$ for some $k < +\infty$, then, $\Lambda_k = \Lambda_\infty$, and the system is stable or marginally stable.*
(2) *If there is a finite number k such that Λ_k is interior to set Λ_0, then $\Lambda_\infty = \{0\}$, and the switched system is unstable.*

Proof For the former statement, let $x(\cdot)$ be a state trajectory of the switched system. The exponential stability implies the existence of a time T with $x(t) \in \Lambda_0$ for $t \ge T$ whenever $x(0) \in \Lambda_0$. For any $k = 0, 1, \dots, T$, Λ_k has the property that $x(0) \in \Lambda_k$ implies that $x(k) \in \Lambda_0$, and visa versa. This means that $\Lambda_{T+1} = \Lambda_T$. Indeed, if this is not true, then, there is a state trajectory $x(\cdot)$ with $x(0) \in \Lambda_T / \Lambda_{T+1}$, which implies that $x(T+1) \notin \Lambda_0$, which is a contradiction. Conversely, if $\Lambda_k = \Lambda_{k-1}$ for some $k < +\infty$, then, it is clear that $\Lambda_k = \Lambda_\infty$ is a polyhedral C-set that is an invariant set for the system. Thus the system is stable or marginally stable.

For the latter statement, we only need to show that $\Lambda_\infty = \{0\}$. As Λ_∞ is an invariant set for the system, it can be seen that $\lambda\Lambda_\infty$ is also an invariant set for any $\lambda \ge 0$. As the set Λ_∞ is interior to the set Λ_0, there is $\lambda > 1$ such that $\lambda\Lambda_\infty \subset \Lambda_0$. This in turn implies that $\lambda\Lambda_\infty \subset \Lambda_k$ for all $k \in \mathbf{N}_+$, which yields $\lambda\Lambda_\infty \subset \Lambda_\infty$. As a result, we have $\Lambda_\infty = \{0\}$. □

It is interesting to notice that $\Lambda_k = \Lambda_\infty$ does not necessarily implies the (exponential) stability. A particular example is the case that $\mathbf{A} = \{I_n\}$, which produces $\Lambda_k = \Lambda_\infty = \Lambda_0$ for any $k \in \mathbf{N}^+$. In fact, if we add the identity matrix to any stable system, we have a marginally stable system which produces the same polyhedral set sequence $\Lambda_0, \Lambda_1, \dots$. To further distinguish between stability and marginal stability, one possible way is to test through the recursive procedure by setting $\mathbf{A} = \{\lambda A_1, \dots, \lambda A_m\}$, where $\lambda > 1$, but $\lambda - 1$ is sufficiently small. If $\Lambda_k = \Lambda_{k-1}$ for a finite k, then, the original system must be stable. Otherwise, the original system is either marginally stable or stable but "nearly marginally stable".

Based on the above discussion, we can develop a computational algorithm that verifies the stability of the discrete-time switched linear system.

Algorithm for calculating the largest invariant set (ACLIS)

Initiation. Set $F^0 := F_{\Lambda_0}$, $X^0 := X_{\Lambda_0}$, $flag := 0$, and $k := 0$. Prespecify a natural number $kmax$.

Step 1. Set $F := F^k$, $X := X^k$, and compute the matrix

$$G = \left[(F_k A_1)^T, \ldots, (F_k A_m)^T\right]^T.$$

Step 2. For each row G^i of G, check if $\|G^i X\|_\infty \leq 1$. If yes, remove the row from matrix G.

Step 3. If $flag = 0$ and G is the vacant matrix, that is, it is 0×0, then output the matrix X and the "stability or marginal stability" message, set $flag := 1$ and go to Step 7. If $flag = 1$ and G is the vacant matrix, then terminate with the message "stability".

Step 4. Set $H := [F^T, G^T]^T$. Compute $\Lambda^{k+1} = \{x \in \mathbf{R}^n : |Hx|_\infty \leq 1\}$ and set $F^{k+1} := F^{k+1}_\Lambda$ and $X^{k+1} := X_{\Lambda^{k+1}}$.

Step 5. If $flag = 0$ and $\|F^0 X^{k+1}\|_\infty < 1$, then, terminate with the message "instability." If $flag = 1$ and $\|F^0 X^{k+1}\|_\infty < 1$, then, terminate with the message "marginal stability or nearly marginal stability."

Step 6. Set $k := k + 1$. If $k \geq kmax$, terminate with message "time is out." Otherwise, go to Step 1.

Step 7. Set a sufficiently small positive number λ, and $A_i = (1 + \lambda)A_i$ for $i = 1, \ldots, m$. Set $F^0 := F$, and go to Step 1.

It can be seen that the algorithm is not efficient as the number of extreme points of polyhedra Λ_k may grow exponentially. Within a recursive loop, the main computation load is in Step 4, which computes the various representations of the new polyhedron. Fortunately, this can be implemented by commercial softwares (e.g. Matlab Geometric Bounding Toolbox [252]).

For continuous-time switched linear systems, it was established in Lemma 2.19 that, if the system is asymptotically stable, then, its *Euler approximating system*

$$x(t+1) = (I_n + \tau A_{\sigma(t)})x(t) \tag{2.80}$$

is also asymptotically stable for sufficiently small τ. A verification procedure can thus been outlined as follows.

Step 1. Choose a sufficiently small positive real number τ.

Step 2. Run the Algorithm ACLIS for system (2.80). If the algorithm terminate with stability, then, set $\Lambda_\infty = \Lambda_k$ and go to Step 4. Otherwise, go to Step 3.

Step 3. Set $\tau = \tau/2$ and go to Step 2.

Step 4. Check if Λ_∞ is positively invariant for the original system. If yes, then terminate with the message "the continuous-time system is stable". Otherwise, go to Step 3.

Note that the fourth step can be conducted with the help of Theorem 2.31. Generally, there is no guarantee that the procedure terminates in a finite time even for stable switched systems. As a matter of fact, the stability verification for continuous-time switched systems is still an open problem for further investigation.

2.5 Notes and References

This chapter introduced the fundamental issues which are relevant to the development of guaranteed stability theory for switched dynamical systems. As a matter of fact, the development of stability theory for switched systems is not isolated. On the contrary, the progress actively interacted with stability and robustness issues for several different system frameworks of various backgrounds, as briefly discussed in Sect. 2.3.1. Indeed, when the switching path is taken as a perturbation variable, the stability of a switched system is in fact robustness against a class of time-varying uncertainties. This can be seen from expression (2.1) where the switching signal $\sigma(t)$ is an unknown time-varying perturbation. A unique feature of the perturbation is that its image set is finite and thus isolated. By taking convex linear combinations as in (2.20) and (2.21), the switched system is naturally connected to the polytopic uncertain system and the relaxed differential inclusion. As these classes of dynamical systems share the same stability properties, it is natural that the stability theory for switched systems has been deeply interacted with that of other system frameworks. To fully understand the major progress in the stability analysis of switched systems, it is important to highlight the various sources of literature from the related disciplines.

The first source of literature is the absolute stability analysis for Lur'e systems. A Lur'e system is a linear plant with a sector-bounded nonlinear output feedback. Specifically, a SISO Lur'e system is mathematically described by

$$\begin{aligned} \dot{x}(t) &= Ax(t) + b\varphi(y), \quad x \in \mathbf{R}^n, \ y \in \mathbf{R}, \\ y(t) &= cx(t), \quad k_1 y^2 \le y\varphi(y) \le k_2 y^2. \end{aligned} \tag{2.81}$$

The problem of absolute stability is to determine the exact bound of k_1 and k_2 that guarantee global asymptotic stability of the system. The research on this problem could be traced back to the 1940s, and the early pioneers were mainly from the Russian applied mathematics community (Lur'e [157], Aizerman [4], Yakubovich [271], etc.). Aizerman [3] conjectured that, if for each $k \in [k_1, k_2]$, the matrix $A + kbc$ is Hurwitz, then the Lur'e system is $[k_1, k_2]$-absolutely stable. This is equivalent to the statement that the stability of each convex combination of $A + k_1bc$ and $A + k_2bc$ implies the stability of the switched linear system $\mathbf{A} = \{A + k_1bc, A + k_2bc\}$. While this conjecture was disproved by counterexamples, it did greatly stimulate the study on the problem of absolute stability. Quadratic Lyapunov functions (sometimes plus an integral of nonlinearity) were sought by Lur'e himself and many other researchers. A quadratic Lyapunov function for the

Lur'e system is in fact a common Lyapunov function for the extreme systems. That is, the existence of a common Lyapunov function for $A + k_1 bc$ and $A + k_2 bc$ implies $[k_1, k_2]$-absolute stability of the Lur'e system. In the 1970–1980s, many researchers realized that the quadratic Lyapunov functions are not universal for absolute stability, and they turned to more general nonquadratic Lyapunov functions. Piecewise quadratic Lyapunov functions were proved to be universal for absolute stability [170–172].

The second source of literature is the boundedness analysis for the infinite products of a set of (complex or real) matrices $\mathbf{A} = \{A_1, \dots, A_m\}$. This topic has been quite active since the 1990s, and the literature can be traced back to the 1950s [260]. It was widely recognized that the joint spectral radius is an index that is closely related to the boundedness of the matrix semigroup [199]. The joint spectral radius and the generalized spectral radius were proved to be equal for any finite set of matrices [25, 69]. Berger and Wang [25] established that a discrete-time switched system is asymptotically stable iff the spectral radius is less than one. As the spectral radius is equal to the least possible induced norms [69], the asymptotic stability is equivalent to the existence of a contractive norm. This norm is in fact a common Lyapunov function for the matrix set. To verify the boundedness, Lagarias and Wang [138] conjectured the existence of a finite k such that the generalized spectral radius $\bar{\rho}(\mathbf{A}) = \bar{\rho}_k(\mathbf{A})^{1/k}$ for any given finite matrix set $\mathbf{A}$. This well-known Finiteness Conjecture was finally disproved [39, 45], which indicates that the exact computation of the spectral radius might be very involved. On the other hand, Brayton and Tong [47, 48] developed constructive procedures for stability verification. The procedures search a piecewise linear common Lyapunov function by recursively approximating its level set. While the computational algorithms are not efficient, the method itself is valuable as it clarifies some useful properties that benefit the forthcoming investigations. Besides the mathematical characteristics, there has been much effort to establish the connections among various notions from the dynamical system's viewpoints. In particular, the notions of vanishing-step(VS)-stability, bounded-variation(BV)-stability, and para-contractility were introduced, and their relationships with the left-convergent-products (LCP) property (all left-infinite products converge) were established [255]. There were also a few works focusing on the more subtle situation that the spectral radius is one, which corresponds to either marginal stability or marginal instability. Among these, the notion of defectivity and its properties were investigated [69, 96].

The third source of literature is the robustness analysis for a class of polytopic linear uncertain systems. When a linear nominal system

$$\dot{x}(t) = A_0 x(t) + B_0 u(t)$$

is perturbed by a time-varying uncertainty with a polyhedral bound

$$d(t) \in \mathrm{co}\big\{A_1 x(t), \dots, A_m x(t)\big\},$$

the perturbed system can be described by

$$\dot{x}(t) = A_0 x(t) + B_0 u(t) + d(t) = A\big(\omega(t)\big) + B_0 u(t),$$

where $\omega(t) \in \{w \in \mathbf{R}^m : w_i \geq 0, \sum_{i=1}^m w_i = 1\}$ and $A(w) = A_0 + \sum_{i=1}^m w_i A_i$. A more general description is

$$\dot{x}(t) = A\big(\omega(t)\big) + B\big(\omega(t)\big)u(t),$$

where the input gain matrix is also perturbed. For this class of systems, the main problems are of various (gain-scheduling, robust) stabilizing design and robustness analysis. Blanchini and his coworkers [30, 32, 34] developed a nonquadratic Lyapunov scheme for stabilizing design of polytopic linear uncertain systems. By developing a Brayton–Tong-like recursive procedure, it was possible to evaluate both the transient performance and the asymptotic behavior of the linear uncertain systems [33].

The fourth source of literature is the stability of differential inclusions. Differential inclusions provide a unified representation of a wide class of dynamical systems. For a differential inclusion

$$\dot{x}(t) \in F\big(x(t)\big)$$

and its relaxed convex system

$$\dot{x}(t) \in \operatorname{co} F\big(x(t)\big),$$

the solution sets for both systems admit a common closure (w.r.t. an appropriate normed functional space) [13, 83]. This implies the fact that the two systems share same stability properties. This observation bridges the stability theories for the switched linear system, the polytopic linear uncertain system (without control input), and the linear convex differential inclusion, as the last one can be seen as the relaxed system for the former two. On the other hand, it has long been established [135, 188] that an asymptotically stable differential inclusion admits a strictly convex and homogeneous Lyapunov function. This paves the way for finding more universal sets of Lyapunov functions. Indeed, as a strictly convex level set can be arbitrarily approximated via a polyhedral set or an intersection of a set of ellipsoids, both sets of piecewise linear and piecewise quadratic functions are universal.

Finally, the literature on switched and hybrid systems has grown rapidly since the 1990s. The common Lyapunov function approach was proposed based on the fact that, if all the subsystems share a common Lyapunov function, then the switched system is stable under arbitrary switching. Much effort was paid to find criteria for the existence of common quadratic Lyapunov functions for switched linear systems. Narendra and Balakrishnan [182] found that, if all linear subsystems are stable with commuting A-matrices, then they share a common quadratic Lyapunov function. The commutation condition in fact implies the simultaneous triangularizability, and for this, the commutation condition can be further relaxed [1, 164]. The commuting criterion was recently extended to switched nonlinear systems [256]. For planar switched linear systems, complete criteria for common quadratic stability were established [53, 204]. There were a few works reporting various converse Lyapunov theorems [64, 159, 168], which in fact can be seen as special cases of the earlier results in different contexts [34, 152, 173].

It should be noticed that, while many early studies focused on one-system framework, more and more researchers took advantage of the tight connections among the schemes [146, 149, 173, 234]. A notable example is that most researchers from various backgrounds realized the limitation of quadratic Lyapunov functions in tackling the stability and robustness problems for linear and quasi-linear systems, which leads to the active research into the nonquadratic Lyapunov approach. Another example is that the constructive criterion for the stability of planar switched linear systems [112] also provides a solution for the absolute stability of planar Lur'e systems [163].

While the above review provides a brief survey on the relevant literature, it only mentioned a small fraction of the existing results, methods, and literature. The reader is referred to [31, 62, 63, 146, 148] and references therein for more details.

In this chapter, we tried to integrate novel ideas, fresh methods, and rigorous results from various schemes into a systematic framework. The richness of the relevant material enables us to highlight the most notable progress within a unified framework.

The notational preliminaries in Sect. 2.1 were adapted from the books [146, 234]. The common Lyapunov function approach presented in Sect. 2.2 is a combination of several works including [31, 152, 224]. In particular, the notion of the common strong Lyapunov function is a mixture of the smooth version [159] and the locally Lipschitz version [30], and the notion of common weak Lyapunov function was taken from [224].

In Sect. 2.3, Lemma 2.11 can be found in [117], and Proposition 2.13 was adapted from [221]. As for Theorem 2.15, the fact of the equivalence between asymptotic stability and exponential stability was reported in [82, 139] for differential inclusions. The equivalence between local attractivity and global exponential stability was established in [8] for switched systems. The context of universal Lyapunov functions in Sect. 2.3.2 was mainly adapted from [173]. The continuous-time version of Propositions 2.21, 2.33, and 2.35 and Lemma 2.34 were adapted from [224], where their discrete-counterparts can be found in [22, 25, 33, 69], respectively. A simplified proof was presented in [156] for Proposition 2.35 in discrete time. Proposition 2.25 was reported in [69] for discrete-time systems and in [223] for continuous-time systems. The important algebraic criterion, Theorem 2.31, was adapted from [32, 173]. The context in Sect. 2.3.5 on triangular systems was adapted from [239]. While conceptually simple, the research on simultaneous triangularization has been quite active, which impacts greatly on the common Lyapunov function approach.

The computational issues presented in Sect. 2.4 highlight some progress in calculating the spectral radius of a matrix set and verifying the stability of the corresponding switched system. It has been revealed that the computation of the joint spectral radius is NP-hard, and the stability verification "$\rho(\mathbf{A}) \leq 1$" is undecidable [41, 251]. This reveals that effective approximating of the joint spectral radius is difficult in general. Nevertheless, much effort has been paid in investigating the approximation issues, and the reader is referred to [7, 37, 95, 158, 190, 249] and the references therein. Proposition 2.46 and Example 2.48 were taken from the thesis [249]. The accuracy of the ellipsoidal norm approximation, Proposition 2.49, could be found in [7, 38]. The approximation by means sum-of-squares homogeneous polynomial Lyapunov functions, which forms the main content of Sect. 2.4.1, was largely borrowed from [190]. The recursive procedure for calculating the largest invariant set contained in a polyhedron, which forms the main content of Sect. 2.4.2, was adapted from [33]. It is also possible to investigate the stability through the largest invariant set containing a polyhedron [47, 48, 193].

Chapter 3
Constrained Switching

3.1 Introduction

In the previous chapter, we addressed the stability issues for switched systems under arbitrary switching. In many practical situations, however, the switching signals are subject to various constraints. For instance, the switching is governed by a random process, that is, the switching times form a random process with known stochastic distribution. Another example is that the switching is autonomous in that a switch occurs when the state enters into a preassigned region. In either case, the set of possible switching signals is only a fraction of the set of arbitrary switching signals, as discussed in the previous chapter. For these constrained switched systems, while the stability criteria for arbitrary switching is still applicable to the stability analysis, the criteria are no doubt too conservative. In this chapter, we present new approaches that lead to less conservative stability analysis for the constrained switched systems. We focus on three types of constrained switching, namely, the random switching, the autonomous switching, and the dwell-time switching.

3.2 Stochastic Stability

3.2.1 Introduction

A stochastic switched system is a hybrid system that consists of a set of deterministic subsystems and a random switching law that coordinates the switching among the subsystems. The random nature of the switching mechanism enables such systems to be widely representative in many real-world processes with abrupt variable structure and/or unpredictable component failure. For the special case where the subsystems are linear and the switching is governed by a Markov process, the switched system is known to be a *jump linear system*. A jump linear system is mathematically described as

$$\dot{x}(t) = A_{\sigma(t)}x(t), \tag{3.1}$$

Z. Sun, S.S. Ge, *Stability Theory of Switched Dynamical Systems*, Communications and Control Engineering,
DOI 10.1007/978-0-85729-256-8_3,

where $x(t) \in \mathbf{R}^n$ is the continuous state, $\sigma(t)$ is a time-homogeneous irreducible Markov stochastic process taking value in the finite discrete state space $M \stackrel{\text{def}}{=} \{1, \ldots, m\}$, and $A_i \in \mathbf{R}^{n\times n}$, $i \in M$, are known real constant matrices. Suppose that $\Psi = (\psi_{ij})_{m\times m}$ is the infinitesimal matrix of the process $\sigma(t)$. Therefore, the *stationary transition probability* is

$$\Pr\bigl(\sigma(t+h) = j|\sigma(t) = i\bigr) = \begin{cases} \psi_{ij}h + o(h), & i \neq j, \\ 1 + \psi_{ii}h + o(h), & i = j, \end{cases} \quad h > 0.$$

3.2.2 Definitions and Preliminaries

Denote the underlying probability space by $(\Omega, \mathcal{F}, \Pr)$, where Ω is the space of elementary events, $\mathcal{F}$ is a σ-algebra, and Pr is a probability measure. Let Ξ be the set of probability measures on M. Let $x_0 \in \mathbf{R}^n$ and $\rho \in \Xi$ be the initial state and initial distribution of $\sigma(t)$, respectively. Given an event $\omega \in \Omega$, the continuous state evolution of system (3.1) is denoted by $\phi(\cdot; x_0, \omega)$ or $x(\cdot)$ in short. The state transition matrix over time interval $[s_1, s_2]$ is denoted by $\Phi(s_2, s_1, \omega)$, which is a random matrix.

Given an initial probability distribution $p_i^0 = \Pr\{\sigma(0) = i\}$ for $i \in M$, the probability distribution $p(t) = [\Pr\{\sigma(t) = 1\}, \ldots, \Pr\{\sigma(t) = m\}]^T$ satisfies

$$\dot{p}(t) = \Psi^T p(t). \tag{3.2}$$

The irreducibility assumption implies the existence of an invariant distribution $\mathbf{p} = [p_1, \ldots, p_m]^T$ that is globally attractive, i.e., $\Psi^T\mathbf{p} = 0$ and $\lim_{t\to+\infty} p(t) = \mathbf{p}$.

Let $0 = t_0 < t_1 < t_2 < \cdots$ be the sequence of switching times. Denote $\tau_{j-1} = t_j - t_{j-1}$ for $j = 1, 2, \ldots$. It is clear that $\tau_0, \tau_1, \tau_2, \ldots$ are successive duration (sojourn) times between jumps. The duration at a discrete state $i \in M$ is exponentially distributed with parameter $-\psi_{ii}$. The corresponding switching index sequence $\sigma(t_0), \sigma(t_1), \sigma(t_2), \ldots$ is said to be the *embedded Markov chain* of the switching signal. This discrete Markov chain is with transition probability

$$\Pr\bigl\{\sigma(t_{k+1}) = j|\sigma(t_k) = i\bigr\} = \begin{cases} -\frac{\psi_{ij}}{\psi_{ii}}, & i \neq j, \\ 0, & i = j. \end{cases}$$

It is clear that the joint sequence $(\tau_j, \sigma(t_j))$, $j = 0, 1, 2, \ldots$, also is a Markov chain. It can be seen that the joint sequence uniquely determines the switching signal, and vice versa. The stationary (invariant) probability distribution for the embedded Markov chain can be computed to be

$$\vartheta = \left[\frac{p_1\psi_{11}}{\sum_{j=1}^m p_j\psi_{jj}}, \ldots, \frac{p_m\psi_{mm}}{\sum_{j=1}^m p_j\psi_{jj}}\right]^T. \tag{3.3}$$

The transition matrix over $[t_0, t_k]$ can thus be expressed by the switching sequence as

$$\Phi(t_k, t_0) = e^{A_{\sigma(t_{k-1})}\tau_{k-1}} \cdots e^{A_{\sigma(t_1)}\tau_1} e^{A_{\sigma(t_0)}\tau_0}. \tag{3.4}$$

Definition 3.1 The jump linear system is said to be

- (*asymptotically*) *mean square stable* if for any initial state x_0 and initial distribution ρ, we have

$$\lim_{t\to+\infty} E_\rho\{|\phi(t; x_0, \omega)|^2\} = 0$$

- *exponentially mean square stable* if for any initial state x_0 and initial distribution ρ, we have

$$E_\rho\{|\phi(t; x_0, \omega)|^2\} \le \beta e^{-\alpha t}|x_0|^2$$

  for some positive real numbers α and β independent of the initial condition
- *stochastically* (*mean square*) *stable* if for any initial state x_0 and initial distribution ρ, we have

$$\int_0^{+\infty} E_\rho\{|\phi(t; x_0, \omega)|^2\}\,dt < +\infty$$

- (*asymptotically*) *almost surely stable* if for any initial state x_0 and initial distribution ρ, we have

$$\Pr\Big\{\lim_{t\to+\infty}|\phi(t; x_0, \omega)| = 0\Big\} = 1$$

Remark 3.2 It is clear that the first three stabilities are moment stabilities, while the last one is sample stability. The mean square stabilities can be extended to δ-moment stabilities (where δ is a positive real number) by simply substituting the square power by δ-power in the definition. Another simple observation is that exponential mean square stability implies both mean square stability and stochastic stability.

3.2.3 Stability Criteria

Recall that a Hermitian matrix $P \in \mathbf{C}^{n\times n}$ is said to be positive definite if $z^* M z > 0$ for all nonzero complex vectors z, where z^* denotes the conjugate transpose of z. Let $P_1, \ldots, P_k$ be a sequence of complex or real matrices. The sequence $P = (P_1, \ldots, P_k)$ is said to be (*semi-*)*positive definite*, denoted $P > 0$ ($P \ge 0$), if each matrix is Hermitian and positive (semi-)definite. For two real $n \times n$ matrices A and B, denote by $A \otimes B$ their Kronecker product, and by $A \oplus B$ their Kronecker sum. Taking $n = 2$ for example, we have

$$A \otimes B = \begin{bmatrix} a_{11}B & a_{12}B \\ a_{21}B & a_{22}B \end{bmatrix} = \begin{bmatrix} a_{11}b_{11} & a_{11}b_{12} & a_{12}b_{11} & a_{12}b_{12} \\ a_{11}b_{21} & a_{11}b_{22} & a_{12}b_{21} & a_{12}b_{22} \\ a_{21}b_{11} & a_{21}b_{12} & a_{22}b_{11} & a_{22}b_{12} \\ a_{21}b_{21} & a_{21}b_{22} & a_{22}b_{21} & a_{22}b_{22} \end{bmatrix}$$

and

$$A \oplus B = A \otimes I_n + I_n \otimes B = \begin{bmatrix} a_{11} + b_{11} & b_{12} & a_{12} & 0 \\ b_{21} & a_{11} + b_{22} & 0 & a_{12} \\ a_{21} & 0 & a_{22} + b_{11} & b_{12} \\ 0 & a_{21} & b_{21} & a_{22} + b_{22} \end{bmatrix}.$$

Note that both $A \otimes B$ and $A \oplus B$ are of $n^2 \times n^2$.

Theorem 3.3 *For the jump linear system, the following statements are equivalent*:

(1) *The system is mean square stable.*
(2) *The system is exponentially mean square stable.*
(3) *The system is stochastically stable.*
(4) *The matrix* $\mathrm{diag}(A_1 \oplus A_1, \dots, A_m \oplus A_m) + \Psi^T \otimes I_{n^2}$ *is Hurwitz.*
(5) *For any real* $S = (S_1, \dots, S_m) > 0$, *the coupled* (*algebraic*) *Lyapunov equations*

$$A_i^T P_i + P_i A_i + \sum_{j=1}^{m} \psi_{ji} P_j = -S_i, \quad i = 1, \dots, m, \tag{3.5}$$

admit a unique solution $P = (P_1, \dots, P_m) > 0$.

To prove the theorem, we need to introduce some auxiliary preliminaries. For a matrix $X \in \mathbf{R}^{n \times n}$, denote its columns by $X_1, \dots, X_n$ and define its *column stacking form* to be $V(X) = [X_1^T, \dots, X_n^T]^T$. Similarly, for a matrix sequence $Y = (Y_1, \dots, Y_k)$ with $Y_i \in \mathbf{R}^{n \times n}$, $i = 1, \dots, k$, define its column stacking form to be $V(Y) = [V^T(Y_1), \dots, V^T(Y_k)]^T$. It is clear that

$$\|V(Y)\|_1 \stackrel{\text{def}}{=} \sum_{i=1}^{k} \|V(Y_i)\|_1 \le n \sum_{i=1}^{k} \|Y_i\|_1. \tag{3.6}$$

For the jump linear system, let

$$Z_i(t) = E\{x(t)x^T(t)1_{\{\sigma(t)=i\}}\}, \quad i = 1, \dots, m,$$
$$Z(t) = (Z_1(t), \dots, Z_m(t)),$$

where $1_{\{\cdot\}}$ is the Dirac measure. It can be seen that

$$\sum_{i=1}^{m} \|Z_i(t)\|_1 \le \sum_{i=1}^{m} E\{x^T(t)x(t)1_{\{\sigma(t)=i\}}\} = E\{x^T(t)x(t)\}. \tag{3.7}$$

Moreover, simple calculation gives

$$\begin{aligned} dZ_i(t) &= \big(A_i Z_i(t) + Z_i(t)A_i^T\big)\,dt + E\big\{x^T(t)x(t)\,d(1_{\{\sigma(t)=i\}})\big\} \\ &= \big(A_i Z_i(t) + Z_i(t)A_i^T\big)\,dt \\ &\quad + \sum_{j=1}^{m} E\big\{E\big\{x^T(t)x(t)1_{\{\sigma(t+dt)=i\}}1_{\{\sigma(t)=j\}}\big\}\big|\mathcal{F}_t\big\} \\ &\quad - E\big\{x^T(t)x(t)1_{\{\sigma(t)=i\}}\,dt\big\} \\ &= \big(A_i Z_i(t) + Z_i(t)A_i^T\big)\,dt \\ &\quad + \sum_{j=1}^{m} \Pr\big(\sigma(t+dt)=i|\sigma(t)=j\big)Z_j(t) - Z_i(t)\,dt, \end{aligned}$$

which further implies that

$$\frac{d}{dt}Z_i(t) = A_i Z_i(t) + Z_i(t)A_i^T + \sum_{j=1}^{m} \psi_{ji} Z_j(t). \tag{3.8}$$

Define the linear mapping

$$\Gamma(P) = \big(\Gamma_1(P), \ldots, \Gamma_m(P)\big),$$

where $P = (P_1, \ldots, P_m)$ is positive definite, and

$$\Gamma_i(P) = A_i P_i + P_i A_i^T + \sum_{j=1}^{m} \psi_{ji} P_j, \quad i = 1, \ldots, m.$$

Similarly, define the linear mapping Υ to be

$$\Upsilon(X_1, \ldots, X_m) = (\Upsilon_1, \ldots, \Upsilon_m)$$

with $\Upsilon_i = A_i^T X_i + X_i A_i + \sum_{j=1}^{m} \psi_{ji} X_j$ for $i = 1, \ldots, m$. It follows that we can rewrite (3.8) in the more compact form

$$\dot{Z}(t) = \Gamma\big(Z(t)\big). \tag{3.9}$$

Finally, define the matrix

$$\Lambda = \operatorname{diag}(A_1 \oplus A_1, \ldots, A_m \oplus A_m) + \Psi^T \otimes I_{n^2}.$$

It can be verified that

$$\begin{aligned} V\big(\Gamma(P)\big) &= \Lambda V(P), \\ V\big(\Upsilon(P)\big) &= \Lambda^T V(P) \end{aligned} \tag{3.10}$$

for any matrix sequence $P = (P_1, \ldots, P_m)$.

Define the exponential of Γ by $e^{\Gamma t}(P) = \sum_{l=0}^{+\infty} \frac{t^l}{l!} \Gamma^l(P)$. The exponential of Υ can be defined in the same way. Then, we have the following lemma.

Lemma 3.4 *For any complex matrix* $P > 0$, $e^{\Gamma t}(P) > 0$ *and* $e^{\Upsilon t}(P) > 0$.

For a proof of the lemma, the reader is referred to [160].

With the above preparations, we are ready to prove the main theorem.

Proof of Theorem 3.3 We proceed to establish (1)→(2), (3)→(4), (4)→(5), (5)→(2), respectively. These relationships, together with the straightforward implications (2)→(1) and (2)→(3), lead to the conclusion.

(1)→(2): Suppose that, for any $\rho \in \Xi$, we have

$$\lim_{t \to +\infty} E_\rho\{\Phi^T(t, 0, \omega)\Phi(t, 0, \omega)\} = 0. \tag{3.11}$$

Let $0 = t_0 < t_1 < \cdots$ be the switching time sequence. It is clear that there exist positive real numbers γ_1 and γ_2 such that

$$\gamma_1 \le \lim_{k \to +\infty} \frac{t_k}{k} \le \gamma_2,$$

which further implies the existence of a sufficiently large number κ_1 such that

$$\frac{\gamma_1}{2} k \le t_k \le 2\gamma_2 k \quad \forall k \ge \kappa_1. \tag{3.12}$$

It follows from (3.11) and (3.12) that, for any initial distribution ρ, there exists a sufficiently large natural number $\kappa_2 \ge \kappa_1$ such that

$$E_\rho\{\Phi^T(t, 0, \omega)\Phi(t, 0, \omega)\} \le \eta < 1 \quad \forall t \ge t_{\kappa_2}, \tag{3.13}$$

where η is a real constant. Let κ be the largest κ_2 when ρ spreads over all the unit vectors. It follows that

$$E_\xi\{\Phi^T(t, 0, \omega)\Phi(t, 0, \omega)\} \le \eta \quad \forall \xi \in \Xi,\ t \ge t_\kappa. \tag{3.14}$$

Simple calculation gives

$$E_\xi\{\Phi^T(t, 0, \omega)\Phi(t, 0, \omega)\} \le \eta^j e^{4\eta\kappa\nu} \le \beta e^{-\alpha t} \quad \forall \xi \tag{3.15}$$

for all $t \in [t_{(j-1)\kappa}, t_{j\kappa}]$ with $j = 1, 2, \ldots$, where $\nu = \max\{\|A_1\|, \ldots, \|A_m\|\}$, $\beta = \eta^{-1} e^{4\eta\kappa\nu}$, and $\alpha = -\frac{\ln \eta}{2\kappa\gamma_2}$. This directly leads to the conclusion that the system is exponentially mean square stable.

(3)→(4): Suppose that the system is stochastically stable. It follows from (3.9) that, for any initial configuration, we have

$$\int_0^{+\infty} \|e^{\Gamma t}(Z(0))\|_1 \, dt = \int_0^{+\infty} \|Z(t)\|_1 \, dt$$

$$\leq \sum_{i=1}^{m} \int_0^{+\infty} \|Z_i(t)\|_1 \, dt$$

$$\leq \sum_{i=1}^{m} \int_0^{+\infty} E\{nx^T(t)x(t)1_{\{\sigma(t)=i\}}\} \, dt$$

$$\leq n \int_0^{+\infty} E\{x^T(t)x(t)\} \, dt < +\infty.$$

Therefore, we have

$$\int_0^{+\infty} \|e^{\Gamma t}(X)\| \, dt < +\infty \tag{3.16}$$

for any real $X = (X_1, \ldots, X_m) > 0$. For complex $X = (X_1, \ldots, X_m) > 0$, it is clear that $X \leq (\lambda_{\max}(X_1)I, \ldots, \lambda_{\max}(X_m)I)$. It follows from Lemma 3.4 that relationship (3.16) also holds for $X = (X_1, \ldots, X_m) > 0$. Furthermore, a complex matrix H can be decomposed into two Hermitian matrices as

$$H = \frac{1}{2}(H + H^*) + \frac{\sqrt{-1}}{2}(\bar{H} + \bar{H}^*),$$

where $\bar{H}$ and H^* denote the complex conjugate and conjugate transpose, respectively, and a Hermitian matrix can be expressed to be a substraction of two positive definite matrices. As a result, we have the decomposition

$$H = H_1 - H_2 + \sqrt{-1}(H_3 - H_4),$$

where $H_1, \ldots, H_4$ are positive definite. This means that (3.16) still holds for any complex $X = (X_1, \ldots, X_m)$ due to the linearity of $e^{\Gamma t}$. On the other hand, it can be routinely verified that

$$\int_0^{+\infty} \|e^{\Lambda t}y\|_1 \, dt \leq \int_0^{+\infty} \|V(e^{\Gamma t}(V^{-1}(y)))\|_1 \, dt$$

$$\leq n^2 \int_0^{+\infty} \|e^{\Gamma t}(V^{-1}(y))\|_1 \, dt \quad \forall y \in \mathbf{R}^{mn^2}, \tag{3.17}$$

where the former inequality comes from (3.10), and the latter from (3.6). Combining (3.16) with (3.17) yields

$$\int_0^{+\infty} \|e^{\Lambda t}y\|_1 \, dt < +\infty \quad \forall y \in \mathbf{R}^{mn^2},$$

which implies that the matrix Λ is Hurwitz.

(4)→(5): Let the mappings V_i^{-1} be the inverse of the ith matrix in V with

$$V(V_1^{-1}(y), \ldots, V_m^{-1}(y)) = y \quad \forall y \in \mathbf{R}^{mn^2}.$$

Define

$$P_i = -V_i^{-1}\big(\Lambda^{-T} V(S)\big), \quad i = 1, \ldots, m, \tag{3.18}$$

and $P = (P_1, \ldots, P_m)$. It is clear that

$$\Lambda^T V(P_1, \ldots, P_m) = -V(S_1, \ldots, S_m).$$

It follows from (3.10) that

$$V\big(\Upsilon(P_1, \ldots, P_m)\big) = -V(S_1, \ldots, S_m),$$

which implies that

$$\Upsilon(P_1, \ldots, P_m) + (S_1, \ldots, S_m) = 0,$$

which is exactly (3.5). Suppose that there is another $P^* = (P_1^*, \ldots, P_m^*)$ satisfying (3.5). Then, it can be seen that

$$\Lambda^T V(P_1, \ldots, P_m) = \Lambda^T V\big(P_1^*, \ldots, P_m^*\big),$$

which means that $V(P) = V(P^*)$ and further $P^* = P$, due to the nonsingularity of Λ. It is clear that

$$\Upsilon\big(P_1^T, \ldots, P_m^T\big) = \big(\Upsilon(P_1, \ldots, P_m)\big)^T = -(S_1, \ldots, S_m).$$

This means that $(P_1^T, \ldots, P_m^T)$ is a solution to (3.5). By the uniqueness of solution, P_i are symmetric for $i = 1, \ldots, m$. Furthermore, utilizing the facts that $\frac{d}{dt}(e^{\Lambda^T t}) = \Lambda^T e^{\Lambda^T t}$ and that Λ^T is Hurwitz, we have

$$\begin{aligned} \int_0^{+\infty} e^{\Lambda^T t} V(S)\, dt &= \Lambda^{-T} \int_0^{+\infty} \frac{d}{dt} e^{\Lambda^T t} V(S)\, dt \\ &= -\Lambda^{-T} V(S) = V(P). \end{aligned} \tag{3.19}$$

On the other hand, it follows from (3.10) that

$$\int_0^{+\infty} e^{\Lambda^T t} V(S)\, dt = \int_0^{+\infty} V\big(e^{\Upsilon t}(S)\big)\, dt = V\bigg(\int_0^{+\infty} e^{\Upsilon t}(S)\, dt\bigg). \tag{3.20}$$

Combining (3.19) with (3.20) gives

$$P = \int_0^{+\infty} e^{\Upsilon t}(S)\, dt \tag{3.21}$$

whose positive definiteness follows from Lemma 3.4.

(5)→ (2): Note that the joint process $\{(x(t), \sigma(t)) : t \geq 0\}$ is a time-homogeneous Markov process with the infinitesimal generator

$$\Theta = \Psi + \operatorname{diag}\bigg(x^T A_1^T \frac{\partial}{\partial x}, \ldots, x^T A_m^T \frac{\partial}{\partial x}\bigg).$$

Define a stochastic Lyapunov function to be

$$V(x,\sigma) = x^T P_\sigma x, \tag{3.22}$$

where $P = (P_1, \ldots, P_m) > 0$ is the solution to (3.5) with, say, $S = (I, \ldots, I)$. Then, we have

$$\begin{aligned}(\Theta V)(x,i) &= \sum_{j=1}^{m} \psi_{ij} V(x,j) + x^T A_i^T \frac{\partial}{\partial} V(x,i) \\ &= x^T \left(\sum_{j=1}^{m} \psi_{ij} P_j + A^T P + PA \right) x = -x^T x.\end{aligned}$$

This implies that

$$(\Theta V)(x,i) \le -\alpha V(x,i) \quad \forall i = 1, \ldots, m,$$

where $\alpha = 1/\max_{i=1}^{m} \lambda_{\max}(P_i) > 0$. It follows that

$$E\{V(x(t),\sigma(t))\} \le e^{-\alpha t} V(x_0,\sigma(0)) \quad \forall t \ge 0,$$

which clearly implies the exponential mean square stability. □

Remark 3.5 The theorem reveals important and rich information about mean square stabilities. First, the stabilities are all equivalent to each other, and thus we do not distinguish them any more. Second, the stability admits Lyapunov characteristics that extend the standard Lyapunov theory for linear systems. Third, the stability is verifiable in terms of the subsystem matrices and the switching transition distribution.

Remark 3.6 From the proof, statement (4) in Theorem 3.3 is further equivalent to either

(4a) Matrix Γ is Hurwitz

or

(4b) Matrix Υ is Hurwitz.

Note that Matrix Λ is of mn^2-dimension. As the Lyapunov operators Γ and Υ are defined over the symmetric matrix group, they are of $\frac{m(n(n+1))}{2}$-dimension, which is approximately half that of the Λ.

Remark 3.7 It can be proven that statement (5) in Theorem 3.3 is equivalent to the following statement

(5a) There is a sequence $P = (P_1, \ldots, P_m) > 0$, such that $\Gamma(P) < 0$.

Besides, statements (5) and (5a) still hold if Γ is replaced by Υ.

Remark 3.8 It is interesting to observe that a mean square stable jump linear system always admits a component-wise quadratic Lyapunov function as in (3.22). In contrast, for a deterministic switched linear system, the existence of such a Lyapunov function is a sufficient condition for guaranteed stability, which may be far from necessity. In this sense, the mean square stability is easier to tackle than the guaranteed stability.

It follows from Theorem 3.3 that any mean square stable system is also structurally mean square stable, as stated in the following corollary.

Corollary 3.9 *For any mean square stable jump linear system* $(\mathbf{A}, \Psi)$, *there exist positive real numbers* ϵ_1 *and* ϵ_2 *such that any perturbed jump linear system* $(\bar{\mathbf{A}}, \bar{\Psi})$ *is mean square stable when* $\|\bar{\mathbf{A}} - \mathbf{A}\| < \epsilon_1$ *and* $\|\bar{\Psi} - \Psi\| < \epsilon_2$.

Example 3.10 For the third-order jump linear system with two subsystems

$$A_1 = \begin{bmatrix} 0 & 0 & -2 \\ 1 & -1 & -2 \\ 1 & 1 & 0 \end{bmatrix}, \qquad A_2 = \begin{bmatrix} 0 & 0 & 1 \\ -1 & -1 & 0 \\ -1 & 1 & 0 \end{bmatrix}$$

and transition probability

$$\Psi = \begin{bmatrix} -1 & 1 \\ 2 & -2 \end{bmatrix},$$

it is clear that the first subsystem is marginally stable while the second is unstable. Routine calculation gives

$$\Lambda = \operatorname{diag}(A_1 \oplus A_1, A_2 \oplus A_2) + \Psi^T I$$

$$= \begin{bmatrix}
-1 & 0 & -2 & 0 & 0 & 0 & -2 & 0 & 0 & 2 & 0 & 0 & 0 & 0 & 0 & 0 & 0 & 0 \\
1 & -2 & -2 & 0 & 0 & 0 & 0 & -2 & 0 & 0 & 2 & 0 & 0 & 0 & 0 & 0 & 0 & 0 \\
1 & 1 & -1 & 0 & 0 & 0 & 0 & 0 & -2 & 0 & 0 & 2 & 0 & 0 & 0 & 0 & 0 & 0 \\
1 & 0 & 0 & -2 & 0 & -2 & -2 & 0 & 0 & 0 & 0 & 0 & 2 & 0 & 0 & 0 & 0 & 0 \\
0 & 1 & 0 & 1 & -3 & -2 & 0 & -2 & 0 & 0 & 0 & 0 & 0 & 2 & 0 & 0 & 0 & 0 \\
0 & 0 & 1 & 1 & 1 & -2 & 0 & 0 & -2 & 0 & 0 & 0 & 0 & 0 & 2 & 0 & 0 & 0 \\
1 & 0 & 0 & 1 & 0 & 0 & -1 & 0 & -2 & 0 & 0 & 0 & 0 & 0 & 0 & 2 & 0 & 0 \\
0 & 1 & 0 & 0 & 1 & 0 & 1 & -2 & -2 & 0 & 0 & 0 & 0 & 0 & 0 & 0 & 2 & 0 \\
0 & 0 & 1 & 0 & 0 & 1 & 1 & 1 & -1 & 0 & 0 & 0 & 0 & 0 & 0 & 0 & 0 & 2 \\
1 & 0 & 0 & 0 & 0 & 0 & 0 & 0 & 0 & -2 & 0 & 1 & 0 & 0 & 0 & 1 & 0 & 0 \\
0 & 1 & 0 & 0 & 0 & 0 & 0 & 0 & 0 & -1 & -3 & 0 & 0 & 0 & 0 & 0 & 1 & 0 \\
0 & 0 & 1 & 0 & 0 & 0 & 0 & 0 & 0 & -1 & 1 & -2 & 0 & 0 & 0 & 0 & 0 & 1 \\
0 & 0 & 0 & 1 & 0 & 0 & 0 & 0 & 0 & -1 & 0 & 0 & -3 & 0 & 1 & 0 & 0 & 0 \\
0 & 0 & 0 & 0 & 1 & 0 & 0 & 0 & 0 & 0 & -1 & 0 & -1 & -4 & 0 & 0 & 0 & 0 \\
0 & 0 & 0 & 0 & 0 & 1 & 0 & 0 & 0 & 0 & 0 & -1 & -1 & 1 & -3 & 0 & 0 & 0 \\
0 & 0 & 0 & 0 & 0 & 0 & 1 & 0 & 0 & -1 & 0 & 0 & 1 & 0 & 0 & -2 & 0 & 1 \\
0 & 0 & 0 & 0 & 0 & 0 & 0 & 1 & 0 & 0 & -1 & 0 & 0 & 1 & 0 & -1 & -3 & 0 \\
0 & 0 & 0 & 0 & 0 & 0 & 0 & 0 & 1 & 0 & 0 & -1 & 0 & 0 & 1 & -1 & 1 & -2
\end{bmatrix},$$

which can be verified to be a Hurwitz matrix. According to Theorem 3.3, the jump linear system is mean square stable. On the other hand, under the natural basis, the

Lyapunov operator $\varUpsilon$ can be computed to be

$$\varUpsilon = \begin{bmatrix} -1 & 0 & -2 & 0 & 0 & 0 & 1 & 0 & 0 & 0 & 0 & 0 \\ 2 & -2 & -2 & 0 & -2 & 0 & 0 & 1 & 0 & 0 & 0 & 0 \\ 2 & 1 & -1 & 0 & 0 & -4 & 0 & 0 & 1 & 0 & 0 & 0 \\ 0 & 1 & 0 & -3 & -2 & 0 & 0 & 0 & 0 & 1 & 0 & 0 \\ 0 & 1 & 1 & 2 & -2 & -4 & 0 & 0 & 0 & 0 & 1 & 0 \\ 0 & 0 & 1 & 0 & 1 & -1 & 0 & 0 & 0 & 0 & 0 & 1 \\ 2 & 0 & 0 & 0 & 0 & 0 & -2 & 0 & 1 & 0 & 0 & 0 \\ 0 & 2 & 0 & 0 & 0 & 0 & -2 & -3 & 0 & 0 & 1 & 0 \\ 0 & 0 & 2 & 0 & 0 & 0 & -2 & 1 & -2 & 0 & 0 & 2 \\ 0 & 0 & 0 & 2 & 0 & 0 & 0 & -1 & 0 & -4 & 0 & 0 \\ 0 & 0 & 0 & 0 & 2 & 0 & 0 & -1 & -1 & 2 & -3 & 0 \\ 0 & 0 & 0 & 0 & 0 & 2 & 0 & 0 & -1 & 0 & 1 & -2 \end{bmatrix},$$

which can also be verified to be Hurwitz. It is clear that $\varLambda$ is of eighteenth order, while $\varUpsilon$ is of twelfth order. Therefore, the latter is more convenient in calculating and expressing. An interesting observation is that all eigenvalues of $\varUpsilon$ are also of $\varLambda$, and they admit a common largest real part. Another interesting point is to calculate the transition probability set that makes the jump linear system stable. For this, define the region

$$\Omega = \left\{ \begin{bmatrix} \alpha \\ \beta \end{bmatrix} \in \mathbf{R}_+^2 : \Psi = \begin{bmatrix} -\alpha & \alpha \\ \beta & -\beta \end{bmatrix}, (A_1, A_2, \Psi) \text{ is mean square stable} \right\}.$$

Figure 3.1 depicts the region within $[0, 20] \times [0, 20]$. It is a little surprising that, though neither subsystem is exponentially stable, the region of stable transition probabilities looks larger than unstable transition probability region.

The next theorem shows that the mean square stability implies almost sure stability.

Theorem 3.11 *Any mean square stable jump linear system is almost surely stable.*

Proof Suppose that the system is exponentially mean square stable. Then, there are positive real numbers α and β such that

$$E_\rho\left\{ \left\| \Phi(t, 0, \omega) \right\|^2 \right\} \le \beta e^{-\alpha t}.$$

Let $\eta = \limsup_{k \to +\infty} \|\Phi(k, 0, \omega)\|$. It can be seen that, for any $\epsilon > 0$, we have

$$\begin{aligned} \Pr\{\eta > \epsilon\} &= \Pr\left\{ \bigcap_{j \ge 0} \bigcup_{k \ge j} \left\{ \left\| \Phi(k, 0, \omega) \right\| > \epsilon \right\} \right\} \\ &\le \lim_{j \to +\infty} \Pr\left\{ \bigcup_{k \ge j} \left\{ \left\| \Phi(k, 0, \omega) \right\| > \epsilon \right\} \right\} \end{aligned}$$

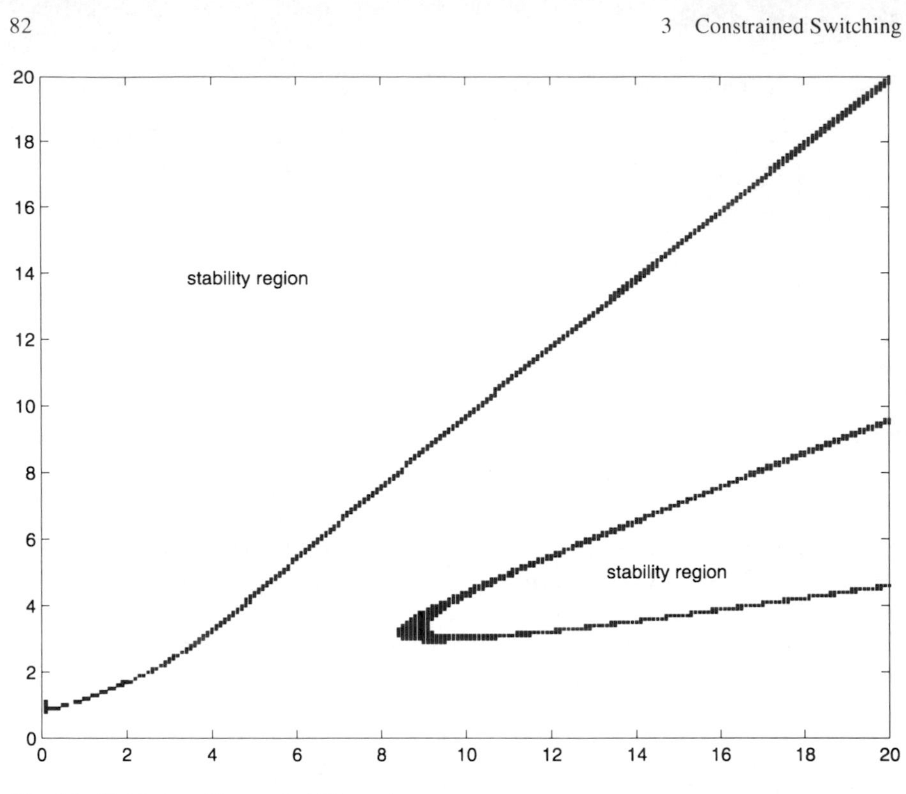

Fig. 3.1 Stable transition probability region

$$
\begin{aligned}
&\leq \lim_{j\to+\infty}\sum_{k=j}^{+\infty}\Pr\bigl\{\bigl\|\Phi(k,0,\omega)\bigr\|>\epsilon\bigr\} \\
&\leq \lim_{j\to+\infty}\sum_{k=j}^{+\infty}\frac{1}{\epsilon^2}E_\rho\bigl\{\bigl\|\Phi(k,0,\omega)\bigr\|^2\bigr\} \\
&\leq \frac{1}{\epsilon^2}\lim_{j\to+\infty}\sum_{k=j}^{+\infty}\beta e^{-\alpha k}=0.
\end{aligned}
$$

As a result, $\Pr\{\eta=0\}=1$, and the theorem follows. □

According to the theorem, almost sure stability is weaker than mean square stability. Therefore, the mean square stability criteria in Theorem 3.3 provide sufficient conditions for almost sure stability. In general, almost sure stability does not imply mean square stability, as illustrated by the following example.

Example 3.12 Let us examine the first-order jump linear system with two subsystems and transition probability matrix as

$$A_1 = a_1, \qquad A_2 = a_2, \qquad \Psi = \begin{bmatrix} -1 & 1 \\ 1 & -1 \end{bmatrix}.$$

First, applying Theorem 3.3 yields the following mean square stability conditions:

$$a_1 + a_2 < 1, \qquad a_1 + a_2 < 2a_1a_2.$$

The stability region in terms of a_1 and a_2 is depicted in Fig. 3.2.

Next, note that, for almost sure stability, the unique stationary distribution is $(\frac{1}{2}, \frac{1}{2})$. The state solution can be expressed by

$$\phi(t, x_0, \omega) = e^{a_1 t_1 + a_2 t_2} x_0,$$

where t_1 and t_2 are the lengths of durations over $[0, t]$ at the first and second subsystems, respectively. Applying the law of large numbers, we have

$$\lim_{t \to +\infty} \frac{t_1}{t} = \lim_{t \to +\infty} \frac{t_2}{t} = \frac{1}{2} \quad \text{a.s.}$$

It follows that

$$\lim_{t \to +\infty} \left|\phi(t, x_0, \omega)\right| = 0 \quad \text{a.s.}$$

when $a_1 + a_2 < 0$. When $a_1 + a_2 = 0$, it can be shown that

$$\Pr\Big\{\lim_{\tau \to +\infty} \sup_{t \geq \tau} \left|\phi(t, t_0, \omega)\right| = 0\Big\} > 0.$$

The above reasonale shows that the system is almost surely stable iff $a_1 + a_2 < 0$. It is clear from Fig. 3.2 that the stability region is larger than that of mean square stability.

The example exhibits that mean square stability can be more conservative than almost sure stability, even for scalar systems. In practical applications, almost sure stability is more interesting as it means that the sample state trajectory is convergent almost surely. For scalar systems, using similar techniques as in the example, it can be proven that almost sure stability is equivalent to

$$p_1 A_1 + p_2 A_2 + \cdots + p_m A_m < 0,$$

where $[p_1, \ldots, p_m]^T$ is an invariant distribution. As a result, the linear convex combination w.r.t. the stationary distribution is stable. Recall that the condition is also equivalent to the existence of a (high-frequency) periodic (deterministic) switching signal that drives the system stable in the deterministic sense. This means that, for scalar switched systems, guaranteed stability is equivalent to almost sure stability w.r.t. arbitrary probability transition. For higher-order systems, the above equivalences still hold when the subsystems are simultaneously triangularizable, that is, the subsystem matrices are upper triangular w.r.t. a common coordinate change. For more discussion on this issue, the reader is referred to [70, 239].

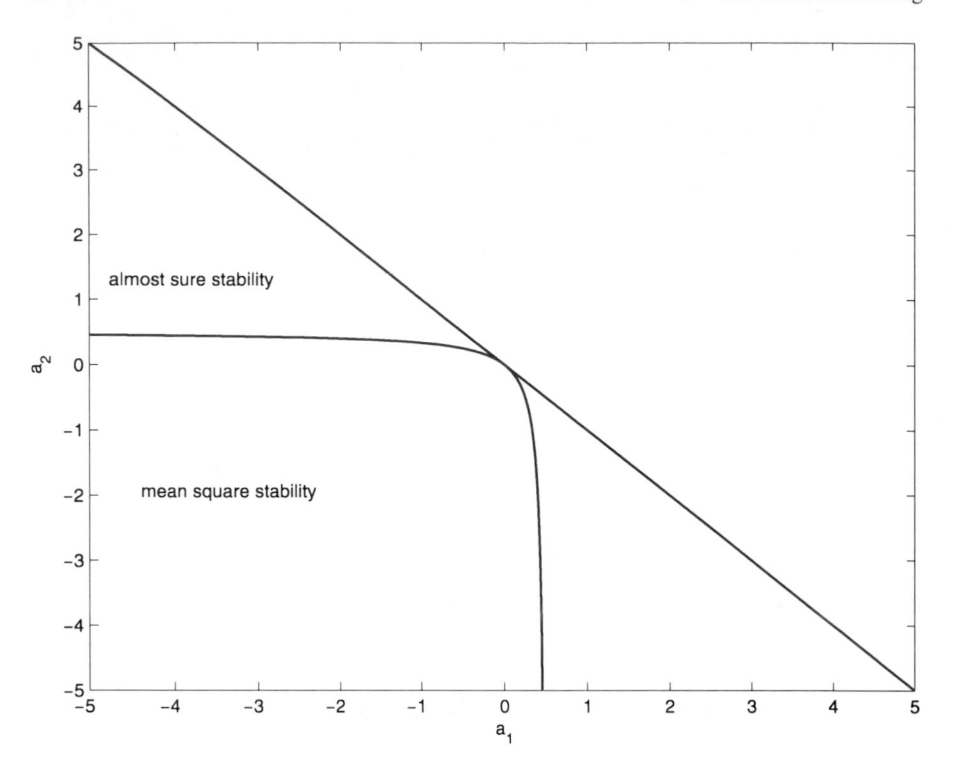

Fig. 3.2 Almost sure stability region vs mean square stability region

To further address almost sure stability, we need to introduce the concept of *top Lyapunov exponent*, which is well known in the mathematics and control literature.

Definition 3.13 For any $\rho \in \Xi$, the *top Lyapunov exponent* is defined to be

$$\lambda_\rho(\omega) = \limsup_{t\to+\infty,|x|=1} \frac{\ln|\phi(t;x,\omega)|}{t}. \tag{3.23}$$

It is clear that the top Lyapunov exponent is well defined and bounded:

$$-\max\{\|A_1\|,\dots,\|A_m\|\} \le \lambda_\rho(\omega) \le \max\{\|A_1\|,\dots,\|A_m\|\}.$$

It captures the rate of divergence (or convergence) of the state trajectory. To be more precise, for any positive real ϵ, there is a positive real δ such that

$$|\phi(t;x,\omega)| \le \delta e^{(\lambda_\rho(\omega)+\epsilon)t}|x| \quad \forall t \ge 0,\ x \in \mathbf{R}^n.$$

While the top Lyapunov exponent is a random variable, it is proved to be almost surely a constant [189], as summarized in the following lemma.

Lemma 3.14 (1) *For any* $\rho \in \Xi$,

$$\lambda_\rho = \lim_{t\to+\infty} E_{\mathbf{p}}\{\ln \|\Phi(t,0,\omega)\|\} = \bar{\lambda} \quad a.s.$$

(2)
$$\bar{\lambda} = \frac{1}{d}\lim_{k\to+\infty}\frac{1}{k}E_\vartheta\{\ln\|\Phi(t_k,0)\|\},$$

where $0 = t_0 < t_1 < \cdots$ *is the switching time sequence with respect to the invariant transition* $\mathbf{p}$, $d = \lim_{k\to+\infty}\frac{t_k}{k}$, *and* ϑ *and* $\Phi(t_k,0)$ *are as in* (3.3) *and* (3.4), *respectively.*

Definition 3.15 The jump linear system is said to be *exponentially almost surely stable* if there exists a positive real number μ such that for any initial state with unit norm and initial distribution of $\sigma(t)$, we have

$$\Pr\left\{\limsup_{t\to+\infty}\frac{\ln|x(t)|}{t} \le -\mu\right\} = 1.$$

It is clear that exponential almost sure stability implies (asymptotic) almost sure stability. It is not clear whether the converse is true. That is, whether asymptotic almost sure stability implies exponential almost sure stability or not is still an open problem. Also it can be seen that exponential almost sure stability is equivalent to $\bar{\lambda} < 0$.

The following theorem presents a necessary and sufficient condition for exponential almost sure stability with the help of Lemma 3.14.

Theorem 3.16 *The jump linear system is exponentially almost surely stable if and only if* $E_\vartheta\{\|\Phi(t_N,0)\|\} < 1$ *for some natural number* N.

Proof Suppose that the system is exponentially almost surely stable. Then, the top Lyapunov exponent $\bar{\lambda} < 0$. It follows from Lemma 3.14 that

$$\lim_{k\to+\infty} E_\vartheta\{\ln\|\Phi(t_k,0)\|\} < 0.$$

As a result, there is a natural number N such that $E_\vartheta\{\|\Phi(t_N,0)\|\} < 1$.

Conversely, the existence of such N implies that

$$\psi \stackrel{\text{def}}{=} \frac{1}{N}E_\vartheta\{\ln\|\Phi(t_N,0)\|\} < 0.$$

For an arbitrarily given natural number k, there exist integers μ_1 and μ_2 with $0 \le \mu_2 \le N-1$ such that $k = \mu_1 N + \mu_2$. We have

$$\begin{aligned}\frac{1}{k}E_\vartheta\{\ln\|\Phi(t_k,0)\|\} \le{}& \frac{1}{\mu_1 N+\mu_2}E_\vartheta\{\ln\|\Phi(t_{\mu_1 N+\mu_2}, t_{\mu_1 N})\|\}\\ &+ \frac{1}{\mu_1 N+\mu_2}E_\vartheta\left\{\sum_{j=1}^{\mu_1}\ln\|\Phi(t_{jN}, t_{(j-1)N})\|\right\}\end{aligned}$$

$$= \frac{1}{\mu_1 N + \mu_2} E_\vartheta \left\{ \ln \left\| \Phi(t_{\mu_1 N + \mu_2}, t_{\mu_1 N}) \right\| \right\}$$
$$+ \frac{\mu_1 \psi}{\mu_1 N + \mu_2},$$

where the last equality follows from stationarity. It can be seen that, as $\mu_1 \to +\infty$,

$$\bar{\lambda} \leq \frac{\psi}{dN} < 0, \tag{3.24}$$

which implies that the system is exponentially almost surely stable. □

Remark 3.17 While the theorem is not generally verifiable, it reveals an essential property for exponential almost sure stability. Indeed, a system is exponentially almost surely stable only if the state transition matrix is norm contractive within a finite number of switches. On the other hand, the norm contractility over a period implies that the transition matrix is also contractive over any other period with the same length. In this way, we can further estimate the rate of convergence in terms of ψ, N, and d, as showed in (3.24).

Remark 3.18 It is interesting to compare the criterion with [216, Prop. 3], which provides a necessary and sufficient condition for consistent stabilizability for deterministic switched systems. By the proposition, for the deterministic switched linear system, the state transition matrix is norm contractive within a finite number of switches iff the system is exponentially stable under a periodic switching signal. Theorem 3.16 provides a stochastic counterpart for this. However, it should be noted that the computation of the stochastic transition matrix is much harder than that of the deterministic case.

Suppose that $E_\vartheta\{\|\Phi(t_N, 0)\|\} < 1$ as in Theorem 3.16. Then, by the continuous dependence of the parameters, the contraction relationship still holds true even if the parameters are slightly perturbed. Hence we have the following robustness property.

Corollary 3.19 *For an exponentially almost surely stable jump linear system* $(\mathbf{A}, \Psi)$, *there exist positive real numbers* ϵ_1 *and* ϵ_2 *such that any perturbed jump linear system* $(\bar{\mathbf{A}}, \bar{\Psi})$ *is exponentially almost surely stable when* $\|\bar{\mathbf{A}} - \mathbf{A}\| < \epsilon_1$ *and* $\|\bar{\Psi} - \Psi\| < \epsilon_2$.

3.3 Piecewise Linear Systems

3.3.1 Introduction

Piecewise linear systems are switched linear systems with autonomous switching. Precisely, suppose that $\Omega_1, \ldots, \Omega_m$ form a nondegenerate polytopic partition of the state space, that is, each region Ω_i is a (convex) polyhedron with nonempty interior, $\bigcup_{i=1}^{m} \Omega_i = \mathbf{R}^n$, and $\Omega_i \cap \Omega_j^o = \emptyset$ for $i \neq j$, where Ω^o denotes the interior of Ω w.r.t. $\mathbf{R}^n$. Then, a piecewise linear system is mathematically described by

$$\dot{x}(t) = A_i x(t) + a_i, \quad x(t) \in \Omega_i \tag{3.25}$$

in continuous time and by

$$x(t+1) = A_i x(t) + a_i, \quad x(t) \in \Omega_i \tag{3.26}$$

in discrete time, where $A_i \in \mathbf{R}^{n \times n}$, and $a_i \in \mathbf{R}^n$. The systems are known as piecewise affine systems in the literature due to the existence of the affine terms. When the affine terms vanish, the system is said to be a piecewise linear system. Here we abuse the notation as a piecewise system can be converted into a piecewise linear system via expanding the system dimension by one. Indeed, by letting

$$\bar{x} = \begin{bmatrix} x \\ 1 \end{bmatrix}, \quad \bar{A}_i = \begin{bmatrix} A_i & a_i \\ 0 & c \end{bmatrix}, \quad i = 1, \ldots, m,$$

where $c = 0$ in continuous time, and $c = 1$ in discrete time, we can reexpress the system as

$$\bar{x}^+(t) = \bar{A}_i \bar{x}(t), \quad x(t) \in \Omega_i, \tag{3.27}$$

which is piecewise linear.

It is clear that the evolution of the continuous state relies only on the initial configuration and is denoted by $\phi(t; t_0, x_0)$.

Piecewise linear systems are very important in representing and approximating many practical systems. First, it is a common practice that, for effectively analyzing and controlling a complex nonlinear dynamical system, the system is linearized around certain operating regions, and the linearized mode is used to approximate the system dynamics in the corresponding region. In this way, the original system is approximately represented by a piecewise linear system as a whole. Second, some well-known phenomena, for instance, saturations, relays, dead zones, etc., are all of piecewise linear forms. Third, when a linear system is controlled by state-feedback switching controllers, the overall system is exactly a piecewise linear system.

A well-known problem of piecewise linear system is the well-posedness problem caused by possible right-hand discontinuities of the system equation when the state crosses the boundaries. As a result, chattering may occur along the boundaries, which implies that a solution in the standard sense may not always exist. The following example exhibits this.

Example 3.20 For the planar two-form piecewise linear system

$$\dot{x} = \begin{cases} \left[\begin{smallmatrix} 0 & -1 \\ 0 & -1 \end{smallmatrix}\right] x & \text{when } [1\ 0]x \geq 0, \\ \left[\begin{smallmatrix} 0 & 1 \\ 0 & -1 \end{smallmatrix}\right] x & \text{when } [1\ 0]x < 0, \end{cases} \tag{3.28}$$

it can be seen that the semi-line

$$\Omega_+ = \{[0, y]^T : y > 0\}$$

is a (stable) sliding mode. Therefore, for any initial state lies on the semi-line, the system does not admit a solution in the standard sense.

The ill-posedness illustrates that piecewise linear systems might exhibit complex system behaviors. To address this issue, more general concepts of solutions, such as the well-known Filippov and Caratheodory definitions, should be exploited. For the above example, it can be seen that the system is always well-posed by Filippov's definition, that is, it admits a unique Filippov solution over $[0, +\infty)$ for any initial state. However, it can be proven that the system is not well-posed by Caratheodory's definition. Furthermore, if we replace A_1 in (3.28) by $\begin{bmatrix} 0 & -1 \\ 0 & 1 \end{bmatrix}$, then the system is ill-posed by either definition, due to the fact that it admits no Caratheodory solution for initial states on the semi-line Ω_+, and it admits more than one (in fact, infinitely many) Filippov solution on Ω_+.

In this section, we present several approaches for analyzing stability of general piecewise linear systems, including the piecewise quadratic Lyapunov approach, the surface Lyapunov approach, and the transition graph approach. In the last subsection we focus on stability analysis of piecewise conic systems. To avoid ill-posedness, we assume that the right-hand side is always continuous over the boundaries. Under this assumption, we further assume without loss of generality that partition cells Ω_i are closed polyhedra. To guarantee that the origin is an equilibrium, we assume that $a_i = 0$ when $0 \in \Omega_i$. For notational convenience, we divide the index set M into M_1 and M_2 such that $i \in M_1$ iff $0 \in \Omega_i$.

3.3.2 *Piecewise Quadratic Lyapunov Function Approach*

Note that, piecewise linear systems with origin equilibrium are locally (around the origin) piecewise linear with autonomous switching. As a result, any sufficient condition for guaranteed stability is also sufficient for the autonomous stability. In particular, if the subsystems admit a common quadratic Lyapunov function, then the system is autonomously stable. While many stability criteria were given by exploiting this idea, the criteria are doomed to be conservative in general. A less conservative idea is to exploit the piecewise quadratic Lyapunov functions, which result in bilinear matrix inequalities or even linear matrix inequalities that are efficiently solvable numerically, as discussed below.

A piecewise quadratic Lyapunov function for the piecewise linear system is of the form

$$V(x) = x^T P_i x + 2q_i x + r_i = \bar{x}^T \bar{P} \bar{x}, \quad x \in \Omega_i, \tag{3.29}$$

where $\bar{x} = \begin{bmatrix} x \\ 1 \end{bmatrix}$, $\bar{P}_i = \begin{bmatrix} P_i & q_i^T \\ q_i & r_i \end{bmatrix}$. When $0 \in \Omega_i$, we require that $q_i = 0$ and $r_i = 0$, which means that $V(x) = x^T P_i x$ for $x \in \Omega_i$. For consistency, we also redefine $\bar{x}$ to be $[x^T, 0]^T$ when $i \in M_1$. To apply the Lyapunov approach, we need to figure out the continuity of the function over the cell boundaries and the definite positiveness of the function.

Fix a natural number k. Let $\bar{F}_i \in \mathbf{R}^{k\times(n+1)}$, $i = 1, \ldots, m$, be a sequence of matrices. The matrix sequence is said to be a *continuity matrix sequence* w.r.t. the partition sequence $\Omega_1, \ldots, \Omega_m$ if

$$\bar{F}_i \bar{x} = \bar{F}_j \bar{x} \quad \forall x \in \Omega_i \cap \Omega_j,\ i \neq j.$$

Similarly, let $\bar{E}_i \in \mathbf{R}^{k\times(n+1)}$, $i = 1, \ldots, m$ be a sequence of matrices. The matrix sequence is said to be *polyhedral cell bounding* w.r.t. the partition sequence $\Omega_1, \ldots, \Omega_m$ if

$$\bar{E}_i \bar{x} \succeq 0 \quad \forall x \in \Omega_i.$$

Lemma 3.21 *Suppose that $\bar{F}_i$, $i = 1, \ldots, m$, is a continuity matrix sequence. Then, for any symmetric matrix $T \in \mathbf{R}^{k\times k}$, the piecewise quadratic scalar function V in* (3.29) *with $\bar{P}_i = \bar{F}_i^T T \bar{F}_i$ is continuous everywhere. Moreover, there exist real numbers β_1 and β_2 such that*

$$\beta_1 x^T x \leq V(x) \leq \beta_2 x^T x.$$

Proof The continuity of V follows straightforwardly from the continuity of $\bar{F}_i x$ over the cell boundaries. For the latter part, note that

$$\lambda_{\min}(\bar{P}_i) x^T x \leq V(x) \leq \lambda_{\max}(\bar{P}_i) x^T x \quad \forall x \in \Omega_i$$

when $i \in M_1$ and

$$\lambda_{\min}(\bar{P}_i) \varpi_1 x^T x \leq V(x) \leq \lambda_{\max}(\bar{P}_i) \varpi_2 x^T x \quad \forall x \in \Omega_i$$

when $i \in M_2$, where $\varpi_1 = 1 + \frac{1}{\min\{x^T x: x\in\Omega_i\}}$, and $\varpi_2 = 1 + \frac{\min\{\mathrm{sgn}(\lambda_{\min}(\bar{P}_i)), 0\}}{\min\{x^T x: x\in\Omega_i\}}$. This clearly leads to the conclusion due to the finiteness of the index set. □

Lemma 3.22 *Let $\bar{E}_i = [E_i, e_i]$, $i = 1, \ldots, m$, be a polyhedral cell bounding matrix sequence. Suppose that symmetric matrices $W_1, \ldots, W_m \in \mathbf{R}_+^{k\times k}$ satisfy*

$$P_i - E_i^T W_i E_i > 0, \quad i \in M_1,$$
$$\bar{P}_i - \bar{E}_i^T W_i \bar{E}_i > 0, \quad i \in M_2.$$

Then, there is $\alpha > 0$ such that the piecewise quadratic function V satisfies

$$V(x) \geq \alpha x^T x \quad \forall x \in \mathbf{R}^n.$$

Proof Note that

$$x^T P_i x > 0 \quad \forall x \in \Omega_i,\ x \neq 0$$

for $i \in M_1$. Clearly the function $\frac{x^T P_i x}{x^T x}$ is radially constant and continuous everywhere in Ω_i except possibly at the origin. Due to the closedness of cells, we have

$$\min_{x \in \Omega_i}^{x \neq 0} \frac{x^T P_i x}{x^T x} > 0, \quad i \in M_1.$$

The same argument also holds for $i \in M_2$. The conclusion follows from the finiteness of the index set. □

With the help of the above lemmas, we are ready to prove the main result on piecewise quadratic stability.

Theorem 3.23 *Let k be a natural number, $\bar{E}_i = [E_i, e_i]$ and $\bar{F}_i = [F_i, f_i]$, $i = 1, \ldots, m$, be polyhedral cell bounding matrix sequence and continuity matrix sequence, respectively, and T, U_i, W_i be symmetric matrices with $T \in \mathbf{R}^{k \times k}$, $U_i, W_i \in \mathbf{R}_+^{k \times k}$, $i = 1, \ldots, m$. Suppose that*

$$P_i = F_i^T T F_i, \quad i \in M_1,$$
$$\bar{P}_i = \bar{F}_i^T T \bar{F}_i, \quad i \in M_2,$$

satisfy

$$\begin{aligned} & A_i^T P_i + P_i A_i + E_i^T U_i E_i < 0, \\ & P_i - E_i^T W_i E_i > 0, \end{aligned} \qquad i \in M_1, \tag{3.30}$$

and

$$\begin{aligned} & \bar{A}_i^T \bar{P}_i + \bar{P}_i \bar{A}_i + \bar{E}_i^T U_i \bar{E}_i < 0, \\ & \bar{P}_i - \bar{E}_i^T W_i \bar{E}_i > 0, \end{aligned} \qquad i \in M_2. \tag{3.31}$$

Then, the piecewise linear system is globally exponentially stable.

Proof It follows straightforwardly from Lemmas 3.21 and 3.22 that the function V is continuous everywhere, $\alpha x^T x \leq V(x) \leq \beta x^T x$ for some positive real numbers α and β, and $\dot{V} \leq -\gamma x^T x$ almost everywhere for some $\gamma > 0$. As a result, the system is globally exponentially stable. □

Remark 3.24 The main advantages of the piecewise quadratic Lyapunov function approach include: (1) the criterion is much less conservative than the existence of a common quadratic Lyapunov function; (2) the searching of piecewise quadratic Lyapunov function is reduced to a set of linear matrix inequalities (LMIs), which admits efficient numerical verification [46]; and (3) software packages are available for the searching of a piecewise quadratic Lyapunov function numerically, see, e.g., [100]. However, to find qualified piecewise quadratic Lyapunov functions, we usually need to further partition the cells, which makes the computation inefficient for higher-dimensional systems.

Example 3.25 For the two-form piecewise linear system

$$\dot{x} = \begin{cases} A_1 x & \text{when } x_1 x_2 \geq 0, \\ A_2 x & \text{when } x_1 x_2 < 0, \end{cases}$$

with

$$A_1 = \begin{bmatrix} -4.6 & 5.5 \\ -5.5 & 4.4 \end{bmatrix} \qquad A_2 = \begin{bmatrix} 4.4 & 5.5 \\ -5.5 & -4.6 \end{bmatrix},$$

it can be verified that both subsystems are stable. While the cones $\{x : x_1 x_2 \geq 0\}$ and $\{x : x_1 x_2 \leq 0\}$ are not convex, each is the union of two quadrants. Therefore, let $A_3 = A_1$, $A_4 = A_2$, and let Ω_i be the ith quadrant for $i = 1, \ldots, 4$. Rewrite the piecewise linear system as

$$\dot{x} = A_i x, \qquad x \in \Omega_i. \tag{3.32}$$

As all the subsystem phase portraits are rotating clockwise, the system is always well defined, though the right side is not continuous over the boundaries. To apply Theorem 3.23, choose the continuity matrix sequence to be

$$F_1 = -F_3 = I_2, \qquad F_2 = -F_4 = \begin{bmatrix} -1 & 0 \\ 0 & 1 \end{bmatrix},$$

and polyhedral cell bounding matrix sequence to be

$$E_i = \begin{bmatrix} F_i \\ I_2 \end{bmatrix}, \qquad i = 1, \ldots, 4.$$

Furthermore, let

$$T = \begin{bmatrix} 1.1 & -0.9 & 0 & 0 \\ -0.9 & 1.1 & 0 & 0 \\ 0 & 0 & 0 & 0 \\ 0 & 0 & 0 & 0 \end{bmatrix}, \qquad U_i = 0, \qquad W_i = 0, \quad i = 1, \ldots, 4.$$

Then, it can be verified that the conditions of Theorem 3.23 are satisfied. As a result, the piecewise linear system is exponentially stable, and a piecewise quadratic Lyapunov function is

$$V(x) = x^T P_i x, \qquad x \in \Omega_i,$$

with $P_1 = P_3 = \left[\begin{smallmatrix} 1.1 & -0.9 \\ -0.9 & 1.1 \end{smallmatrix}\right]$ and $P_2 = P_4 = \left[\begin{smallmatrix} 1.1 & 0.9 \\ 0.9 & 1.1 \end{smallmatrix}\right]$.

Figure 3.3 depicts the phase portrait of the state trajectory and the level set of the Lyapunov function. The flower-like trajectory indicates that there does not exist a quadratic Lyapunov function whose level sets are invariant. In fact, it can be verified that the convex combination $0.4A_1 + 0.6A_2^{-1}$ has an eigenvalue with positive real part. As a result, the switched system does not admit any common quadratic Lyapunov function [205].

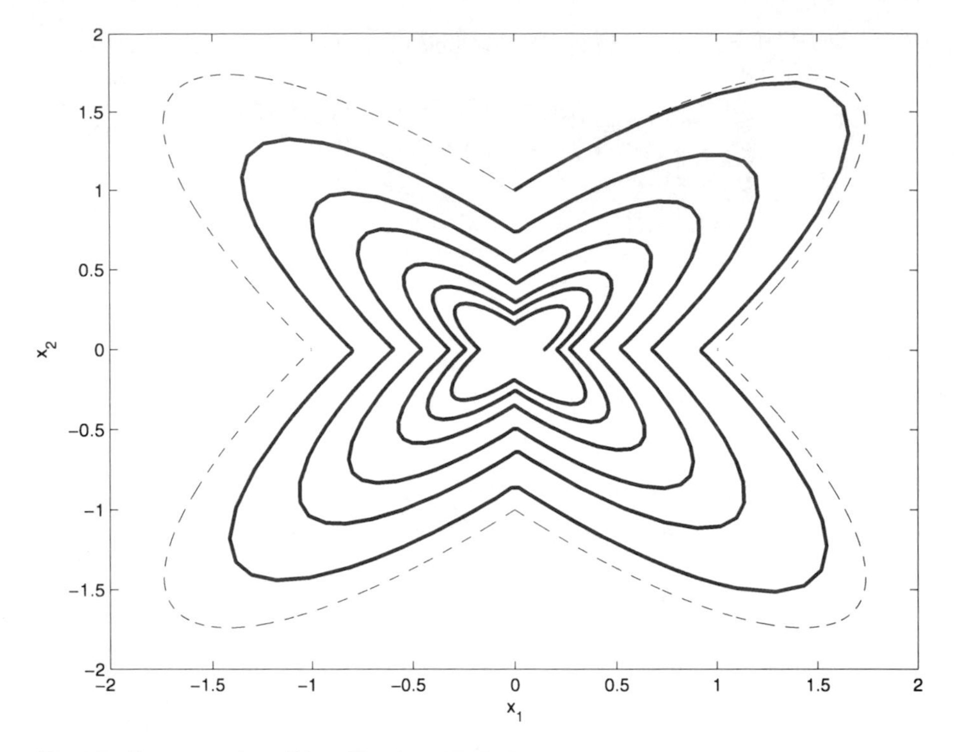

Fig. 3.3 Phase portrait (*solid*) and level set (*dashed*)

3.3.3 Surface Lyapunov Approach

While the piecewise quadratic Lyapunov approach introduced in the previous subsection provides much less conservative criteria than the conventional quadratic Lyapunov approach, the approach may admit very heavy computational load. Indeed, for many piecewise linear systems, it is necessary to seek more refined partitions (than the natural partitions) so that a valid piecewise quadratic function could be found. For continuous-time piecewise linear systems, any state trajectory has to cross cell boundaries or otherwise stay within a cell forever. It is thus heuristic to analyze the stability through the system dynamics across the switching surfaces. This is clearly illustrated by the following example.

Example 3.26 For the continuous-time two-form piecewise linear system

$$\dot{x} = \begin{cases} \begin{bmatrix} 0.9 & 1 \\ -1 & -1 \end{bmatrix} x & \text{if } x_1 \geq 0, \\ \begin{bmatrix} 1.1 & 3 \\ -1 & -1 \end{bmatrix} x & \text{if } x_1 \leq 0, \end{cases} \tag{3.33}$$

it is clear that the first subsystem is unstable. Therefore, the piecewise system does not admit any global Lyapunov function, nor it admits any piecewise quadratic Lyapunov function w.r.t. the natural cell partition (cf. [100]). However, the system is

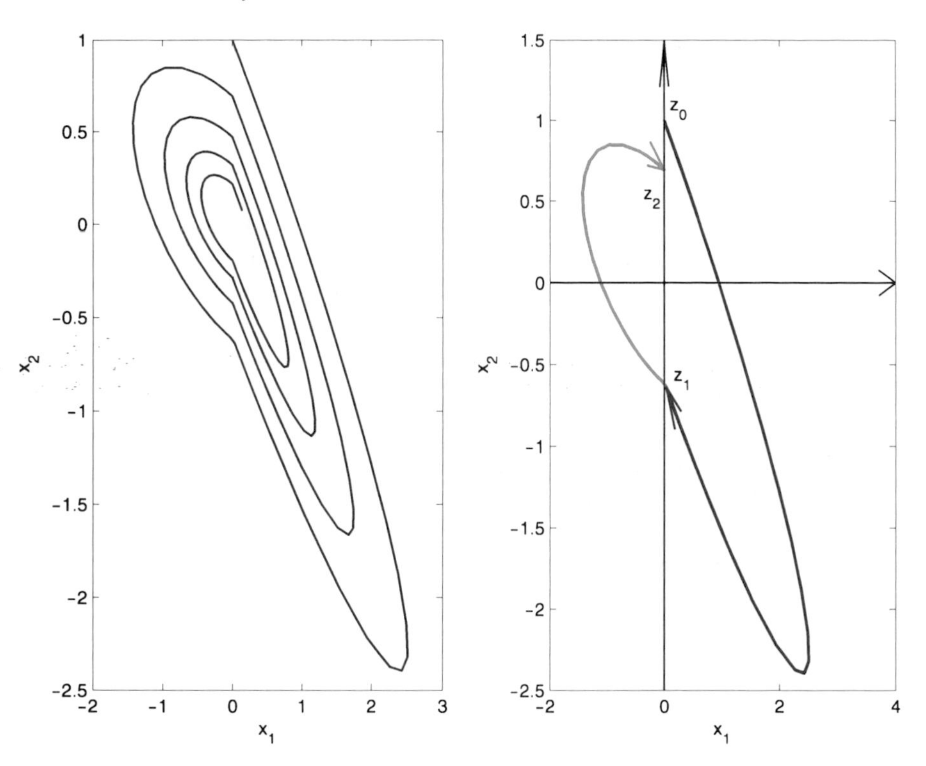

Fig. 3.4 Phase portrait (*left*) and transition map (*right*)

exponentially stable, as illustrated from the phase portrait (the left of Fig. 3.4). In fact, starting from the unit vector $z_0 = [0\ 1]^T$ on x_2-axis, the state comes back to the axis at $z_1 \approx [0\ -0.62]^T$ and $z_2 \approx [0\ 0.69]^T$. It is clear that the map at the switching surface (from z_0 to z_2) over a period is contractive w.r.t. the Euclidean norm, which is shown on the right of Fig. 3.4. The exponential convergence follows from the homogeneity of the switched system.

The example indicates a way to analyze the stability of piecewise linear systems through the contractility of surface (impact) maps, which are defined on the switching surfaces instead of the total state space. More accurately, if each map from one switching surface to another is contractive w.r.t. an energy function, the piecewise linear system is stable if no unstable state trajectory exists within a cell. In this way, the searching of a Lyapunov function is reduced to the searching of a surface Lyapunov function, which is required to decrease along the impact maps. The major steps toward the approach include the computation of the impact maps over possible switching transitions and the searching of a proper surface Lyapunov function for the impact maps. Let us define the impact maps first.

Consider the affine linear system given by

$$\dot{x} = Ax + B, \tag{3.34}$$

where $x \in \mathbf{R}^n$, $A \in \mathbf{R}^{n\times n}$, and $B \in \mathbf{R}^n$. Suppose that the system is defined over a polyhedron $X \subset \mathbf{R}^n$ and that S_0 and S_1 are two hyperplanes in the boundary of X with

$$S_0 = \{x \in \mathbf{R}^n : C_0 x = d_0\}, \qquad S_1 = \{x \in \mathbf{R}^n : C_1 x = d_1\},$$

where C_0, C_1, and d_0, d_1 are matrices and column vectors with compatible dimensions. Let $\bar{X}$ be the closure of X, S_0^d be a polytopic subset of S_0 with the property that any trajectory $x(\cdot)$ starting at S_0^d satisfies $x(t_1) \in S_1$ for some time t_1 and $x(t) \in \bar{X}$ for $t \in [0, t_1]$, and S_1^a be the subset of S_1 which is the collection of the above $x(t_1)$'s. The impact map from S_0^d to S_1^a is the map from $x_0 = x(0) \in S_0^d$ to $x_1 = x(t_1) \in S_1^a$ through the state trajectory $x(\cdot)$. Note that the map is not necessarily injective. Indeed, when a state trajectory starting from S_0 reaches S_1 more than once within region $\bar{X}$, there is a sequence of time constants $t_1 < t_2 < \cdots < t_k$, $k \geq 2$, such that $x(t_j) \in S_1$, $j = 1, \ldots, k$, and $x(t) \in \bar{X}$ for $t \in [0, t_k]$. As a result, the impact map may be multivalued and discontinuous.

Suppose that x_0^* and x_1^* are two states belonging to the hyperplanes S_0 and S_1, respectively. A state x_0 in S_0^d can be parameterized by $x_0 = x_0^* + \Delta_0$ with $C_0\Delta_0 = 0$. Similarly, a state x_1 in S_1^d can be parameterized by $x_1 = x_1^* + \Delta_1$ with $C_1\Delta_1 = 0$. In this way, the impact map from x_0 to x_1 reduces to the map from Δ_0 to Δ_1. Let $x(t; x_0)$ be the state trajectory with initial state $x(0) = x_0$, and by $x^0(t)$ and $x^1(t)$ we shortly denote $x(t; x_0^*)$ and $x(t; x_1^*)$, respectively.

Definition 3.27 Let $x_0 = x_0^* + \Delta_0 \in S_0^d$. Define the set of switching times of the impact map at Δ_0 to be

$$t_{\Delta_0} = \{s \in \mathbf{R}^+ : x(s) \in S_1, x(t; x_0) \in \bar{X}\ \forall t \in [0, s]\}.$$

Define the set of switching times of the impact map over $S_0^d - x_0^*$ to be

$$T = \bigcup_{\Delta \in S_0^d - x_0^*} t_\Delta.$$

In general, the sets of switching times may be unbounded, and it is hard to express the sets in a closed form. However, by defining the functions

$$w(t) = \frac{C_1 e^{At}}{d_1 - C_1 x^0(t)}$$

and

$$H(t) = e^{At} + \big(x^0(t) - x_1^*\big)w(t),$$

the impact map can be characterized as follows.

Theorem 3.28 *Assume that $C_1 x^0(t) \neq 0$ for $t \in T$. Then, for any $\Delta_0 \in S_0^d - x_0^*$, there exists $t \in T$ such that the impact map is given by*

$$\Delta_1 = H(t)\Delta_0. \tag{3.35}$$

Proof Suppose that $x_0 = x_0^* + \Delta_0 \in S_0^d$ and $x_1 = x_1^* + \Delta_1 \in S_1^a$. Integrating the system equation (3.34) yields

$$x_1 = e^{At} x_0 + \int_0^t e^{A(t-\tau)} B \, d\tau$$

for some $t \in T$. This further implies that

$$\Delta_1 = e^{At} \Delta_0 + x^0(t) - x_1^*. \tag{3.36}$$

On the other hand, it follows from $C_1 \Delta_1 = 0$ and $C_1 x_1^* = d_1$ that

$$C_1 e^{At} \Delta_0 = d_1 - C_1 x^0(t),$$

which leads to

$$w(t)\Delta_0 = 1. \tag{3.37}$$

Combining (3.36) with (3.37) yields

$$\Delta_1 = e^{At} \Delta_0 + \big(x^0(t) - x_1^*\big) w(t) \Delta_0,$$

which is exactly (3.35). □

Remark 3.29 The theorem asserts that the impact map from Δ_0 to Δ_1 is linear, provided that the switching time is fixed. To utilize this fact, it is necessary to examine the submanifold of S_0^d that corresponds to the same switching time. For a time $t \in T$, let S_t be the set of states $x_0^* + \Delta_0$ with $t \in t_{\Delta_0}$. It is clear that S_t is nonempty and that $\bigcup_{t \in T} S_t = S_0^d$. Furthermore, as $w(t)\Delta_0 = 1$ and $C_0 \Delta_0 = 0$, S_t is a subset of a linear manifold of dimension $n - 2$.

Based on the above preparations, we are ready to introduce the *surface Lyapunov functions* for analyzing the global asymptotic stability of piecewise linear systems. Let V_0 and V_1 be two smooth positive definite functions defined over $S_0^d - x_0^*$ and $S_1^a - x_1^*$, respectively. Clearly, if

$$V_1(\Delta_1) < V_0(\Delta_0) \quad \forall \Delta_0 \in S_0^d - x_0^*, \tag{3.38}$$

then the impact map from S_0^d to S_1^a is a contraction. In this case, V_0 and V_1 are called surface Lyapunov functions defined over S_0^d and S_1^a, respectively. Usually, it is very hard to verify the contractility for general surface Lyapunov functions due to the nonlinear nature of the impact map. So we restrict our attention to piecewise quadratic surface Lyapunov functions of the form

$$V_i(x) = x^T P_i x - 2g_i^T x + \alpha_i, \quad i = 0, 1, \tag{3.39}$$

where $0 < P_i \in \mathbf{R}^{n\times n}$, $g_i \in \mathbf{R}^n$, and $\alpha_i \in \mathbf{R}$, $i = 0, 1$. For a symmetric matrix $M \in \mathbf{R}^{n\times n}$ and a region $Y \subseteq \mathbf{R}^n$, by $M > 0$ on Y we mean $x^T M x > 0$ for any $0 \neq x \in Y$.

Theorem 3.30 *Suppose that* $V_i, i = 0, 1$, *are piecewise quadratic surface Lyapunov functions as in* (3.39) *and define*

$$R(t) = P_0 - H^T(t)P_1H(t) - 2\big(g_0 - H^T(t)g_1\big)w(t) + w^T(t)\alpha w(t), \tag{3.40}$$

where $\alpha = \alpha_0 - \alpha_1$. *Then, inequality* (3.38) *holds iff* $R(t) > 0$ *on* $S_t - x_0^*$ *for all* $t \in T$.

Proof Rewrite (3.38) as

$$\Delta_1^T P_1 \Delta_1 - 2g_1^T \Delta_1 + \alpha_1 < \Delta_0^T P_0 \Delta_0 - 2g_0^T \Delta_0 + \alpha_0.$$

Applying Theorem 3.28 gives

$$\Delta_0^T\big(P_0 - H^T(t)P_1H(t)\big)\Delta_0 - 2\big(g_0^T - g_1^T H(t)\big)\Delta_0 + \alpha > 0.$$

Utilizing the fact that $w(t)\Delta_0 = 1$, we obtain

$$\Delta_0^T\big(P_0 - H^T(t)P_1H(t) - 2w^T(t)\big(g_0^T - g_1^T H(t)\big) + w^T(t)\alpha w(t)\big)\Delta_0 > 0,$$

which verifies the theorem. □

Corollary 3.31 *The impact map from* $S_0^d - x_0^*$ *to* $S_1^a - x_1^*$ *is a contraction if there exist* P_0, $P_1 > 0$ *and* g_0, g_1, α *such that* $R(t) > 0$ *on* $S_t - x_0^*$ *for all* $t \in T$.

This corollary provides a criterion for contractility of the impact map. However, the computation of the manifold S_t is usually untractable. One way to cope with this difficulty is to further relax the condition as follows.

Corollary 3.32 *The impact map from* $S_0^d - x_0^*$ *to* $S_1^a - x_1^*$ *is a contraction if there exist* P_0, $P_1 > 0$ *and* g_0, g_1, α *such that*

$$R(t) > 0 \quad \text{on } S_0 - x_0^* \quad \forall t \in T. \tag{3.41}$$

Condition (3.41) is an infinite set of linear matrix inequalities (LMIs). Computationally, the set should be gridded into a finite subset of LMIs, which can be efficiently solved using available softwares. More details could be found in [92, 93].

To summarize, for a piecewise linear system, the main steps toward global asymptotic stability analysis include: (1) identifying all impact maps associated with the piecewise linear system; (2) defining all surface quadratic functions on the respective domains of impact maps; and (3) adjusting the parameters of the surface Lyapunov function so that condition (3.41) holds. In contrast to the piecewise quadratic Lyapunov function approach introduced in the previous subsection, the surface Lyapunov approach has several advantages including: (1) the analysis is based on the original cell partition, and no refinement of the partition is needed, and thus the computational burden is usually much lower; (2) the approach scales better with the dimension of the system due to the fact that the Lyapunov function is only needed to be sought on the switching surfaces instead of the total state space; and (3) the approach can prove global asymptotic stability, while the piecewise quadratic Lyapunov function approach can only prove exponential stability. However, it is clear that the approach applies only to continuous-time systems, while the piecewise quadratic Lyapunov function approach applies to both continuous- and discrete-time systems. Besides, the identification of impact maps is usually not an easy task, especially when there are three or more switching surfaces.

Next, we apply the above analysis approach to a class of saturated systems. A single-input single-output linear system with a saturated unitary output feedback could be described by

$$\begin{cases} \dot{x} = Ax + bu, \\ y = c^T x, \\ u = \operatorname{sat}(y), \end{cases} \tag{3.42}$$

where sat$(\cdot)$ is the standard saturation function

$$\operatorname{sat}(y) = \begin{cases} -1 & \text{if } y < -1, \\ y & \text{if } |y| \le 1, \\ 1 & \text{if } y > 1. \end{cases}$$

Note that the saturated system always admits a unique solution due to the fact that $Ax + b\operatorname{sat}(c^T x)$ is globally Lipschitz continuous.

It is clear that there are two switching surfaces, i.e.,

$$S^+ = \{x \in \mathbf{R}^N : c^T x = 1\}$$

and

$$S^- = \{x \in \mathbf{R}^N : c^T x = -1\}.$$

The surfaces are origin-symmetric and separated. Denote $A_1 = A + bc^T$. To ensure that the origin is the unique equilibrium with local stability, we must have that A_1 is Hurwitz, and A does not admit any eigenvalue with positive real part. Besides, when A is nonsingular, it must satisfy $-c^T A^{-1} b < 1$. Define

$$S_+^+ = \{x \in S^+ : c^T A_1 x \ge 0\}, \qquad S_-^+ = \{x \in S^+ : c^T A_1 x \le 0\}$$

and

$$S_+^- = \{x \in S^- : c^T A_1 x \ge 0\}, \qquad S_-^- = \{x \in S^- : c^T A_1 x \le 0\}.$$

As A_1 is Hurwitz, there exists a nonempty subset of S_-^+, denoted S^*, such that any state trajectory starting from S^* will not switch again and will converge to the origin asymptotically. To see this, let $P > 0$ be a solution of $A_1^T P + PA_1 = -I$, and define $\varpi = \max_{x^T Px=1} c^T x$; then the intersection $S^+ \cap \{x : x^T Px = \varpi^2\}$ is inside S^*. It can be seen that the subset S^* is a convex and closed subset of S_-^+.

Taking advantage of the origin-symmetry of S^+ and S^-, we need to take care of three impact maps: (1) the map from S_+^+ to S_-^+; (2) the map from $S_-^+ - S^*$ to S_+^+; and (3) the map from $S_-^+ - S^*$ to S_-^-. Let T_1, T_2, T_3 denote the sets of the switching times for the impact maps, respectively. For a state trajectory starting from $x_0 \in S_+^+$, it switches at $x_1 \in S_-^+$. If $x_1 \in S^*$, then the trajectory converges to the origin without any further switching. Otherwise, $x_1 \in S_-^+ - S^*$, and it may switch at either $x_{2a} \in S_+^+$ or $x_{2b} \in S^-$. The next switching occurs at $x_{3a} \in S_-^+$ and $x_{3b} \in S_-^-$, respectively. The idea is to check whether sequence $x_1, x_{3a}/-x_{3b}, \ldots$ is contractive with respect to a quadratic surface Lyapunov function. If so, then the sequence will eventually enter into set S^*, and global asymptotic stability of the piecewise linear system is established.

Choose $x_1^* = P^{-1}c/(c^T P^{-1}c) \in S^*$, where $P > 0$ satisfies $A_1^T P + PA_1 = -I$. $x_0^* \in S_+^+$ can be chosen similarly in a subtler manner. Let $x^0(\cdot)$ and $x^1(\cdot)$ be the state trajectories starting from x_0^* and x_1^*, respectively.

Define the functions

$$w_1(t) = c^T e^{At}/\left(1 - c^T x^0(t)\right), \qquad w_2(t) = c^T e^{A_1 t}/\left(1 - c^T x^1(t)\right),$$
$$w_3(t) = c^T e^{A_1 t}/\left(-1 - c^T x^1(t)\right),$$

and further the impact maps

$$H_1(t) = e^{At} + \left(x^0(t) - x_1^*\right)w_1(t), \qquad H_2(t) = e^{A_1 t} + \left(x^1(t) - x_0^*\right)w_2(t),$$
$$H_3(t) = e^{A_1 t} + \left(x^1(t) + x_0^*\right)w_3(t).$$

By applying Corollary 3.32, we have the following stability criterion for saturated systems.

Proposition 3.33 *Saturated system* (3.42) *is globally asymptotically stable if there exist matrices* P_1, $P_2 > 0$, *vectors* g_1 *and* g_2, *and a real number* α *such that the inequalities*

$$P_1 - H_1^T(t)P_2 H_1(t) - 2\left(g_1 - H_1^T(t)g_2\right)w_1(t) + w_1^T(t)\alpha w_1(t) > 0 \quad \text{on } S^+ - x_0^*,$$
$$P_2 - H_2^T(t)P_1 H_2(t) - 2\left(g_2 - H_2^T(t)g_1\right)w_2(t) + w_2^T(t)\alpha w_2(t) > 0 \quad \text{on } S^+ - x_1^*,$$
$$P_2 - H_3^T(t)P_1 H_3(t) - 2\left(g_2 - H_3^T(t)g_1\right)w_3(t) + w_3^T(t)\alpha w_3(t) > 0 \quad \text{on } S^+ - x_1^*$$

hold for $t \in T_1$, $t \in T_2$, *and* $T \in T_3$, *respectively.*

Finally, we present a numerical example to illustrate the effectiveness of the surface Lyapunov function approach for saturated systems.

Example 3.34 For the planar linear system with saturated output feedback

$$\begin{cases} \dot{x} = \begin{bmatrix} -1 & 0 \\ 0 & 0 \end{bmatrix} x + \begin{bmatrix} 2 \\ -0.4 \end{bmatrix} u, \\ y = [0.6 \quad 1] x, \\ u = \operatorname{sat}(y), \end{cases} \tag{3.43}$$

it is clear that the system is locally stable at the origin equilibrium. An elementary analysis shows that any state trajectory starting from the saturated area ($|c^T x(0)| > 1$) crosses the switching surfaces. There are three impact maps: the map from S_+^+ to S^*, the map from $S_-^+ - S^*$ to S_-^-, and the map from S_-^- to $-S^*$, which is symmetric to the first map. Note that any state trajectory starting from the switching surfaces enters into $\pm S^*$ by two or less switches. As any state trajectory starting from $\pm S^*$ asymptotically converges to the origin, global asymptotic stability is guaranteed. Figures 3.5 and 3.6 show sample phase portraits starting from S^+ and the corresponding impact maps, respectively. In this simple case, no explicit surface Lyapunov function is needed for stability analysis.

An interesting point here is that the system is not globally exponentially stable. Indeed, let $x(t; x_0)$ be the state trajectory with $x_0 = [a \; 1 - 0.6a]^T \in S^+$, where a is

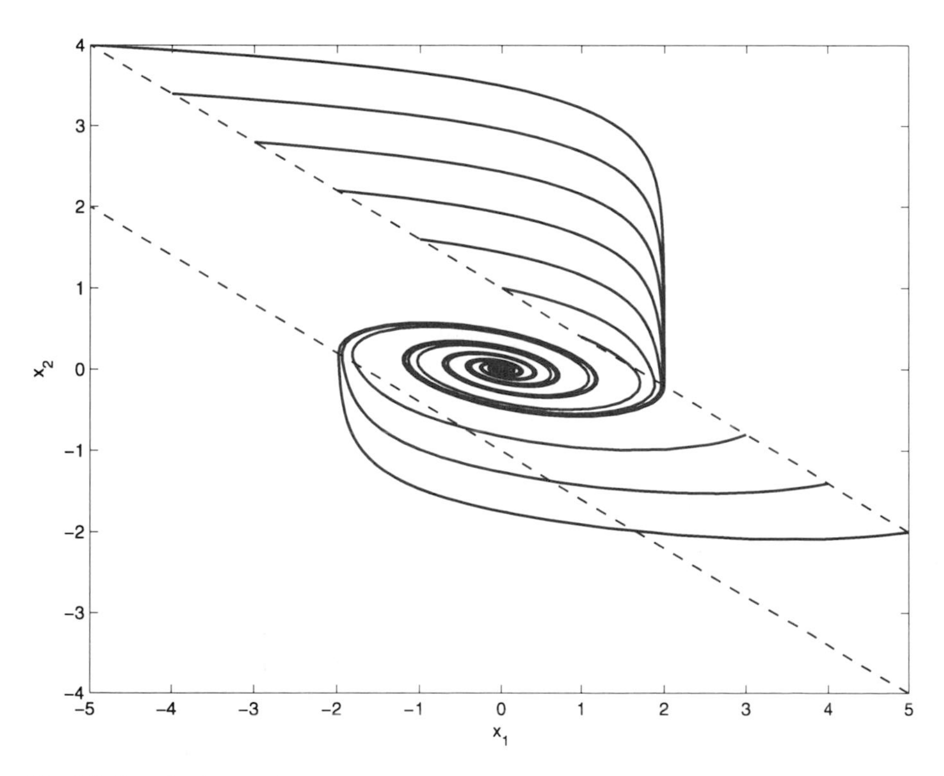

Fig. 3.5 Sample phase portraits

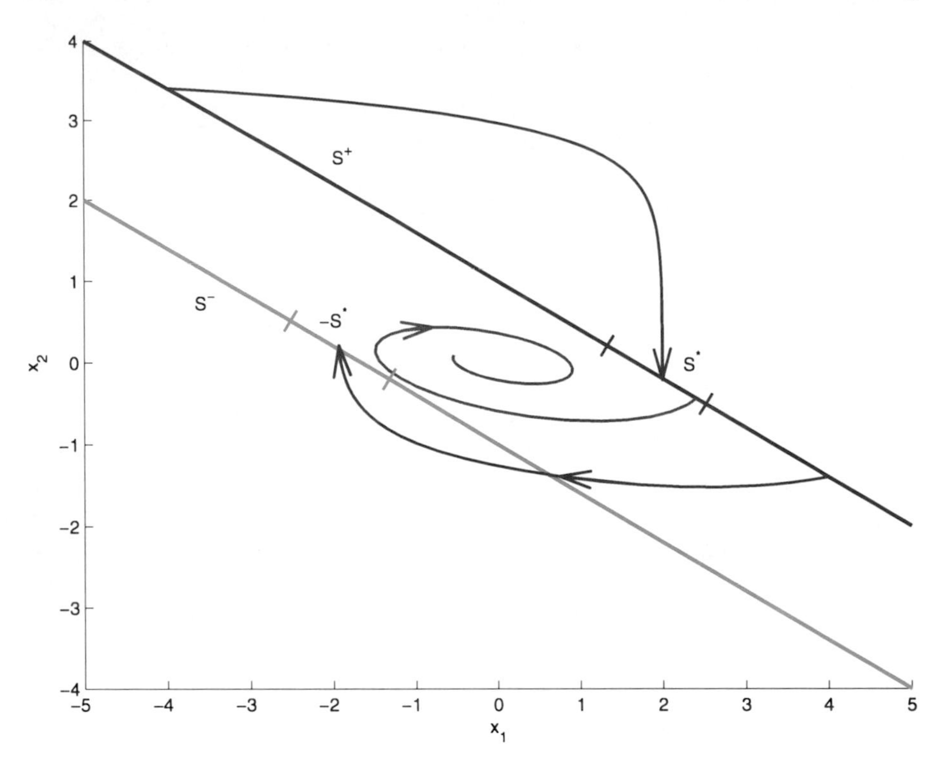

Fig. 3.6 Illustration of impact maps

a real parameter. Furthermore, let t_1 be the first switching time with $x(t_1; x_0) \in S^+$. Then, it can be shown that

$$\lim_{a \to -\infty} \frac{t_1}{a} = -\frac{3}{2},$$

which excludes the possibility of global exponential stability of the piecewise linear system. This means that the piecewise Lyapunov function approach does not apply to the system.

3.3.4 Transition Analysis: A Graphic Approach

An intrinsic difficulty of stability analysis for piecewise linear systems lies in the fact that switching transitions among the cells are autonomous. Therefore, the autonomous stability is neither robust as in guaranteed stability nor flexible as in stabilizing switching design. The lack of both robustness and flexibility in the switching mechanism makes autonomous stability very hard to understand.

In this subsection, rather than taking the autonomous switching as a negative factor, we try to take advantage of the switching transitions and present a graphic approach for autonomous stability analysis.

Definition 3.35 Discrete state $i \in M$ is said to be *weakly transitive* if for all $x_0 \in \Omega_i - \{0\}$, we have either

$$\exists t > 0, \quad \phi(t; 0, x_0) \notin \Omega_i$$

or

$$\forall t > 0, \quad \phi(t; 0, x_0) \in \Omega_i, \qquad \lim_{t \to +\infty} \phi(t; 0, x_0) = 0.$$

It is clear that, when $0 \notin \Omega_i$, then state i is weakly transitive if for all $x_0 \in \Omega_i$, there exists $t > 0$ such that $\phi(t; 0, x_0) \notin \Omega_i$. That is to say, there does not exist a whole (continuous) state trajectory that stays within the cell. In this case, state i is said to be *transitive*. Another observation is that a necessary condition for global asymptotic stability of the origin is the weak transitivity of all discrete states.

Based on the transitive relationship, we could formulate a graphic approach for addressing the stability of the piecewise linear system.

Let (V, Ξ) be a *directed graph* with $V = M$ being the *node set*, $(i, j) \in \Xi$ if $i \neq j$, and there exist a state $x_0 \in \Omega_i$ and a time $t > 0$ such that $\phi(t; 0, x_0) \in \Omega_j$. When $(i, j) \in \Xi$, (i, j) is said to be an *edge* of the graph, and Ξ is the edge set. For a node i, its neighbor set is

$$\Gamma_i = \{ j \in V : (i, j) \in \Xi \}.$$

Definition 3.36 A collection of nodes $V_0 \subset V$ is said to be *invariant* if $\Gamma_i \subset V_0$ for all $i \in V_0$.

As a trivial fact, the node set V itself is invariant.

Definition 3.37 A collection of nodes $V_0 \subset V$ is said to be attractive if it is invariant and the set $V - V_0$ does not contain any loop.

It is clear that, a node set is attractive if each path with sufficiently large length contains a node in this set and all the successive nodes stay in the set forever.

An attractive set is said to be *minimal* if any strict subset is not attractive. A minimal set is denoted $V_{\min}$.

For the piecewise linear system, any invariant node set $V_0 \subset V$ induces a subdynamics described by

$$x^+(t) = A_i x + a_i, \quad x \in \Omega_i, \ i \in V_0 \tag{3.44}$$

whose dynamics is part of that of the original system, and the stability is simpler to analyze. For simplicity, system (3.44) is denoted by V_0-induced piecewise linear system.

Theorem 3.38 *The piecewise linear system is globally asymptotically stable if the following conditions hold*:

(1) *each discrete state is weakly transitive*
(2) *there is an index* $i \in M$ *such that* $0 \in \Omega_i^o$ *and* $i \in V_{\min}$, *and*
(3) *the* $V_{\min}$*-induced system is globally asymptotically stable*

Proof The proof is straightforward. Any state trajectory either converges to the origin without going through the region $\bigcup_{i\in V_{\min}} \Omega_i$ due to the weak transitivity or enters into the region in a finite time and then converges to the origin within the region. □

Remark 3.39 The merits of the theorem include simplifying the stability verification with the aid of graphic decomposition and applying to both discrete-time and continuous-time systems. More importantly, the graphic approach provides an effective method for attractivity analysis. Indeed, when V_0 is attractive, then $\bigcup_{i\in V_0} \Omega_i$ is an invariant domain.

Example 3.40 H–K model of opinion dynamics.

Krause–Hegselmann [103] proposed a model of opinion dynamics with a uniform confident level, which is mathematically described by

$$x_i(t+1) = \frac{1}{|N_i|} \sum_{j\in N_i} x_j(t), \quad i = 1, \dots, \kappa, \tag{3.45}$$

where $x_i \in \mathbf{R}$ is the opinion value of agent i, $x = [x_1, \dots, x_\kappa]^T$, r is the level of confidence, N_i is the neighbor set of agent i defined by $N_i(t) = \{j : |x_j(t) - x_i(t)| \le r\}$, and $|N_i|$ denotes the cardinality of set N_i. The model is called *K–H model* or *H–K model* in the literature.

The *problem of asymptotic consensus* is to determine the region of initial states that lead the agents' opinions to a common value. For this, define the region

$$X = \Big\{x_0 \in \mathbf{R}^\kappa : \lim_{t\to+\infty} \big(x_i(t, x_0) - x_j(t, x_0)\big) = 0 \quad \forall i \ne j\Big\}, \tag{3.46}$$

where $x_i(t, x_0)$ denotes the state x_i at time t with initial condition $x_i(0) = x_{0i}$. It has been established in [103] that any consensus process is finite-time convergent, that is,

$$x_i(t, x_0) = x_j(t, x_0) \quad \forall i \ne j,\ x_0 \in X,\ t \ge T,$$

where T is a natural number that relies on κ. From this and from the complete orderedness of the opinions we can further prove that each process reaches its final dynamics in a finite time, that is, there is a natural number T^* such that

$$x_i\big(t + T^*, x_0\big) = x_i\big(T^*, x_0\big) \quad \forall t \in \mathcal{T}_0,\ x_0 \in \mathbf{R}^\kappa.$$

As a result, the limit in (3.46) always exists. Furthermore, when $x_i(0) \le x_j(0)$, then we have $x_i(t) \le x_j(t)$ for all $t > 0$. As a result, the order of the opinion values always keeps unchanged. Without loss of generality, we assume that

$x_1 \le x_2 \le \cdots \le x_\kappa$. Another useful observation is that if $x_{i+1}(t) - x_i(t) > r$ for some i and t, then the inequality holds for all forward time. Therefore, a necessary condition for consensus is

$$x_{i+1}(t) - x_i(t) \le r, \quad \forall i = 1, \ldots, \kappa - 1,\ t \in \mathcal{T}_0. \tag{3.47}$$

To apply the graphic approach, one way is to redefine the state vector that takes $x_i - x_j$ as elements, and the consensus problem reduces to the attractivity of the piecewise linear system. However, for the sake of simplicity, here we still use x as the state vector, and the graphic approach is utilized in a heuristic manner.

First, define

$$X_1 = \{x \in \mathbf{R}^\kappa : x_\kappa - x_1 \le r\}.$$

This implies that $|N_i| = \kappa$ for all i. It is clear that any opinion process initiated from X_1 achieves consensus in one step, and any consensus process must enter into X_1 before the final consensus is achieved. As a result, $X_1 \subset X$, and X_1 is the minimal attractive region w.r.t. X. That is to say, any other cell within X must be transitive to X_1 in a finite time. Note that when $\kappa = 2$, $X = X_1$, and the consensus is equivalent to initial connectedness of the opinion dynamics network.

Next, for $\kappa \ge 3$, define

$$X_2 = \{x \in \mathbf{R}^\kappa : x_\kappa - x_2 \le r,\ x_{\kappa-1} - x_1 \le r,\ x_\kappa - x_1 > r\}.$$

This corresponds to $|N_1| = |N_\kappa| = \kappa - 1$ and $|N_i| = \kappa$ otherwise. For any $x(0) \in X_2$, we have

$$\begin{aligned} x_\kappa(1, x(0)) - x_1(1, x(0)) &= \frac{1}{\kappa - 1}\sum_{i=2}^{\kappa} x_i(0) - \frac{1}{\kappa - 1}\sum_{i=1}^{\kappa-1} x_i(0) \\ &= \frac{1}{\kappa - 1}\big(x_\kappa(0) - x_1(0)\big) \\ &\le \frac{2}{\kappa - 1} r, \end{aligned}$$

which means that $x(1, x(0)) \in X_1$. That is, X_2 is transitive to X_1 in one step. Note that when $\kappa = 3$, $X = X_1 \cup X_2$, and the consensus is equivalent to initial connectedness of the opinion dynamics network.

Then, continue the analysis process and define

$$\begin{aligned} X_{l_1} = \{x \in \mathbf{R}^\kappa : \ & x_\kappa - x_{l_1} \le r,\ x_{\kappa - l_2} - x_1 \le r, \\ & x_\kappa - x_{l_1 - 1} > r,\ x_{\kappa - l_2 + 1} - x_1 > r\}, \end{aligned}$$

where $l_1 = \lfloor \frac{\kappa+1}{2} \rfloor$ is the largest integer less than or equal to $\frac{\kappa+1}{2}$, and $l_2 = \lceil \frac{\kappa+1}{2} \rceil$ is the smallest integer larger than or equal to $\frac{\kappa+1}{2}$. It can be verified that, for any $x(0) \in X_{l_1}$, we have

$$x_\kappa(1, x(0)) - x_{l_1}(1, x(0)) \le r$$

and

$$x_{l_2}\big(1, x(0)\big) - x_1\big(1, x(0)\big) \le r.$$

This means that $x(1, x(0)) \in \bigcup_{j=1}^{l_1} X_j$, which further implies that $x(t, x(0)) \in \bigcup_{j=1}^{l_1} X_j$ for all $t \in \mathcal{T}_0$. It follows from the opinion dynamics equation that both $x_\kappa(t, x(0)) - x_{l_1}(t, x(0))$ and $x_{l_2}(t, x(0)) - x_1(t, x(0))$ are contractive as t increases, which further implies that region X_{l_1} is transitive to X_1 in a finite time.

To summarize, the regions $X_2, \dots, X_{l_1}$ are transitive to X_1 in a finite time, and any state in X_1 achieves consensus in one step. By applying Theorem 3.38, we obtain the following consensus criterion.

Proposition 3.41 *The opinion dynamics reaches a consensus if* $|N_i(0)| \ge \frac{\kappa+1}{2}$ *for any* $i = 1, \dots, \kappa$.

While conservative, the proposition provides an easily verifiable sufficient condition, which only relies on the initial network graph. Besides, it tells us that, when each person in a society finds that more than half of the population (nearly) agree with him, then a consensus could be finally achieved.

To illustrate the effectiveness of the above analysis, we take $\kappa = 5$. Let $r = 1$ and $x_0 = [-1, -0.9, 0, 0.9, 1]^T$. It can be verified that the condition in Proposition 3.41

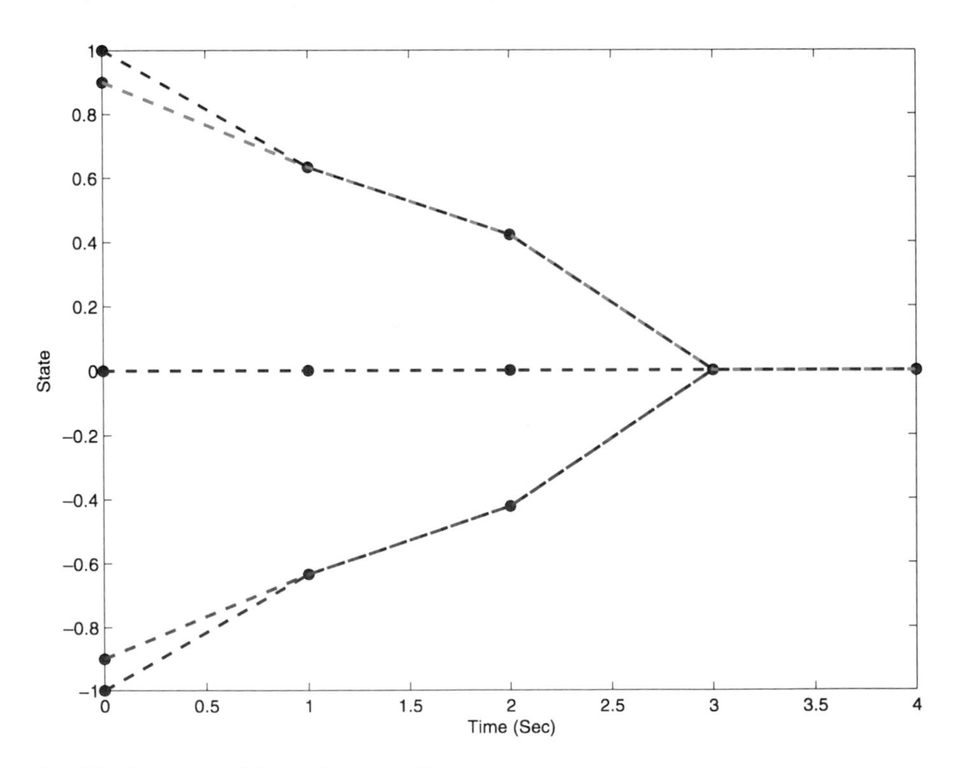

Fig. 3.7 Consensus of the opinion dynamics

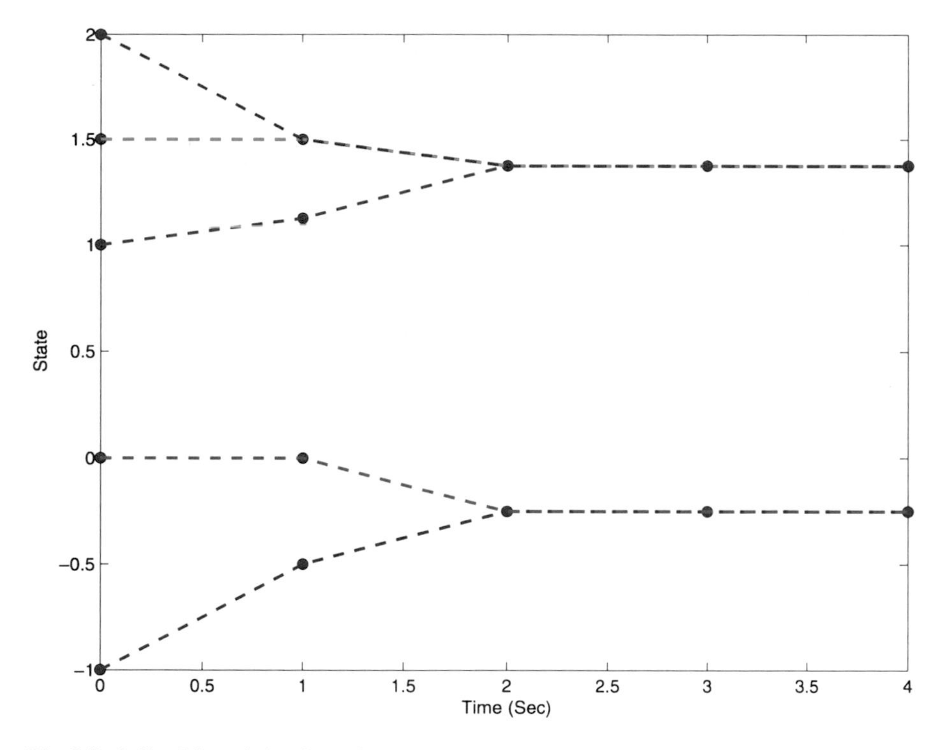

Fig. 3.8 Split of the opinion dynamics

is satisfied. Therefore, consensus can be achieved. Figure 3.7 shows the evolution of the opinion dynamics, which achieves consensus in three steps.

Finally, take the initial state to be $x_0 = [-1, 0, 1, 1.5, 2]^T$. It can be verified that $|N_1(0)| = 2$, which means that the condition in Proposition 3.41 is violated. Figure 3.8 depicts the evolution of the opinion dynamics, and it is clear that the agents split into two parties within two steps.

3.3.5 Conewise Linear Systems

A continuous function $f : \mathbf{R}^n \to \mathbf{R}^n$ is said to be *conewise linear* if there exist a finite set of convex polyhedral cones $\{\mathcal{X}_1, \ldots, \mathcal{X}_m\}$ with $\bigcup_{i=1}^m \mathcal{X}_i = \mathbf{R}^n$ and $n \times n$ matrices $\{L_1, \ldots, L_m\}$ such that $f(x) = L_i x$ for $x \in \mathcal{X}_i$.

A conewise linear system is a differential or difference equation with a conewise linear function as the right-hand side:

$$x^+(t) = A_i x, \quad x \in \mathcal{X}_i, \tag{3.48}$$

where $A_1, \ldots, A_m$ are $n \times n$ real constant matrices, and $\{\mathcal{X}_1, \ldots, \mathcal{X}_m\}$ is a set of convex polyhedral cones with $\bigcup_{i=1}^m \mathcal{X}_i = \mathbf{R}^n$ and $\mathcal{X}_i^o \cap \mathcal{X}_j^o = \emptyset$ for $i \neq j$.

It is clear that conewise linear systems are piecewise linear systems with conic state partitions and without affine terms. Other important features include that the system is homogeneous with degree one and that the right-hand side is globally Lipschitz continuous. As a result, the system always admits a unique solution over the time space from any initial state, and local asymptotic stability implies global asymptotic stability. Furthermore, we can prove the equivalence among the stability notions, as stated in the following proposition.

Proposition 3.42 *For the conewise linear system, the following statements are equivalent*:

(1) *The system is attractive.*
(2) *The system is asymptotically stable.*
(3) *The system is exponentially stable.*

Proof We only need to prove that attractivity implies exponential stability. For this, suppose that the conewise system is attractive. This means that, for any state x on the unit sphere, there is a time $t^x > 0$ such that

$$\left|\phi\left(t^x; 0, x\right)\right| \le \frac{1}{4}.$$

As the vector field of the system is globally Lipschitz continuous, there is a positive real number r such that

$$\left|\phi\left(t^x; 0, x\right) - \phi\left(t^x; 0, y\right)\right| \le \frac{1}{4} \quad \forall y \in \mathbf{B}_r(x).$$

Combining the above inequalities yields

$$\left|\phi\left(t^x; 0, y\right)\right| \le \frac{1}{2} \quad \forall y \in \mathbf{B}_r(x).$$

Letting x vary along the unit sphere, we have

$$\bigcup_{x\in\mathbf{H}_1} \mathbf{B}_r(x) \supset \mathbf{H}_1.$$

As the unit sphere is compact, by the Finite Covering Theorem, there is a finite set of states $x_1, \dots, x_l$ on the unit sphere such that

$$\bigcup_{j=1}^{l} \mathbf{B}_r(x_j) \supset \mathbf{H}_1.$$

As a result, we can partition the unit sphere into l regions $R_1, \dots, R_l$ such that $R_j \subset \mathbf{B}_r(x_j)$, $j = 1, \dots, l$, $\bigcup_{j=1}^{l} R_j = \mathbf{H}_1$, and $R_j \cap R_k = \emptyset\ \forall j \neq k$. Define accordingly

$$\Omega_j = \{x : \exists\lambda \ge 0,\ \lambda x \in R_j\}, \qquad s_j = t^{x_j}, \quad j = 1, \dots, l.$$

For any initial state x_0, define recursively

$$\begin{aligned} i_k &= \#\{x \in \Omega_j\}, \\ t_{k+1} &= t_k + s_{i_k}, \\ z_{k+1} &= \phi(t_{k+1}; t_k, z_k), \quad k = 0, 1, 2, \ldots, \end{aligned}$$

with $t_0 = 0$ and $z_0 = x_0$. It can be seen that

$$z_k = \phi(t_k; 0, x_0), \qquad |z_{k+1}| \le \frac{|z_k|}{2}, \quad k = 0, 1, 2, \ldots,$$

which further implies that

$$\left|\phi(t; 0, x_0)\right| \le \beta e^{-\alpha t}|x_0| \quad \forall t \in \mathcal{T}_0,$$

where $\beta = e^{\max\{\|A_1\|,\ldots,\|A_m\|\}\max\{s_1,\ldots,s_l\}}$, and $\alpha = \frac{\ln 2}{\max\{s_1,\ldots,s_l\}}$. The conclusion follows due to the arbitrariness of x_0. □

Note that the stabilities do not always coincide for general piecewise linear systems, as exhibited by Example 3.34.

Next, we are to present a verifiable criterion for stability of conewise linear systems. For this, we introduce the notion of unit-sphere contractility as defined below.

Definition 3.43 For a given norm $|\cdot|$, the conewise linear system is *unit-sphere contractive* (w.r.t. norm $|\cdot|$) if for any state x with unit norm, there is a time $T > 0$ such that $|\phi(T; 0, x)| < 1$.

Theorem 3.44 *The following statements are equivalent to each other*:

(1) *The conewise linear system is exponentially stable.*
(2) *The conewise linear system is unit-sphere contractive w.r.t. any given norm.*
(3) *For a given norm, the conewise linear system is unit-sphere contractive.*

Proof It is clear that (1) $\Longrightarrow$ (2) $\Longrightarrow$ (3). So we need only to prove (3) $\Longrightarrow$ (1).

For any arbitrarily given but fixed state x with unit norm, let T be such that $|\phi(T; 0, x)| = \mu_x < 1$. Define

$$T_x = \min\{t : \left|\phi(t; 0, x)\right| = \mu_x\}.$$

By the continuous dependence of initial state, there exists an open neighborhood N_x of x such that

$$\left|\phi(T_x; 0, y) - \phi(T_x; 0, x)\right| \le \frac{1 - \mu_x}{2} \quad \forall y \in N_x,$$

which further implies that

$$\left|\phi(T_x; 0, y)\right| \le \frac{1 + \mu_x}{2} \quad \forall y \in N_x.$$

Letting x vary along the unit sphere, it is obvious that

$$\bigcup_{x \in \mathbf{H}_1} N_x \supseteq \mathbf{H}_1.$$

By the compactness of the unit sphere and the Finite Covering Theorem, there exist a finite number l and a set of states $x_1, \ldots, x_l$ on the unit sphere such that

$$\bigcup_{i=1}^{l} N_{x_i} \supseteq \mathbf{H}_1.$$

Accordingly, we can partition the unit sphere into l regions $R_1, \ldots, R_l$ such that $\bigcup_{i=1}^{l} R_i = \mathbf{H}_1$. In addition, for each i, $1 \le i \le l$, we have

$$\left|\phi(T_{x_i}; 0, y)\right| \le \frac{1 + \mu_{x_i}}{2} \quad \forall y \in R_i.$$

For any state $z \in \mathbf{R}^n$, there are $x \in R_i$ with $1 \le i \le l$ and a positive real number λ such that $z = \lambda x$. Denote $T_z = T_{x_i}$ and $\mu = \max\{\mu_{x_1}, \ldots, \mu_{x_l}\}$. It is clear that

$$\left|\phi(T_z; 0, z)\right| \le \frac{1 + \mu}{2} |z| \quad \forall z \in \mathbf{R}^n. \tag{3.49}$$

Then, for any $x_0 \neq 0$, define recursively the sequence of states

$$\begin{aligned} z_0 &= x_0, \\ z_{k+1} &= \phi(T_{z_k}; 0, z_k), \quad k = 0, 1, \ldots. \end{aligned}$$

Similarly, define recursively a sequence of times

$$\begin{aligned} t_0 &= 0, \\ t_{k+1} &= t_k + T_{z_k}, \quad k = 0, 1, \ldots. \end{aligned}$$

It follows from (3.49) that

$$\left|\phi(t_k; 0, z)\right| \le \left(\frac{1 + \mu}{2}\right)^k |z|, \quad k = 1, 2, \ldots. \tag{3.50}$$

Finally, we prove that each state trajectory under the above switching path is exponentially convergent. To see this, let

$$T = \max_{i=1}^{l} T_{x_i}, \qquad \alpha = \ln 2/T, \qquad \eta = \max_{j=1}^{m} \|A_j\|, \qquad \beta = 2e^{\eta T}.$$

It follows that

$$\left|\phi(t; 0, x_0)\right| \le \beta e^{-\alpha t} |x_0| \quad \forall x_0 \in \mathbf{R}^n, \ t \in \mathcal{T}_0. \tag{3.51}$$

This completes the proof. □

Next, we are to develop a constructive procedure for verifying the unit-sphere contractility, which by the theorem provides a computational verification of exponential stability of the conewise linear system. The idea is to select a set of states with unit norm and determine a neighbor of contractility for each state. When all the neighbors cover the unit sphere, the system is unit-sphere contractive. A key issue is to estimate a contractility neighbor for a given initial state. For this, denote $L = \max\{\|A_1\|, \dots, \|A_m\|\}$ and fix a real number $\delta \in (0, 1)$.

Proposition 3.45 *Let x be a state with unit norm, and $T_x > 0$ be such that $|\phi(T_x; 0, x)| \le \delta$. Then, for any y with $|y - x| < \eta_x(1 - \delta)$, we have*

$$\bigl|\phi(T_x; 0, y)\bigr| < 1, \tag{3.52}$$

where $\eta_x = e^{-LT_x}$ in continuous time, and $\eta_x = \frac{1}{L^{T_x}}$ in discrete time.

Proof Note that L is a Lipschitz constant for the conewise linear system, that is, for any indices $i, j \in \{1, \dots, m\}$ and any states $x \in \mathcal{X}_i$ and $y \in \mathcal{X}_j$, we have

$$|A_i x - A_j y| \le L|x - y|.$$

To see this, denote the ordered set of intermediate states

$$z_0 = x, \ z_1, \ \dots, \ z_k, z_{k+1} = y$$

on the segment (x, y) such that (z_l, z_{l+1}) is within some $\mathcal{X}_{\mu_l}$ for each $l \in \{0, 1, \dots, k\}$. It is clear that

$$\begin{aligned} |A_i x - A_j y| &\le |A_i x - A_i z_1| + \cdots + |A_j z_k - A_j y| \\ &\le L\bigl(|x - z_1| + \cdots + |z_k - y|\bigr) \\ &= L|x - y|. \end{aligned}$$

This implies that, for continuous-time systems, we have

$$\bigl|\phi(T_x; 0, y) - \phi(T_x; 0, x)\bigr| \le |x - y| + L\int_0^{T_x} \bigl|\phi(t; 0, y) - \phi(t; 0, x)\bigr|\,dt.$$

It follows from the Bellman–Gronwall lemma that

$$\bigl|\phi(T_x; 0, y) - \phi(T_x; 0, x)\bigr| \le e^{LT_x}|y - x|,$$

which leads to inequality (3.52). The discrete time case can be established in the same manner. □

Remark 3.46 The proposition presents an estimate of the radius of a ball within the contractility neighbor for any given state with unit norm. The radius is dependent on the contractility ratio, the subsystem matrices, and the time length T_x for contractility. For any exponentially stable conewise linear system with convergence estimate (3.51), it can be seen that T_x can be chosen such that

$$T_x \leq \frac{\ln \beta - \ln \delta}{\alpha} \quad \forall x.$$

This means that a uniform lower bound of the radius could be explicitly computed.

Based on the proposition, we could outline a computational procedure for verifying the unit-sphere contractility. For this, denote the region $N_x = \{y : |x - y| < \eta_x(1 - \delta)\}$ for a state x.

(1) Select a set of initial states $x_1, \ldots, x_k$ that are uniformly distributed on the unit sphere with a preassigned dense.
(2) Determine the times T_{x_i} by simulation.
(3) Check whether $\mathbf{H}_1 \subset \bigcup_{i=1}^k N_{x_i}$ or not. If yes, return "System is exponentially stable." Otherwise, double the dense of the initial states and repeat the process.

Conceptually, the procedure does work as it terminates in a finite time when the system is unit-sphere contractive. However, it does not terminate when the system is not unit-sphere contractive. Technically, it is usually not an easy task for representation of a region and a union of regions, and for verification of subset relationship between two regions. In particular, region N_x is norm-dependent, that is, different norms may correspond to different regions. Note that, for the ℓ_2-norm, the region is generally nonconvex (and not a union of a finite set of convex regions) and nonpolyhedric as well. In this case, it is very hard to verify the subset relation $\mathbf{H}_1 \subset \bigcup_{i=1}^k N_{x_i}$. Accordingly, we use the ℓ_1-norm instead. In this case, each N_x is a convex polytope, and the subset relationship can be verified by means of commercial numerical softwares (for example, MATLAB GBT Toolbox [252]).

To further reduce the computational load, we take the reduced-order approach. For this, first consider the (upper) half unit sphere given by

$$\mathbf{H}_1^+ = \{x = [x_1, \ldots, x_n]^T : x_n \geq 0,\ |x|_1 = 1\}.$$

This sphere can be projected into the $(n-1)$-dimensional unit ball by the map

$$\mathcal{P}_n \colon \mathbf{H}_1^+ \mapsto \mathbf{B}_1^{n-1}, \qquad \mathcal{P}_n x = [x_1, \ldots, x_{n-1}]^T.$$

The inverse map is

$$\mathcal{P}_n^{-1} y = [y_1, \ldots, y_{n-1}, 1 - |y|_1]^T, \quad y \in \mathbf{B}_1^{n-1}.$$

Define $N_x^+ = \{y \in \mathbf{H}_1^+ : |x - y| < \eta_x(1 - \delta)\}$ and $PN_x^+ = \{\mathcal{P}_n y : y \in N_x^+\} \subset \mathbf{B}_1^{n-1}$. Similar reduction can be made for the (lower) half unit sphere given by

$$\mathbf{H}_1^- = \{x = [x_1, \ldots, x_n]^T : x_n \leq 0,\ |x|_1 = 1\}.$$

The subset relation $\mathbf{H}_1 \subset \bigcup_{i=1}^{k} N_{x_i}$ is reduced to

$$\bigcup_{x_k \in \mathbf{H}_1^+} PN_{x_k}^+ = \mathbf{B}_1^{n-1}, \qquad \bigcup_{x_k \in \mathbf{H}_1^-} PN_{x_k}^- = \mathbf{B}_1^{n-1},$$

which could be verified over the $(n-1)$-dimensional unit ball.

Example 3.47 For the planar discrete-time conewise linear system

$$x(t+1) = A_i x(t), \qquad x(t) \in \mathcal{X}_i, \quad t = 0, 1, \ldots,$$

with

$$A_1 = \begin{bmatrix} -1.4078 & 0.1223 \\ 1.3846 & 0.4437 \end{bmatrix},$$

$$A_2 = \begin{bmatrix} -0.5837 & -0.7019 \\ 0.5213 & 1.3070 \end{bmatrix},$$

$$A_3 = \begin{bmatrix} 0.2405 & 0.1223 \\ -0.3420 & 0.4437 \end{bmatrix},$$

$$\mathcal{X}_1 = \left\{ x \in \mathbf{R}^2 : \begin{bmatrix} 1 & 0 \\ 1 & -1 \end{bmatrix} x \succeq 0 \right\},$$

$$\mathcal{X}_2 = \left\{ x \in \mathbf{R}^2 : \begin{bmatrix} 1 & 1 \\ -1 & 1 \end{bmatrix} x \succeq 0 \right\},$$

$$\mathcal{X}_3 = \left\{ x \in \mathbf{R}^2 : \begin{bmatrix} -1 & 0 \\ -1 & -1 \end{bmatrix} x \succeq 0 \right\},$$

it can be verified that the first and the second subsystems are unstable, and the third one is stable.

Set contractility rate $\delta = 0.8$, and dense size $\gamma = 0.01$. Let $y_k = -1 + \gamma k$, $k = 0, 1, \ldots, 2/\gamma$, which are equally distributed on the one-dimensional unit ball $[-1, 1]$. By projecting y_k into the two-dimensional upper and lower half unit spheres, respectively, we have $x_k^u = [y_k, 1 - |y_k|]^T$ and $x_k^l = [y_k, -1 + |y_k|]^T$ for $k = 0, 1, \ldots, 2/\gamma$. The next step is to determine T_x by simulation for each initial state $x \in \{x_k^u, x_k^l, k = 0, 1, \ldots, 2/\gamma\}$. It was found that each initial state is δ-contractive within three steps. Simple calculation shows that the radius of N_x is not less than 0.0092, which corresponds to $r = 0.0065$ when projected into the one-dimensional space. In this way, the interval $(y_k - r, y_k + r)$ is a contractility neighbor of y_k. It is clear that

$$[-1, 1] \subset \bigcup_{k=0}^{2/\gamma} (y_k - r, y_k + r),$$

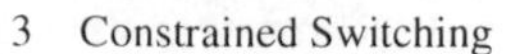

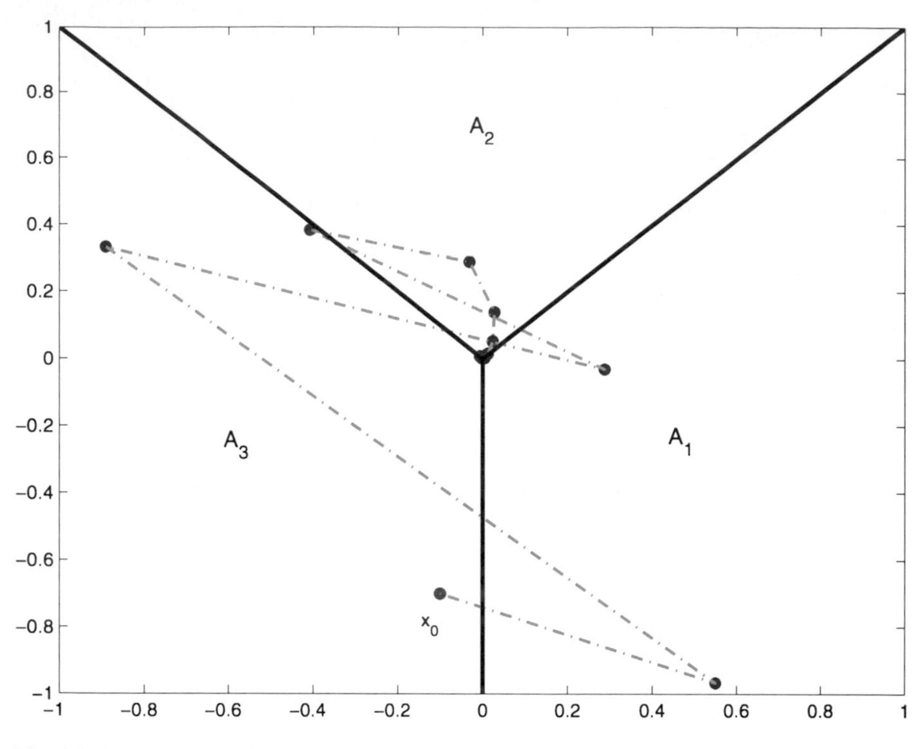

Fig. 3.9 Phase portrait of the conewise linear system

which means that the conewise linear system is exponentially stable. Figure 3.9 depicts a sample phase portrait, which stays and converges within $\mathcal{X}_2$ after several switches between $\mathcal{X}_1$ and $\mathcal{X}_3$.

3.4 Dwell-Time Switching

For switched linear systems, it is well known that the exponential stability of the subsystems does not necessarily imply stability of the switched systems under arbitrary switching. However, under a switching law with sufficiently large dwell time, the exponential stability of the subsystems does imply the asymptotic stability of the switched systems, due to the fact that any exponentially stable system admits a finite transient process. This arises an interesting problem of finding the least dwell time such that the switched system is globally stable under any switching signal with the dwell time. The problem, which we refer to the *least stable dwell time problem*, is both theoretically challenging and practical appealing. Indeed, the least stable dwell time is a measure of fault tolerance for systems with a nominal subsystem and a substitute subsystem which works when the nominal loop fails to work properly. It is clear that the smaller the least stable dwell time, the better ability of fault tolerance. Theoretically, the least stable dwell time captures the subtle property of marginal transition among subsystems, which is a unique and important phenomenon for switched systems.

On the other hand, when the subsystems are unstable, the switched system can still be stabilizable by means of proper switching among the subsystems. A well-known condition for stabilizability is the existence of a stable convex combination of the subdynamics, for which high-frequency switching can stabilize the switched system. However, high-frequency switching is usually harmful and undesirable in most applications. This arises the problem of finding stabilizing switching with the largest possible dwell time, which we refer to the *slow switching (design) problem*.

In this section, we investigate the above problems in a heuristic manner. As the problems are very difficult, we mainly focus on two-form switched linear systems.

3.4.1 Preliminaries

For a time-driven switching signal σ with the switching sequence

$$(t_0, i_0), (t_1, i_1), \dots, (t_l, i_l), \quad l \leq +\infty,$$

its *dwell time* is the least duration of switching, that is,

$$d_\sigma = \inf_{k=1}^{l} \{t_k - t_{k-1}\}.$$

It is clear that the dwell time is nonnegative, and positive dwell time implies well-definedness, but the converse is not true. For a nonnegative real number τ, let $\mathcal{S}_\tau$ be the set of switching signals with dwell time greater than or equal to τ, that is,

$$\mathcal{S}_\tau = \{\sigma \in \mathcal{S} : d_\sigma \geq \tau\}.$$

As an extension, we define the *average dwell time* of switching signal σ as

$$d_\sigma^a = \inf \lim_{k \to l} \frac{t_k - t_0}{k}.$$

For a nonnegative real number τ, the set of switching signals with average dwell time greater than or equal to τ is denoted by $\mathcal{S}_\tau^a$. It is clear that the average dwell time is larger than or equal to the dwell time for any switching signal, and $\mathcal{S}_\tau$ is a subset of $\mathcal{S}_\tau^a$ for any τ.

For the switched linear system

$$\dot{x}(t) = A_{\sigma(t)} x(t), \tag{3.53}$$

let $\mathcal{S}^*$ be the set of switching signals that make the switched system asymptotically stable. A nonnegative real number τ is said to be a *stable dwell time* if each switching signal in $\mathcal{S}_\tau$ makes the switched system asymptotically stable, and the *least stable dwell time* is defined to be

$$\tau_* = \inf\{\tau : \mathcal{S}_\tau \subseteq \mathcal{S}^*\},$$

which is the infimum over the stable dwell times. The least stable average dwell time can be defined in the same manner.

It can be seen that, when all the subsystems are exponentially stable, the least stable (average) dwell time is always finite. Indeed, there exist positive real number pairs (α_j, β_j), $j = 1, 2, \ldots, m$, such that

$$\left\| e^{A_j t} \right\| \le \beta_j e^{-\alpha_j t} \quad \forall t \in \mathcal{T}_0. \tag{3.54}$$

Suppose that the switching signal σ is with switching sequence

$$(t_0, i_0), (t_1, i_1), \ldots, (t_l, i_l).$$

Then the state transition matrix is

$$\Phi(t, t_0, \sigma) = e^{A_{i_k}(t-t_k)} e^{A_{i_{k-1}}(t_k - t_{k-1})} \cdots e^{A_{i_0}(t_1 - t_0)} \quad \forall t \in (t_k, t_{k+1}].$$

Define

$$\tau_1 = \max\left\{ \frac{\ln \beta_1}{\alpha_1}, \ldots, \frac{\ln \beta_m}{\alpha_m} \right\}.$$

It can be seen that any switching signal with dwell time τ_1 makes the state transition matrix norm contractive. As a result, τ_1 is an upper bound of the least stable dwell time. Note that such an upper bound is norm-dependent.

Another approach for approximating the least stable dwell time is the Lyapunov approach. Suppose that $P_1, \ldots, P_m$ is a sequence of symmetric and positive definite matrices. Let τ be a positive real number satisfying

$$A_i^T P_i - P_i A_i < 0 \quad \forall i \in M$$

and

$$e^{A_i^T \tau} P_j e^{A_i \tau} < P_i \quad \forall i, j \in M, \ i \ne j.$$

Then, by taking the piecewise quadratic Lyapunov function

$$V(x, t) = x^T P_{\sigma(t)} x,$$

it can be seen that, for any switching signal with dwell time τ, the Lyapunov function is strictly decreasing along switching instants. Hence, the switched system is asymptotically stable. As a result, τ is an upper bound of the least stable dwell time.

The above idea of using piecewise quadratic functions can be further extended to piecewise norm functions. Indeed, the following result has been established.

Lemma 3.48 (See [261]) *τ is a stable dwell time for the switched linear system iff there exist norms v_i, $i = 1, \ldots, m$, such that*

$$\begin{aligned} &v_i\left(e^{A_i t} x\right) - v_i(x) < 0 \quad \forall x \ne 0, \ t > 0, \\ &v_j\left(e^{A_i \tau} x\right) - v_i(x) < 0 \quad \forall x \ne 0, \ i, j \in M, \ i \ne j. \end{aligned} \tag{3.55}$$

To find a lower bound of the least stable dwell time, we examine the sampled-data system

$$x_{k+1} = D_{\sigma}^{\tau} x_k$$

with sampling period τ, and $D_j^{\tau} = e^{A_j \tau}$, $j = 1, \ldots, m$. It can be seen that a necessary condition for τ to be a stable dwell time is the guaranteed asymptotic stability of the sampled switched system, which can be verified by calculating the spectral radius of the sampled system. When the generalized spectral radius is larger than or equal to one, then τ is a lower bound of the least stable dwell time. Suppose that we have found an upper bound τ_1 for τ_*. Here we provide a random search procedure for computing such a lower bound.

Step 1. Fix an integer K and set $\tau_0 := 0$ and $k := 0$.
Step 2. Randomly choose a sampling period τ from the interval (τ_0, τ_1). Sample the switched system with the period.
Step 3. Calculate the generalized spectral radius $\bar{\lambda}$ of the sampled system. If $\bar{\lambda} \geq 1$, then set $\tau_0 := \tau$. Set $k := k + 1$.
Step 4. If $k \leq K$, then go back to Step 2. Otherwise, return τ_0 as the lower bound estimate of the least stable dwell time.

Example 3.49 For the planar two-form switched linear system with

$$A_1 = \begin{bmatrix} 0.0957 & 1.4148 \\ -0.9812 & -0.3837 \end{bmatrix}, \qquad A_2 = \begin{bmatrix} 0.0517 & -0.5547 \\ 0.7801 & -0.4392 \end{bmatrix},$$

by solving the Lyapunov equations

$$A_i^T P_i + P_i A_i = -I_2, \quad i = 1, 2,$$

we obtain (cf. (3.54))

$$\alpha_i = \frac{1}{2\lambda_{\max}(P_i)}, \qquad \beta_i = \sqrt{\lambda_{\max}(P_i)/\lambda_{\min}(P_i)}, \quad i = 1, 2,$$

which further gives $\tau_1 = 3.4630$. On the other hand, applying the above search procedure (with $K = 1000$) gives $\tau_0 = 1.2425$. Thus the least stable dwell time $\tau_* \in (1.2425, 3.4630)$.

Next, we turn to the problem of slow switching, where the largest possible stabilizing dwell time is to be sought. To be more precise, a nonnegative real number τ is a *stabilizing dwell time* if there exists a switching signal in $\mathcal{S}_\tau$ that steers the switched system asymptotically stable, and the *largest stabilizing dwell time* is defined to be

$$\tau^* = \sup\{\tau : \mathcal{S}_\tau \cap \mathcal{S}^* \neq \emptyset\},$$

which is the supremum over the stabilizing dwell times. The largest stabilizing average dwell-time can be defined in the same manner.

For a two-form switched linear system with at least one unstable subsystem, we define the set

$$\Delta = \left\{(\tau_1, \tau_2) : \tau_1 > 0, \tau_2 > 0, e^{A_1\tau_1} e^{A_2\tau_2} \text{ is Schur}\right\}.$$

It can be seen that, when $(\tau_1, \tau_2) \in \Delta$, the switched system is asymptotically stable if we take the periodic switching signal

$$\sigma(t) = \begin{cases} 1 & \text{if} \quad \mathrm{mod}\,(t, \tau_1 + \tau_2) < \tau_1, \\ 2 & \text{otherwise.} \end{cases} \tag{3.56}$$

As a result, $\min\{\tau_1, \tau_2\}$ is a lower bound of the largest stabilizing dwell time, and $(\tau_1 + \tau_2)/2$ is a lower bound of the largest stabilizing average dwell time. To obtain a tighter estimate, we use the quantities

$$\sup_{(\tau_1,\tau_2)\in\Delta} \min\{\tau_1, \tau_2\} \quad \text{and} \quad \sup_{(\tau_1,\tau_2)\in\Delta} (\tau_1 + \tau_2)/2 \tag{3.57}$$

to serve as lower bounds for the largest stabilizing dwell time and the largest stabilizing average dwell time, respectively.

It is interesting to examine the open set Δ. It is clear that, when A_1 and A_2 admit a stable convex combination, 0 is an accumulating point of the set, and vice versa. Even in this case, the set is not necessarily convex. In fact, the set may contain disconnected subsets, and each connected subset is possibly nonconvex. For example, for the two-form switched system with

$$A_1 = \begin{bmatrix} -2.1 & 1.4 & 5.9 \\ -8.0 & -5.7 & -0.2 \\ 0.6 & 5.8 & 1.6 \end{bmatrix}, \qquad A_2 = \begin{bmatrix} 1.0 & -0.5 & -2.8 \\ 4.8 & -5.0 & 1.1 \\ -1.0 & -6.6 & -2.1 \end{bmatrix}, \tag{3.58}$$

it can be verified that $wA_1 + (1-w)A_2$ is Hurwitz for $0.33 \le w \le 0.75$. Therefore, for any $w \in (0.33, 0.75)$, there exists a positive real number τ such that the segment from $(0, 0)$ to $(w\tau, (1-w)\tau)$ belongs to the set Δ. Figure 3.10 depicts the set that contains several isolated subregions. It is clear that the largest subregions are not convex. The nonconvexity nature makes the effective computation of the set very difficult.

Another observation is that a periodic switching signal with the largest dwell time as defined in (3.57) steers the switched system marginally stable with possible very long transient process and very large overshoot. In fact, even for a dwell time near the boundary of the set Δ, the corresponding transient process may be quite long. For system (3.58), if we take the periodic switching signal with

$$\tau_1 = 2.1009, \qquad \tau_2 = 2.1132,$$

then it can be verified that the matrix $e^{A_1\tau_1} e^{A_2\tau_2}$ admits spectra

$$\{0.6384, -0.7891, -0.0000\},$$

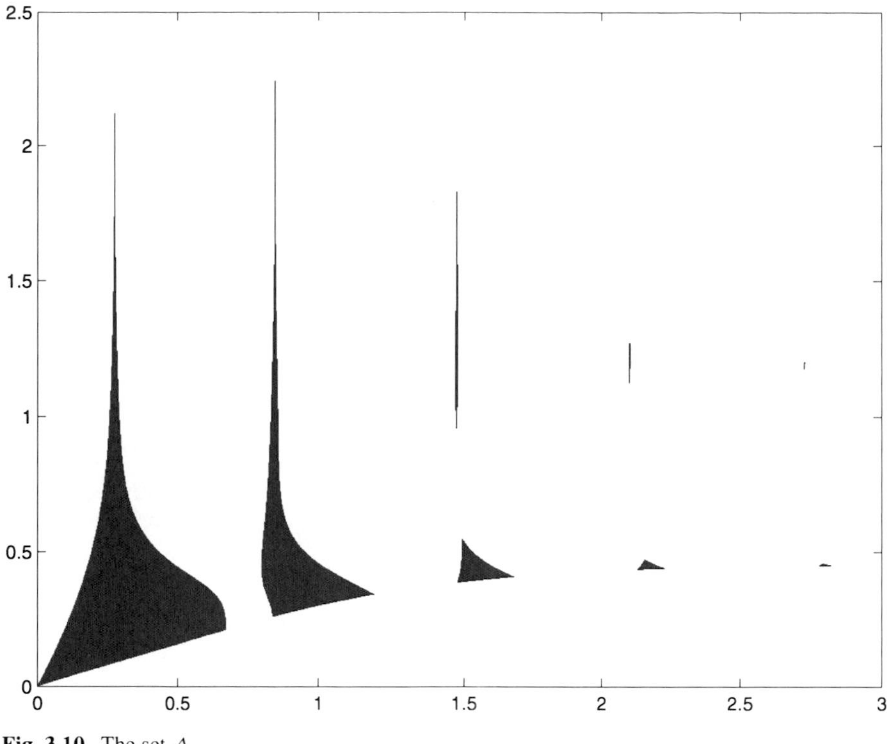

Fig. 3.10 The set Δ

which means that the transition matrix is Schur stable. This implies that $(\tau_1, \tau_2) \in \Delta$. Figure 3.11 shows the state trajectory of the switched system under the periodic switching with initial state $x_0 = [1, -1, 0]^T$. It is clear that the transient process is quite long and the overshoot is very large.

To improve the system performance, we introduce the notion of ϵ-robust stabilizing dwell time pair. Precisely, given $\epsilon > 0$, a pair of positive real numbers (τ_1, τ_2) is said to be an *ϵ-robust stabilizing dwell time pair* if

$$\mathbf{B}_\epsilon(\tau_1, \tau_2) \in \Delta.$$

That is, the ball centered at (τ_1, τ_2) with radius ϵ is inside the stabilizing dwell time set Δ. The set of such pairs, denoted Δ_ϵ, is a strict subset of the set Δ. Instead of computing the largest dwell time as in (3.57), we are to find the largest ϵ-robust stabilizing dwell time defined by

$$\sup_{(\tau_1,\tau_2)\in\Delta_\epsilon} \min\{\tau_1, \tau_2\}. \tag{3.59}$$

The next random searching procedure aims at computing the dwell time.

Initial Setting

Set $\tau_1 := 0$ and $\tau_2 := 0$. Set two searching steps, k_1 and k_2. Set ϵ.

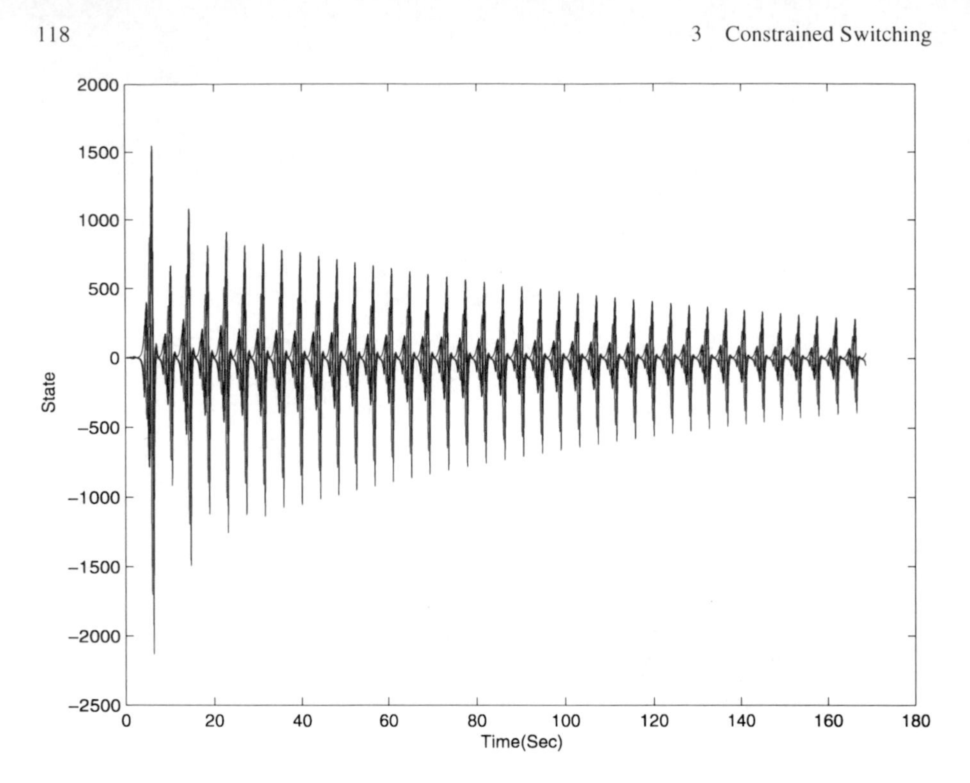

Fig. 3.11 State trajectory under periodic switching

Finding Dwell Times

(1) Set $k_1 := k_1 - 1$ and input two random positive real numbers a and b.
(2) If $k_1 <= 0$, go to Step 4. Otherwise, check the Schur stability of the matrix $e^{A_1(\tau_1+a)}e^{A_2(\tau_2+b)}$. If not, go to Step 1.

Checking ϵ-robustness

(3.1) Set $j := k_2$.
(3.2) If $j <= 0$, go to Step 3.4. Otherwise, input two random numbers c and d between $-\epsilon$ and ϵ.
(3.3) Check the Schur stability of the matrix $e^{A_1(\tau_1+a+c)}e^{A_2(\tau_2+b+d)}$. If not, then go to Step 1. Otherwise, set $j := j - 1$ and go to Step 3.2.
(3.4) Set $\tau_1 := \tau_1 + a$ and $\tau_2 := \tau_2 + b$. Go to Step 1.

Conclusion

(4) Set $\tau := \min(\tau_1, \tau_2)$, which is an estimate of the largest ϵ-robust dwell time.

Finally, let us examine the behavior of system (3.58) under various ϵ's. As we mentioned before, the transient performance is very poor when $\epsilon = 0$ (cf. Fig. 3.11). When $\epsilon = 0.01$, the computed largest ϵ-robust dwell time is 0.8503 with $\tau_1 = 0.8503$ and $\tau_2 = 0.8524$. Figure 3.12 shows the state trajectory under the periodic switching, which admits much better transient process than that of $\epsilon = 0$. When

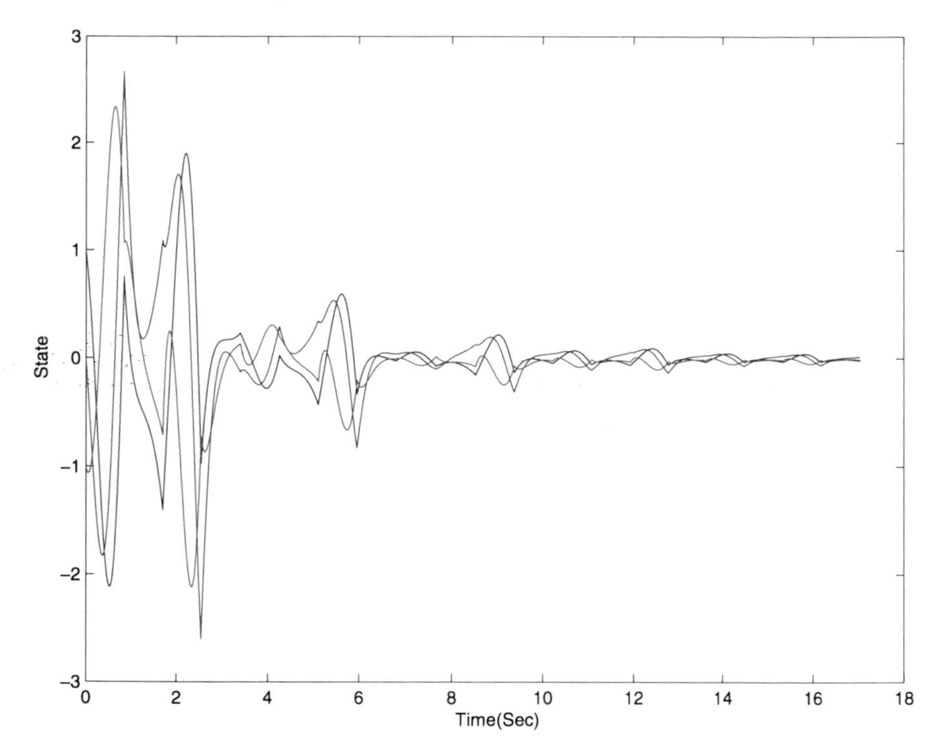

Fig. 3.12 State trajectory of under periodic switching with $\epsilon = 0.01$

$\epsilon = 0.1$, the computed largest ϵ-robust dwell time is 0.3350 with $\tau_1 = 0.3350$ and $\tau_2 = 0.3634$. The corresponding state trajectory, shown in Fig. 3.13, has smaller settling time and overshoot than those with smaller ϵ's. Notice the interesting trade-off between the switching period and the system state performance, which is in fact the trade-off between the discrete state quality and continuous state quality.

3.4.2 Homogeneous Polynomial Lyapunov Approach

As stated in Lemma 3.48, τ is a stable dwell time if inequality (3.55) holds from some norms v_i, $i = 1, \dots, m$. This indicates a way of approximating the least dwell time. In this subsection, we present a computational approach that searches for a least possible stable dwell time when the norms are adopted from homogeneous polynomial functions.

Fix a natural number d. For a positive definite homogeneous polynomial with degree $2d$ given by

$$p(x) = \sum_{\substack{i_1,\dots,i_n \geq 0 \\ i_1+\cdots+i_n=2d}} a_{i_1,\dots,i_n} x_1^{i_1} \cdots x_n^{i_n},$$

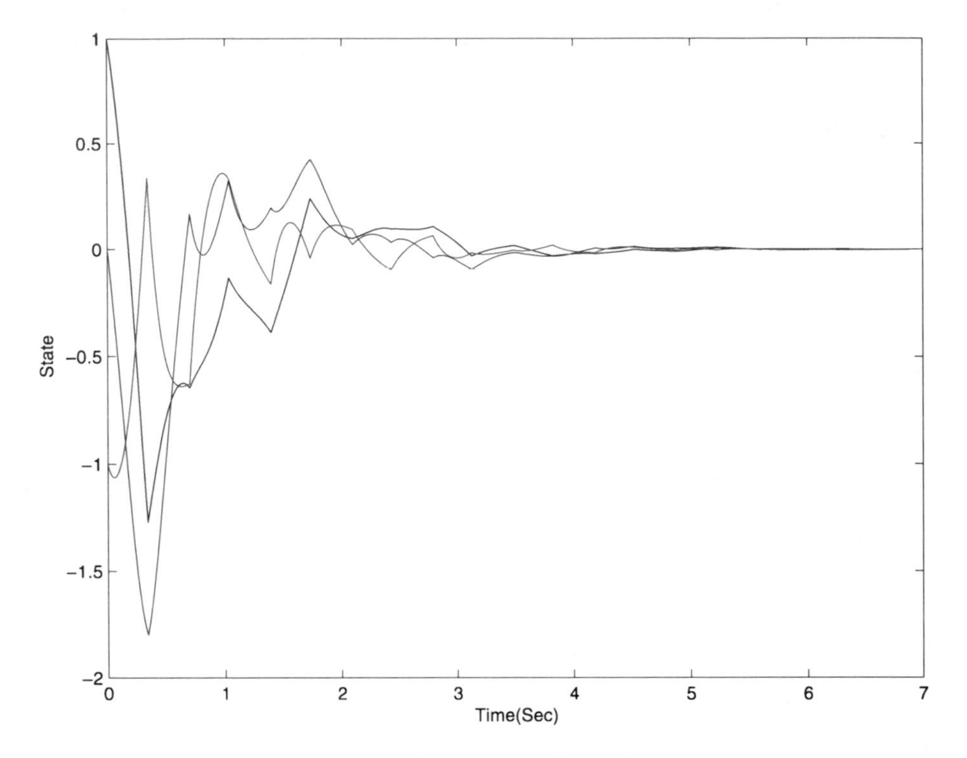

Fig. 3.13 State trajectory of under periodic switching with $\epsilon = 0.1$

a norm on $\mathbf{R}^n$ is induced as $v(x) = p^{\frac{1}{2d}}(x)$. It is clear that, if there exist a sequence of positive definite homogeneous polynomials $p_1, \ldots, p_m$ such that

$$\begin{aligned} &p_i(e^{A_i t}x) - p_i(x) < 0 \quad \forall x \neq 0,\ t > 0, \\ &p_j(e^{A_i \tau}x) - p_i(x) < 0 \quad \forall x \neq 0,\ i, j \in M,\ i \neq j, \end{aligned} \tag{3.60}$$

for some τ, then the induced norms $v_i, i = 1, \ldots, m$ satisfy (3.55), which means that τ is a stable dwell time for the switched linear system. Note that when $d = 1$, the set of Lyapunov functions degenerate into the set of piecewise quadratic Lyapunov functions, and the computation of the quantity can be conducted by means of linear matrix inequalities.

As discussed in Sect. 2.4.1, we take the set of positive definite homogeneous polynomials that admit a sum-of-squares expression as

$$p(x) = \left(x^{[d]}\right)^T P x^{[d]},$$

where P is a positive definite matrix, and $x^{[d]}$ is the d-lift of state x. Denote by $A_i^{[d]}$ the induced matrix such that $(A_i x)^{[d]} = A^{[d]} x^{[d]}$ and

$$\frac{dx^{[d]}(t)}{dt} = A_{\sigma(t)}^{[k]} x^{[d]}(t).$$

Let $N_n^d = \binom{n+d-1}{d}$. It can be seen that $x^{[d]}$ is N_n^d-dimensional and that the matrix $A^{[d]}$ is of $N_n^d \times N_n^d$.

Lemma 3.50 *τ is a stable dwell time of the switched system if there exists a sequence of positive definite matrices $Q_i \in \mathbf{R}^{N_n^d \times N_n^d}$, $i = 1, \dots, m$, such that*

$$\begin{aligned}\left(A_i^{[d]}\right)^T Q_i + Q_i A_i^{[d]} < 0 \quad \forall i \in M,\\ \exp\left(\left(A_i^{[d]}\right)^T \tau\right) Q_j \exp\left(A_i^{[d]} \tau\right) < Q_i \quad \forall i, j \in M,\ i \neq j.\end{aligned} \tag{3.61}$$

Proof It is clear that $\exp(A_i^{[d]}\tau) = (e^{A_i \tau})^{[d]}$. The conclusion follows from (3.60) with $p_i(x) = (x^{[d]})^T Q_i x^{[d]}$. □

Note that, when τ satisfies the second inequality of (3.61), any real number $\varsigma > \tau$ also satisfies the inequality with τ replaced by ς. Indeed, it follows from the first inequality of (3.61) that

$$e^{(A_i^{[d]})^T(\varsigma-\tau)} Q_i e^{A_i^{[d]}(\varsigma-\tau)} \le Q_i, \quad i \in M.$$

Pre- and post-multiplying the second inequality of (3.61) by $e^{(A_i^{[d]})^T(\varsigma-\tau)}$ and $e^{A_i^{[d]}(\varsigma-\tau)}$, respectively, we obtain

$$\exp\left(\left(A_i^{[d]}\right)^T \varsigma\right) Q_j \exp\left(A_i^{[d]} \varsigma\right) < e^{(A_i^{[d]})^T(\varsigma-\tau)} Q_i e^{A_i^{[d]}(\varsigma-\tau)} \le Q_i.$$

This means that, if we take τ as a variable, then inequality (3.61) admits a set of solutions with a unique infimum, which is an upper bound of the least stable dwell time.

The above discussion enable us to formulate the optimization problem

$$\begin{gathered}\inf \tau\\ \exists Q_i > 0 \quad \text{s.t.} \quad (3.61) \text{ holds.}\end{gathered}$$

The problem can be solved by means of semi-definite programming technique. The resultant solution, denoted τ_*^d, is the smallest upper bound of τ_* when the Lyapunov function is in the set of homogeneous sum-of-squares with degree $2d$.

Example 3.51 Consider the planar two-form switched system with

$$A_1 = \begin{bmatrix} 0 & 1 \\ -2 & -1 \end{bmatrix}, \qquad A_2 = \begin{bmatrix} 0 & 1 \\ -9 & -1 \end{bmatrix}.$$

Taking $d = 1, 2, 3, 4$, we obtain the upper bounds τ_*^d presented in the table below.

d	1	2	3	4
τ_*^d	0.6222	0.6079	0.6073	0.6073

It is clear that 0.6073 is an upper bound of the least stable dwell time. On the other hand, we can prove that 0.6073 is also a lower bound of τ_*. Indeed, let $h_1 = 0.8800$ and $h_2 = 0.6073$. It is straightforward to verify that the state transition matrix $e^{A_1h_1}e^{A_2h_2}$ admits the spectra $\{-1.0000, -0.2260\}$, which shows the existence of a periodic switching that steers the switched system marginally stable.

To conclude, we have $\tau_* = 0.6073$, and the homogeneous polynomial Lyapunov function approach provides a nonconservative estimate of the least stable dwell time for this example.

3.4.3 Combined Switching

Based on the largest ϵ-robust stabilizing dwell time, we can further enlarge the dwell time by introducing a combined switching mechanism, that is, the state-feedback switching with fixed dwell time.

Next, suppose that τ_1 and τ_2 are fixed and that the matrix $e^{A_1\tau_1}e^{A_2\tau_2}$ is Schur stable. Clearly, there exist a positive definite matrix P and a real number $\delta \in (0, 1)$ such that

$$\left(e^{A_2\tau_2}e^{A_1\tau_1}\right)^T Pe^{A_2\tau_2}e^{A_1\tau_1} \le (1-\delta)P. \tag{3.62}$$

Assume without loss of generality that $\tau_1 < \tau_2$.

Fix a real number $\eta \in (0, \frac{\delta}{1-\delta})$. Let $Q_i = A_i^T P + PA_i$, $i = 1, 2$. Define

$$\nu = \ln\left(1 + 2\eta\|A_1\|\frac{\lambda_{\min}(P)}{\lambda_{\max}(Q_1)}\right)/(2\|A_1\|),$$

where $\lambda_{\min}(P)$ and $\lambda_{\max}(Q_1)$ are the smallest and the largest eigenvalues of P and Q_1, respectively.

Suppose that x is initialized at $x(t_0) = x_0$. Define the switching sequence

$$\sigma(t_0) = \begin{cases} 1 & \text{if } x_0^T Q_1 x_0 \le x_0^T Q_2 x_0, \\ 2 & \text{otherwise,} \end{cases}$$

$$t_{k+1} = \begin{cases} \inf\{t > t_k + \nu : x(t)^T Q_1 x(t) \ge 0\} + \tau_1 & \text{if } \sigma(t_k) = 1, \\ \inf\{t \ge t_k + \tau_2 : x(t)^T Q_2 x(t) \ge 0\} & \text{if } \sigma(t_k) = 2, \end{cases} \tag{3.63}$$

$$\sigma(t_{k+1}) = \begin{cases} 2 & \text{if } \sigma(t_k) = 1, \\ 1 & \text{if } \sigma(t_k) = 2, \end{cases} \quad k = 0, 1, \ldots.$$

According to this strategy, when the first subsystem is activated, it is first kept active for time ν, and then it must be kept active for the additional dwell time τ_1 after the state-feedback switching time is due. On the other hand, if the second subsystem is activated, it must be kept active for the dwell time τ_2, and then the state-feedback switching law decides the next switching time. In this way, the dwell time of the combined switching is greater than or equal to $\min(\tau_1 + \nu, \tau_2)$, which is larger than the original dwell time τ_1.

The following result states the main property of the proposed switching strategy.

Theorem 3.52 *The switched system is exponentially stable under switching law* (3.63).

Proof Let $V(x) = x^T Px$ be the Lyapunov candidate. Fix an initial state and suppose that the switching time sequence is $t_0, t_1, \dots$. Taking any switching instant t_k with $\sigma(t_k) = 1$, we examine the monotonicity of the Lyapunov candidate along with the time interval $[t_k, t_{k+2})$. According to the switching law, the time-driven period is

$$[t_k, t_k + \nu) \cup [t_{k+1} - \tau_1, t_{k+1} + \tau_2),$$

while the state-driven period is

$$[t_k + \nu, t_{k+1} - \tau_1) \cup [t_{k+1} + \tau_2, t_{k+2}).$$

Note that the Lyapunov candidate function is decreasing during the state-driven period. During the time-driven period $[t_{k+1} - \tau_1, t_{k+1} + \tau_2)$, it follows from (3.62) that

$$V\big(x(t_{k+1} + \tau_2)\big) \le (1 - \delta) V(t_{k+1} - \tau_1).$$

For the time-driven period $[t_k, t_k + \nu)$, we have

$$\int_{t_k}^{t_k+\nu} x^T(t) Q_1 x(t)\, dt \le \eta V\big(x(t_k)\big),$$

which further implies that

$$V\big(x(t_k + \nu)\big) < \frac{1}{1-\delta} V\big(x(t_k)\big).$$

Combining the above facts together yields

$$V\big(x(t_{k+2})\big) < V(x_{t_k}).$$

As the time-driven period is with fixed length $\tau_1 + \tau_2 + \nu$ for each switching cycle, the Lyapunov candidate converges exponentially, and the switched system is exponentially stable.

To illustrate the effectiveness of the proposed switching law, we reexamine the example as in (3.58). We take $\tau_1 = 0.3350$ and $\tau_2 = 0.3634$ that correspond to the largest ϵ-robust stabilizing dwell time with $\epsilon = 0.1$. Furthermore, take

$$P = \begin{bmatrix} 1.1125 & 0.0681 & 0.0878 \\ 0.0681 & 2.1136 & 1.6565 \\ 0.0878 & 1.6565 & 3.4735 \end{bmatrix}, \qquad \eta = \delta = 0.2180.$$

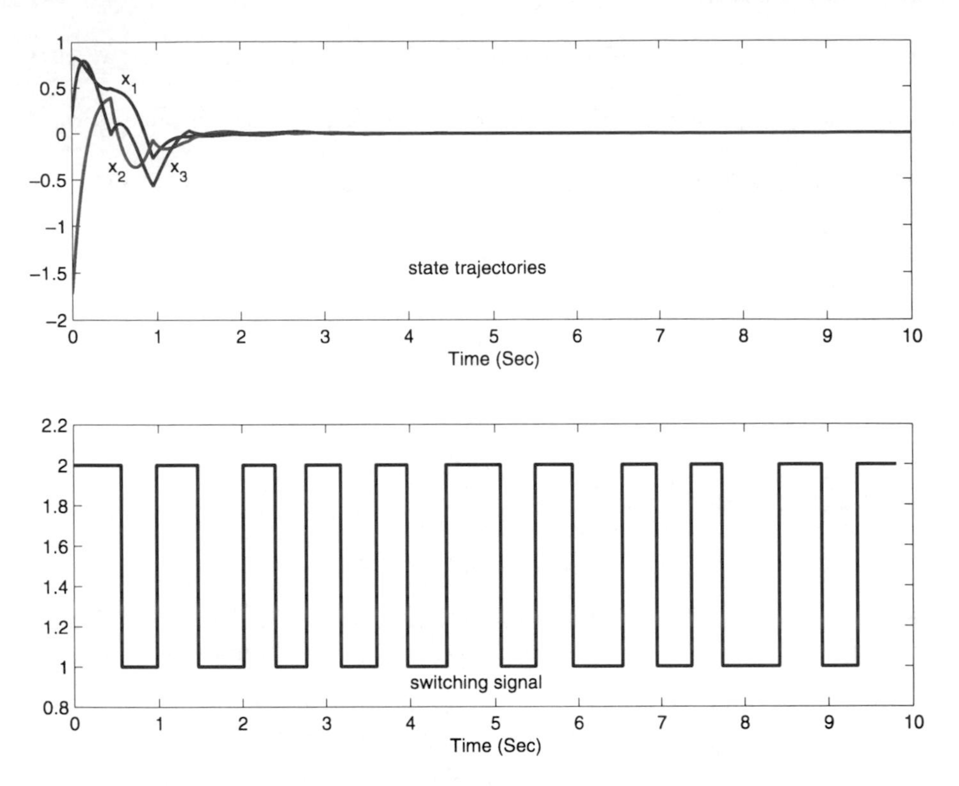

Fig. 3.14 Sample state trajectory (*upper*) and switching signal (*lower*)

As a result, we have $\nu = 0.0249$. It follows that switching law (3.63) admits the dwell time of 0.3599, which is larger than 0.3350 for the periodic switching law (3.56). Moreover, due to the introduction of the state-driven mechanism, the average dwell time can be further enlarged. Figure 3.14 depicts the state trajectory and switching signal of the switched system under switching law (3.63) with initial state $x_0 = [0.8040, -1.7240, 0.1741]^T$. For this piece of sample switching path, the average dwell time is 0.4672 sec, which is one third larger than that of periodic switching (3.56), 0.3492 sec. □

3.5 Notes and References

Stability of stochastic systems is not a new topic, and the history can be traced back to the 1950s. Systematic investigations were made by quite a few scholars such as Bharucha [27], Kozin [133], Kushner [136], and Khasminskii [131]. In particular, Kozin [133] established that exponential mean square stability implies almost sure stability, as stated in Theorem 3.11. In the survey paper [134], Kozin clarified some confusing concepts and explained the relationships among the stability concepts.

Much progress has been made in understanding various stabilities since the 1990s. The reader is referred to [43, 59, 74, 79, 140, 167] and the references therein.

The equivalence among the mean square stabilities, as stated in Items (1)–(3) of Theorem 3.3, was obtained by Feng et al. [79], which was extended to the δ-moment stabilities by Fang [70]. The stability criteria, Items (4) and (5) of Theorem 3.3, were presented in [79] and [160], respectively. The proof of Theorem 3.3 was combined from [70, 79, 160]. Lemma 3.4 was adapted from [160]. For almost sure stability, while many sufficient conditions and necessary conditions were presented in the literature (see, e.g., [70, 71, 144, 166]), it was well recognized that, as the Lyapunov component is notoriously difficult to calculate [10, 251], almost sure stability is much harder to tackle than moment stabilities [61]. It was established, however, that as $\delta \to 0$, the stability region (w.r.t. subsystem matrices) of δ-moment stability shrinks exactly to the stability region of almost sure stability [70, 73]. While this reveals the insightful connection between almost sure stability and moment stability, it provides little help in practical verification of almost sure stability, due to the lack of verifiable moment stability criteria. The necessary and sufficient condition for almost sure stability, Theorem 3.16, was adapted from [42]. Though the criterion is of limited practical value, it does provide a counterpart of the transition contraction criterion for deterministic switched systems, as presented in [216, Prop. 3]. Lemma 3.14 was adapted from [70].

It should be noted that the material here is limited to continuous-time homogeneous jump linear systems. For the discrete-time counterpart and/or jump diffusion counterpart, the reader is referred to [59, 70, 72, 144, 160] and the references therein.

Piecewise linear systems are switched linear systems with state-space-partition-based switching. Theoretically, piecewise linear systems are powerful in approximating highly nonlinear dynamic systems [210], in representing interconnections of linear systems and finite automata [212], and in characterizing control systems with fuzzy logics [76]. From the model point of view, piecewise linear systems provide an equivalent framework to the well-known linear complementary systems [101, 102, 143] and mixed logical dynamical systems [23, 174]. Primary topics in the literature include well-posedness [116, 265], controllability, reachability, and observability [11, 49, 202], stability and stabilization [93, 94, 119, 125, 126], and computational complexity [40], among others. While the study on stability of piecewise linear systems has been attracting increasing attention, the stability problem is found to be notoriously challenging [29, 230]. One reason for this is the fact that switching is autonomous based on cell partitions which usually induce highly nonlinear transition maps that are very difficult to characterize. Nevertheless, remarkable progress has been made during the last decades, with much attention being paid to the development of computational approaches. The piecewise quadratic Lyapunov approach, proposed initially by Johansson and Rantzer in [125, 126], provides a rigorous method for analyzing the stability of general piecewise linear systems. To find qualified piecewise quadratic Lyapunov functions, we usually need to further partition the cells, which makes the computation inefficient for higher-dimensional systems. This motivates the development of a natural-partition-based mechanism to reduce the computational burden. The surface Lyapunov function approach, proposed by Goncalves et al. in [92, 93], is exactly such an approach. It focuses on the contractility of the impact maps defined over the switching surfaces instead of the total state space, which captures and utilizes the subtle properties of the impact maps that are generally nonlinear and multivalued. The approach is powerful in coping with piecewise linear systems whose impact maps could be parameterized. Particular but important examples include a linear plant with a nonlinear relay/on–off/saturated feedback. It is clear that the approach only applies to continuous-time systems. An extended scheme is to analyze the transition relationships among different cells, which could possibly be represented by a digraph. By applying the graph theory, it is possible to reduce global graph analysis to that of subgraphs which are invariant and attractive. While simple, the approach applies to both continuous time and discrete time.

The main results in Sect. 3.3.2 were adapted from [126], and the contents in Sect. 3.3.3 were mainly taken from [93]. The transition analysis presented in Sect. 3.3.4 was proposed in [226], where the concept of (weak) transitivity was borrowed from [119]. As a specific class of piecewise linear systems, conewise linear systems admit simpler behavior; for example, a continuous conewise linear system is always well defined as no Zeno phenomenon occurs [50]. Proposition 3.42 reveals another nice property for the system, which naturally follows from the homogeneity of the switched system [8]. The other part of Sect. 3.3.5 was taken from [229].

Switching with positive dwell time or average dwell time was proposed to address the stability problem by Morse and Hespanha in [110, 178], and numerous works could be found along this line in the literature, see, e.g., [56, 90, 99, 197, 278] and the references therein. While simple in idea and popular in literature, finding the least stable dwell time for stability of switched stable linear systems is very difficult due to the fact that the concatenation of two feasible switching signals is not necessarily feasible, which destroys the semigroup nature of the transition matrices [261]. A closely related problem is to find the largest dwell time for stabilizability of switched unstable linear systems, which is also an open problem. The reader is referred to [234, Problem 7.6] for more discussion on the problems.

The homogeneous polynomial Lyapunov function approach presented in Sect. 3.4.2 was mainly adopted from [56], and the reader is referred to [57] for more detailed background about the approach. The other material of Sect. 3.4 was mostly taken from [227].

Chapter 4
Designed Switching

4.1 Preliminaries

To effectively control complex dynamics with either high nonlinearity or large-scale unknown/uncertain parameters, it is a common practice to use the "divide and conquer" strategy. One approach in light of this strategy is the hybrid control scheme, which amounts to designing a set of candidate controllers, each working around a local support, and a switching mechanism coordinating the switching among the candidate controllers. Indeed, this hybrid control scheme integrates the ideas from several well-known conventional control schemes such as gain scheduling, intelligent control, and adaptive control [78, 107, 183]. One good example is hybrid control of nonholonomic systems which are not stabilizable by means of any individual continuous state feedback controller [106, 132]. Even for simple linear time-invariant (LTI) systems, the performance (e.g., transient response) can be improved through controllers/compensators switching [81, 118, 169].

For a hybrid control system, when the candidate controllers are known or designed, the overall system is the switched dynamical system described by

$$x^+(t) = f_{\sigma(t)}\big(x(t)\big), \tag{4.1}$$

where $x(t) \in \mathbf{R}^n$ is the continuous state, $\sigma(t) \in M \overset{\text{def}}{=} \{1, \ldots, m\}$ is the discrete state or switching signal, $f_i : \mathbf{R}^n \mapsto \mathbf{R}^n$ is a Lipschitz continuous vector field with $f_i(0) = 0$ for any $i \in M$, and x^+ denotes the derivative operator in continuous time and the shift forward operator in discrete time.

The issue of this chapter is the stabilizing switching design of switched dynamical system (4.1). For this, we assume that the switching signal is observable and controllable and that we can freely select the switching mechanism.

For clarity, we denote by $\phi(t; t_0, x_0, \sigma)$ the continuous state of system (4.1) at time t with initial condition $x(t_0) = x_0$ and switching path/signal/law σ. Denote ℓ_p-norm by $|\cdot|_p$ for $p \in [1, +\infty]$. For any positive real number r, let

$$\mathbf{B}_r = \big\{x \in \mathbf{R}^n : |x| \le r\big\}, \qquad \mathbf{H}_r = \big\{x \in \mathbf{R}^n : |x| = r\big\}.$$

Z. Sun, S.S. Ge, *Stability Theory of Switched Dynamical Systems*,
Communications and Control Engineering,
DOI 10.1007/978-0-85729-256-8_4, © Springer-Verlag London Limited 2011

Finally, recall that, for a time interval ϖ, $\mathcal{S}_{\varpi}$ is the set of well-defined switching paths defined over the interval, and $\mathcal{S}$ is the set of well-defined switching signals.

Definition 4.1 Switched system (4.1) is said to be

(1) *(uniformly) stabilizable* if for any $\varepsilon > 0$, there exist $\delta > 0$ and a switching law $\{p^x : x \in \mathbf{B}_\delta\}$ such that

$$\left|\phi\left(t; 0, x, p^x\right)\right| \le \varepsilon \quad \forall x \in \mathbf{B}_\delta,\ t \in \mathcal{T}_0 \tag{4.2}$$

(2) *(globally uniformly) switched attractive* if for any $\epsilon > 0$ and $\gamma > 0$, there exist a switching law $\{p^x : x \in \mathbf{B}_\gamma\}$ and a time $T > 0$ such that

$$\left|\phi\left(t; 0, x, p^x\right)\right| \le \epsilon \quad \forall x \in \mathbf{B}_\gamma,\ t \in \mathcal{T}_T \tag{4.3}$$

(3) *(globally uniformly) asymptotically stabilizable* if for any $\varepsilon > 0$, $\epsilon > 0$, and $\gamma > 0$, there exist $\delta > 0$, $T > 0$, and a switching law $\{p^x : x \in \mathbf{R}^n\}$ such that relationships (4.2) and (4.3) hold simultaneously, and
(4) *(globally uniformly) exponentially stabilizable* if there are positive real numbers α and β and a switching law $\{p^x : x \in \mathbf{R}^n\}$ such that

$$\left|\phi\left(t; 0, x, p^x\right)\right| \le \beta e^{-\alpha t}|x| \quad \forall x \in \mathbf{R}^n,\ t \in \mathcal{T}_0$$

The switching law $\{p^x\}$ in the definition is said to be an *(asymptotically/exponentially) stabilizing/switched-attractive switching law*, respectively.

Note that the uniformity is referred to the switching law rather than the initial time as in the conventional stability notions, and asymptotic stabilizability means stabilizability and switched attractivity w.r.t. a (common) switching law. In particular, even when a switched system is both switched attractive and stabilizable, it is not necessarily asymptotically stabilizable. Indeed, if we take the first subsystem with an attractive but unstable origin equilibrium and the second subsystem with an identity motion (that is, each state is an invariant equilibrium), then, it is clear that the switched system is both switched attractive (by assigning $\sigma = \hat{1}$) and stabilizable (by assigning $\sigma = \hat{2}$), but the switched system is not asymptotically stabilizable.

In this chapter, we address the problem of stabilization by designing various switching mechanisms including time-driven switching, state-feedback switching, and mixed-driven switching. The aim is to reveal the capability and limitation of each switching mechanism, to provide a comprehensive understanding of the stabilization problem and to present computation design procedures.

4.2 Stabilization via Time-Driven Switching

In this section, we examine the possibility of achieving stabilizability by means of time-driven switching. For this, we introduce the notion of consistent stabilizability.

Definition 4.2 Switched system (4.1) is said to be *consistently* (*asymptotically, exponentially*) *stabilizable* if there is a consistent switching signal σ such that the system is well defined and uniformly (asymptotically, exponentially) stable.

It is clear that a consistent stabilizable system is also stabilizable in the sense of Definition 4.1. We are interested in whether the converse is still true or not. Indeed, if any stabilizable system is also consistently stabilizable, then the stabilized system is in fact a linear time-varying system with a piecewise constant system matrix, and the problem of stabilization can be seen as a special case of stability problem for linear time-varying systems. This, however, is not generally true, as shown in the sequel.

Consider the switched linear system given by

$$x^+(t) = A_{\sigma(t)}x(t), \tag{4.4}$$

where A_i, $i \in M$, are real constant matrices.

Lemma 4.3 *Suppose that the switched linear system is consistently stabilizable. Then, there is $k \in M$ such that*

$$\sum_{i=1}^{n} \lambda_i(A_k) \le 0 \tag{4.5}$$

in continuous time and

$$\left|\prod_{i=1}^{n} \lambda_i(A_k)\right| \le 1 \tag{4.6}$$

in discrete time, where $\lambda_i(A)$, $1 \le i \le n$, are the eigenvalues of the matrix A.

Proof We proceed with the continuous-time case, and the discrete-time case could be proven in a similar way. Let σ be a consistent switching signal that stabilizes the switched system. Suppose that the switching duration sequence of σ is

$$DS_\sigma = \{(i_0, h_0), (i_1, h_1), \dots\}.$$

If the sequence is finite, i.e., there involve only finite switches in σ, then, it can be seen that the last active subsystem must be stable, and the theorem follows immediately. If the sequence is infinite, it follows from the well-definedness of σ that there involve only finite switches in any finite time. As a consequence, $\sum_{i=1}^{l} h_i \to \infty$ as $l \to \infty$. According to Definition 4.2, by setting $\varepsilon = 1$, there exists $\delta > 0$ such that

$$\|x_0\| \le \delta \quad \Longrightarrow \quad \|\phi(t; 0, x_0, \sigma)\| \le 1 \quad \forall t \ge t_0.$$

In particular,

$$\|e^{A_{i_s}h_s} \cdots e^{A_{i_1}h_1} e^{A_{i_0}h_0} x_0\| \le 1 \quad \forall x_0 \in \mathbf{B}_\delta,\ s = 0, 1, \dots.$$

As a consequence, all entries of the matrices

$$e^{A_{i_0}h_0}, e^{A_{i_1}h_1}e^{A_{i_0}h_0}, \ldots, e^{A_{i_s}h_s}\cdots e^{A_{i_1}h_1}e^{A_{i_0}h_0}, \ldots \tag{4.7}$$

must be bounded by $\frac{1}{\delta}$. Suppose that

$$\varrho = \min_{k \in M}\left\{\sum_{i=1}^{n}\lambda_i(A_k)\right\} > 0.$$

Then, we have

$$\det e^{A_k h} = \exp\left(h\sum_{i=1}^{n}\lambda_i(A_k)\right) \geq e^{\varrho h}, \quad k \in M,\ h > 0.$$

As a result, we have

$$\det e^{A_{i_s}h_s}\cdots e^{A_{i_1}h_1}e^{A_{i_0}h_0} \geq e^{\varrho\sum_{j=0}^{s}h_j} \to \infty \quad \text{as } s \to \infty.$$

This contradicts the boundedness of entries of the matrices. This establishes the former part of the theorem. The latter part can be proven in a similar manner. □

Recall that a switching signal $\theta_{[0,\infty)}$ is said to be *periodic* if there exists a positive time T such that

$$\theta(t+T) = \theta(t) \quad \forall t \geq 0.$$

Proposition 4.4 *If a switched system is consistently asymptotically stabilizable, then, there is a periodic switching signal that asymptotically stabilizes the switched system.*

Proof If there is a subsystem, say, A_k, that is asymptotically stable, then the constant switching signal $\sigma \equiv k$ works. Otherwise, suppose that a switching signal σ with duration sequence

$$DS_\sigma = \{(i_0, h_0), (i_1, h_1), \ldots\}$$

asymptotically stabilizes the switched system. It is obvious that this switching signal must involve infinite switches. From the proof of Lemma 4.3, matrix sequence (4.7) converges to the zero matrix. Consequently, there is a finite number N such that

$$\left\|e^{A_{i_N}h_N}\cdots e^{A_{i_1}h_1}e^{A_{i_0}h_0}\right\| < 1. \tag{4.8}$$

It can be verified that the periodic and synchronous switching path θ with duration sequence

$$DS_\theta = \{(i_0, s_0), \ldots, (i_N, s_N), (i_0, s_0), \ldots, (i_N, s_N), \ldots\} \tag{4.9}$$

asymptotically stabilizes the switched system. □

Estimation (4.8) is very important in analyzing the convergence of the systems. It establishes the contractibility uniformly for all initial states. Indeed, let $\Phi(T, 0, p)$ be the state transition matrix under switching path p over interval $[0, T]$. Inequality (4.8) is equivalent to the contraction of the transition matrix, i.e., $\|\Phi(T, 0, p)\| < 1$. This motivates us to construct a periodic stabilizing switching signal σ as $\sigma(t) = p(t)$ for $t \in [0, T)$ and $\sigma(t + T) = \sigma(t)$ for all t. Under this periodic switching signal, the resultant switched system is in fact a periodic linear time-varying system whose lifting system is Schur stable, and its dynamics is well understood. Note that the stabilizing switching signal is always well defined as it is self-concatenated of the well-defined switching path p.

Corollary 4.5 *For a switched linear system, the following statements are equivalent*:

(i) *the system is consistently asymptotically stabilizable*
(ii) *the system is consistently exponentially stabilizable*
(iii) *the system is periodically asymptotically stabilizable*
(iv) *there exist a natural number l, an index sequence $i_1, \dots, i_l$, and a positive real number sequence $h_1, \dots, h_l$ such that the matrix $e^{A_{i_l} h_l} \cdots e^{A_{i_1} h_1}$ is Schur, and*
(v) *for any real number $s \in (0, 1)$, there exist a natural number $l = l(s)$, an index sequence $i_1, \dots, i_l$, and a positive real number sequence $h_1, \dots, h_l$ such that*

$$\left\| e^{A_{i_l} h_l} \cdots e^{A_{i_1} h_1} \right\| \le s. \tag{4.10}$$

Proof From the proof of Proposition 4.4, (i) implies that there is a finite number N such that

$$\left\| e^{A_{i_N} h_N} \cdots e^{A_{i_1} h_1} e^{A_{i_1} h_1} \right\| = \gamma < 1$$

for some sequences $i_1, \dots, i_N$ and $h_1, \dots, h_N$. Let $l = kN$, where k will be determined later. Define

$$i_{j+\mu N} = i_j \quad \text{and} \quad h_{j+\mu N} = h_j, \qquad j = 1, \dots, N, \ \mu = 1, \dots, k-1.$$

It can be seen that

$$\left\| e^{A_{i_l} h_l} \cdots e^{A_{i_1} h_1} e^{A_{i_1} h_1} \right\| = \left(\left\| e^{A_{i_N} h_N} \cdots e^{A_{i_1} h_1} e^{A_{i_1} h_1} \right\| \right)^k = \gamma^k.$$

Accordingly, for any $s \in (0, 1)$, by letting $k \ge \frac{\ln s}{\ln \gamma}$, inequality (4.10) holds. This means that (i) $\Longrightarrow$ (v). In the same manner, we can prove that (iv) $\Longrightarrow$ (v). Other implications are trivial, and the corollary follows. □

Finally, utilizing the Lemma 4.3, we present two examples that are stabilizable but not consistently stabilizable.

Example 4.6 For the planar continuous-time two-form switched system with

$$A_1 = \begin{bmatrix} 2 & 0 \\ 0 & -1 \end{bmatrix}, \qquad A_2 = \begin{bmatrix} 1 & 1 \\ -1 & 1 \end{bmatrix},$$

it can be verified that condition (4.5) does not hold, and thus the system is not consistently stabilizable.

On the other hand, the system can be made asymptotically stable by means of the following switching law: activating the second subsystem until the state reaches the x_2-axis and then turning to the first subsystem forever.

Example 4.7 For the planar discrete-time two-form switched system with

$$A_1 = \begin{bmatrix} 3 & 0 \\ 0 & 1/2 \end{bmatrix}, \qquad A_2 = \frac{101}{100} \begin{bmatrix} \cos(\pi/18) & \sin(\pi/18) \\ -\sin(\pi/18) & \cos(\pi/18) \end{bmatrix},$$

it can be verified that condition (4.6) does not hold, and hence the system is not consistently stabilizable. On the other hand, the system is asymptotically stabilizable due to the following facts:

(1) for any state y in the cone $\Lambda = \{x \in \mathbf{R}^2 : |\arctan(x_1/x_2)| \le \pi/36\}$, $|A_1 y| < \kappa |y|$ with $\kappa \approx 0.7596$
(2) for any state y outside the cone, applying A_2 by k_y times ($k_y \le 17$) will steer y into the cone, and
(3) $|A_1 A_2^{k_y} y| < \kappa |A_2^{k_y} y| \le \kappa \|A_2\|^{k_y} |y| \le \kappa \|A_2\|^{17} |y| < \mu |y|$ with $\mu \approx 0.8996$

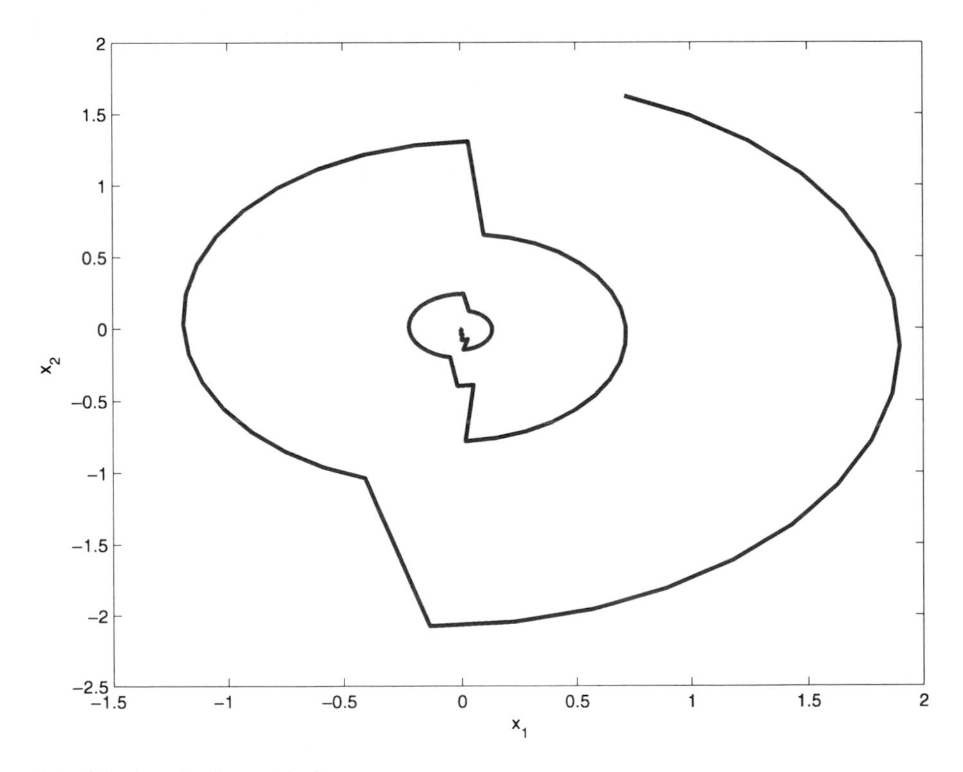

Fig. 4.1 Sample phase portrait

Therefore, a stabilizing switching law is to activate the first subsystem when the state is in cone Λ and to activate the second subsystem when the state is outside the cone. Figure 4.1 shows a sample phase portrait of the switched system.

The above examples exhibit that not all stabilizable systems can be made stable by means of time-driven switching laws. That is, time-driven switching laws are not universal for stabilizing switched systems.

4.3 Stabilization via State-Feedback Switching: The Lyapunov Approach

In this section, we investigate the possibility of using the Lyapunov approach for addressing stabilizability issues. For this, we need to define a proper set of Lyapunov function candidates.

Definition 4.8 Function $V: \mathbf{R}^n \mapsto \mathbf{R}_+$ is said to be a *switched Lyapunov function* for switched system (4.1) if

(1) it is locally Lipschitz continuous
(2) it admits class $\mathcal{K}_\infty$ bounds, that is, there are class $\mathcal{K}_\infty$ functions η_1 and η_2 such that

$$\eta_1\big(|x|\big) \leq V(x) \leq \eta_2\big(|x|\big) \quad \forall x \in \mathbf{R}^n$$

and
(3) the least upper Dini derivative of V along vectors $f_i(x)$, $i \in M$, is negative definite. That is, there is a positive definite continuous function $w: \mathbf{R}^n \mapsto \mathbf{R}_+$ such that

$$\min_{i\in M} \mathcal{D}^+ V(x)|_{f_i} \stackrel{\text{def}}{=} \min_{i\in M} \liminf_{\tau\to 0^+} \frac{V(x+\tau f_i(x)) - V(x)}{\tau} \leq -w(x)$$

in continuous time and

$$\min_{i\in M} \mathcal{D}^+ V(x)|_{f_i} \stackrel{\text{def}}{=} \min_{i\in M} V\big(f_i(x)\big) - V(x) \leq -w(x)$$

in discrete time

Remark 4.9 The local Lipschitz continuity of the Lyapunov function implies that

$$\liminf_{\tau\to 0^+} \frac{V(x+\tau f_i(x)) - V(x)}{\tau} = \liminf_{\tau\to 0^+} \frac{V(\phi(\tau;0,x,\hat{i})) - V(x)}{\tau}$$

in continuous time, where $\hat{i}$ stands for the constant switching signal $\sigma(t) = i\ \forall t$.

Suppose that switched system (4.1) admits a switched Lyapunov function V. Then, the switching law

$$\sigma(t+) = \arg\min_{i\in M}\{\mathcal{D}^+ V(x(t))|_{f_i}\} \tag{4.11}$$

steers the switched system asymptotically stable, provided that the switching law is well defined. Note that chattering might occur when the switching surface

$$\{x \in \mathbf{R}^n : \exists i, j \text{ s.t. } \mathcal{D}^+ V(x)|_{f_i} = \mathcal{D}^+ V(x)|_{f_j}\}$$

is a sliding mode. To avoid such ill-definedness, we can use the hysteresis switching law as follows. Define the regions

$$\Omega_i = \left\{x \in \mathbf{R}^n : \mathcal{D}^+ V(x)|_{f_i} \le \frac{1}{2}\min_{j\in M} \mathcal{D}^+ V(x)|_{f_j}\right\}, \quad i \in M. \tag{4.12}$$

Starting from any given initial state $x(0) = x_0$, select the index of the active subsystem at $t_0 = 0$ as

$$\sigma^{x_0}(t_0) = \arg\min_{i\in M}\{\mathcal{D}^+ V(x_0)|_{f_i}\}.$$

If there are two or more such indices, we simply choose the minimum one. The consecutive switching times/indices can be recursively selected as

$$\begin{aligned} t_k &= \inf\{t > t_{k-1} : \mathcal{D}^+ V(x(t))|_{f_{\sigma(t_{k-1})}} \notin \Omega_{\sigma(t_{k-1})}\}, \\ \sigma^{x_0}(t_k) &= \arg\min_{i\in M}\{\mathcal{D}^+ V(x(t_k))|_{f_i}\}, \quad k = 1, 2, \ldots. \end{aligned} \tag{4.13}$$

It can be proven that this hysteresis switching law is well defined and that it steers the switched system asymptotically stable.

4.3.1 Converse Lyapunov Theorems

Suppose that the switched system is asymptotically stabilizable. We are to prove the existence of a switched Lyapunov function as in Definition 4.8. To this end, we need some technical preliminaries adapted from the proofs for converse Lyapunov theorems of (nonswitched) nonlinear systems (cf. [130, 152]).

Firstly, it follows from the asymptotic stabilizability that there exist a class $\mathcal{K}_\infty$ function α and a class $\mathcal{KL}$ function $\bar{\beta}$ such that for any initial state x, we have a switching signal $\theta^x \in \mathcal{S}$ satisfying

$$|\phi(t; 0, x, \theta^x)| \le \alpha(|x|) \quad \forall t \in \mathcal{T}_0 \tag{4.14}$$

and

$$|\phi(t; 0, x, \theta^x)| \le \bar{\beta}(|x|, t) \quad \forall t \in \mathcal{T}_0.$$

The latter can be equivalently represented by

$$\alpha\big(\big|\phi(t;0,x,\theta^x)\big|\big) \le \beta\big(|x|,t\big) \quad \forall x \in \mathbf{R}^n,\ t \in \mathcal{T}_0, \tag{4.15}$$

with $\beta = \alpha \circ \bar{\beta}$. It is clear that β is a class $\mathcal{KL}$ function.

Secondly, define the function $g: \mathbf{R}^n \mapsto \mathbf{R}_+$ as

$$g(x) = \sup_{t\in\mathcal{T}_0} \inf_{\sigma\in\mathcal{S}_{[0,t]}} \big|\phi(t;0,x,\sigma)\big|.$$

It is clear that the function is well defined and

$$|x| \le g(x) \le \alpha\big(|x|\big) \quad \forall x \in \mathbf{R}^n. \tag{4.16}$$

It can be seen that

$$g(x) \ge \inf_{\sigma\in\mathcal{S}_{[0,t]}} g\big(\phi(t;0,x,\sigma)\big), \quad \forall t \in \mathcal{T}_0,\ x \in \mathbf{R}^n. \tag{4.17}$$

Thirdly, we show that the function g is locally Lipschitz continuous at any nonorigin state. Given $x \neq 0$, let positive real numbers γ_1 and γ_2 be such that $\gamma_2 < |x| < \gamma_1$, and let T be the time constant corresponding to $\gamma = \gamma_1$ and $\epsilon = \gamma_2$ as defined in Definition 4.1. It can be seen that

$$g(x) = \max_{t\in[0,T]} \inf_{\sigma\in\mathcal{S}_{[0,t]}} \big|\phi(t;0,x,\sigma)\big|.$$

As a result, for any $\varepsilon > 0$, there is a switching path $\theta \in \mathcal{S}_{[0,T]}$ such that

$$g(x) \ge \max_{t\in[0,T]} \big|\phi(t;0,x,\theta)\big| - \varepsilon.$$

Let $L^x = \max\{L_1^x, \dots, L_m^x\}$, where L_i^x is the Lipschitz constant of f_i at x for $i \in M$. Choose a sufficiently small positive real number r such that

$$|x| + r < \gamma_1, \qquad |x| - r > \gamma_2,$$

and

$$\big|f_i(y_1) - f_i(y_2)\big| \le L^x |y_1 - y_2| \quad \forall i \in M,\ y_1, y_2 \in \mathbf{B}(x,r).$$

It can be verified that, for any $y \in \mathbf{B}(x,r)$, we have

$$\begin{aligned}
g(y) &\le \max_{t\in[0,T]} \big|\phi(t;0,y,\theta)\big| \\
&\le \max_{t\in[0,T]} \big(\big|\phi(t;0,x,\theta)\big| + e^{L^x t}|x-y|\big) \\
&\le \max_{t\in[0,T]} \big|\phi(t;0,x,\theta)\big| + e^{L^x T}|x-y| \\
&\le g(x) + e^{L^x T}|x-y| + \varepsilon.
\end{aligned}$$

From the arbitrariness of ε and the interchangeability of x and y we have

$$\left|g(x)-g(y)\right| \le e^{L^x T}|x-y| \quad \forall y \in \mathbf{B}(x,r),$$

which shows that g is locally Lipschitz continuous at x. Note that while we do not prove that g is locally Lipschitz continuous at the origin, g is indeed continuous at the origin due to the facts that $g(x) \le \alpha(|x|)$ and α is of class $\mathcal{K}$.

Fourthly, define the function $V: \mathbf{R}^n \mapsto \mathbf{R}_+$ by

$$V(x) = \sup_{t\in\mathcal{T}_0} \inf_{\sigma\in\mathcal{S}} \left\{ g\big(\phi(t;0,x,\sigma)\big)\frac{1+2t}{1+t} \right\}.$$

It follows from relationships (4.16) and (4.17) that

$$|x| \le g(x) \le V(x) \le 2g(x) \le 2\alpha\big(|x|\big). \tag{4.18}$$

By the similar argument used with g, we can show that V is locally Lipschitz continuous at any nonorigin state. It is clear that V is continuous at the origin.

Fifthly, fix $x \neq 0$ and $i \in M$. It follows from inequality (4.15) that there exists a function $\psi: \mathbf{R}^+ \times \mathbf{R}^+ \mapsto \mathbf{R}^+$ such that $\psi(u,\cdot)$ is continuous and decreasing for any fixed u, $\psi(\cdot,v)$ is increasing for any fixed v, and

$$16\beta\big(u,\psi(u,v)\big) \le \alpha^{-1}(v) \quad \forall u,v \in \mathbf{R}^+.$$

Denote $\rho = (1+\psi(|x|,\alpha(|x|)))^2$ and let $\tau > 0$ be a time with

$$\inf_{t\in[0,\tau]} \inf_{\sigma\in\mathcal{S}_{[0,t]}} V\big(\phi(t;0,x,\sigma)\big) \ge \frac{|x|}{2}.$$

For any $h \in [0,\tau]$, we have

$$\begin{aligned} V\big(\phi(h;0,x,\hat{i})\big) &= \sup_{t\in\mathcal{T}_0} \inf_{\sigma\in\mathcal{S}} \left\{ g\big(\phi\big(t;0,\phi(h;0,x,\hat{i}),\sigma\big)\big)\frac{1+2t}{1+t} \right\} \\ &= \sup_{t\in\mathcal{T}_0} \inf_{\sigma\in\mathcal{S}} \left\{ g\big(\phi(t+h;0,x,\sigma\circ_h \hat{i})\big)\frac{1+2t}{1+t} \right\}, \end{aligned}$$

where $\sigma \circ_h \hat{i}$ is the switching path that concatenates $\hat{i}$ and σ at time h,

$$(\sigma \circ_h \hat{i})(t) = \begin{cases} i, & t \in [0,h), \\ \sigma(t-h), & t \in [h,+\infty). \end{cases}$$

It follows that, for any given $\varepsilon \in (0, \frac{|x|}{8\rho})$, there is a time $\zeta \in (0,\tau)$ such that

$$\min_{i\in M} V\big(\phi(h;0,x,\hat{i})\big) \le \sup_{t\in\mathcal{T}_0} \inf_{\sigma\in\mathcal{S}} \left\{ g\big(\phi(t+h;0,x,\sigma)\big)\frac{1+2t}{1+t} \right\} + \varepsilon \tag{4.19}$$

for any $h \in [0, \zeta]$. Utilizing the fact that

$$\inf_{\sigma \in \mathcal{S}} g\big(\phi(t+h; 0, x, \sigma)\big)\frac{1+2t}{1+t} \le 2 \inf_{\sigma \in \mathcal{S}} \alpha\big(\big|\phi(t+h; 0, x, \sigma)\big|\big)$$
$$\le 2\beta\big(|x|, t+h\big),$$

we see that for all $t + h \ge \psi(|x|, \alpha(|x|))$,

$$\inf_{\sigma \in \mathcal{S}} g\big(\phi(t+h; 0, x, \sigma)\big)\frac{1+2t}{1+t} + \varepsilon$$
$$\le \frac{1}{4}|x| \le \inf_{\sigma \in \mathcal{S}} \frac{1}{2} V\big(\phi(h; 0, x, \sigma)\big) \le \min_{i \in M} \frac{1}{2} V\big(\phi(h; 0, x, \hat{i})\big).$$

This, together with relationship (4.19), implies that the supremum in (4.19) is reached at some s with $s + h \le \psi(|x|, \alpha(|x|))$. As a result, we have

$$\begin{aligned}
\min_{i \in M} V\big(\phi(h; 0, x, \hat{i})\big) &\le \inf_{\sigma \in \mathcal{S}} \left\{ g\big(\phi(s+h; 0, x, \sigma)\big)\frac{1+2s}{1+s} \right\} + \varepsilon \\
&\le \inf_{\sigma \in \mathcal{S}} \left\{ g\big(\phi(s+h; 0, x, \sigma)\big)\frac{1+2s+2h}{1+s+h}\left(1 - \frac{h}{2\rho}\right) \right\} + \frac{|x|}{4\rho} \\
&\le V(x)\left(1 - \frac{h}{4\rho}\right) \quad \forall h \in [0, \zeta].
\end{aligned} \tag{4.20}$$

Finally, we show that the function V is strictly decreasing along at least one subsystem. Define the function

$$w(x) = \frac{|x|}{4\rho} = \frac{|x|}{4(1 + \psi(|x|, \alpha(|x|)))^2},$$

which can be verified to be continuous and positive definite. It follows from (4.20) that

$$\min_{i \in M} \liminf_{\tau \to 0^+} \frac{V(\phi(\tau; 0, x, \hat{i})) - V(x)}{\tau} \le -w(x) \tag{4.21}$$

for any $x \ne 0$. It is clear that the inequality still holds at the origin. By Remark 4.9, the local Lipschitz continuity of the Lyapunov function implies that

$$\min_{i \in M} \liminf_{\tau \to 0^+} \frac{V(x + \tau f_i(x)) - V(x)}{\tau} \le -w(x) \quad \forall x \in \mathbf{R}^n.$$

As a result, the function V is a switched Lyapunov function for the switched system.

The above discussion leads to the following conclusion.

Theorem 4.10 *A switched nonlinear system is asymptotically stabilizable iff it admits a (smooth) switched Lyapunov function.*

Remark 4.11 The theorem converts the verification of asymptotic stabilizability of a switched system into the searching of a switched Lyapunov function for the system. The equivalence connection is important as it extends the conventional Lyapunov method to the switched Lyapunov approach for the stabilizability of switched systems.

Next, we restrict our attention to the switched linear system (4.4).

Note that, if the system admits a switched Lyapunov function V which is positively homogeneous with degree one, i.e.,

$$V(\lambda x) = |\lambda| V(x) \quad \forall x \in \mathbf{R}^n,\ \lambda \in \mathbf{R},$$

then the state-space partitions Ω_i, $i \in M$, as defined in (4.12), are 0-symmetric cones, i.e.,

$$x \in \Omega_i \quad \Longleftrightarrow \quad \lambda x \in \Omega_i \quad \forall x,\ i, \lambda \neq 0.$$

In this case, the resultant hysteresis switching law (4.13) is radially invariant in the sense that $\sigma^x = \sigma^{\lambda x}$ for any $x \in \mathbf{R}^n$ and $\lambda \neq 0$. This further implies that the state trajectories are radially linear, i.e.,

$$\phi\big(t; 0, \lambda x, \sigma^{\lambda x}\big) = \lambda \phi\big(t; 0, x, \sigma^x\big) \quad \forall x \in \mathbf{R}^n,\ \lambda \in \mathbf{R}. \tag{4.22}$$

The radial linearity property is important due to the fact that, if we make the dynamical system well behaved locally, then it also well behaves globally.

Theorem 4.12 *Any asymptotically stabilizable switched linear system admits a switched Lyapunov function which is globally Lipschitz continuous and positively homogeneous with degree one.*

Proof We proceed with the continuous-time case, and the discrete-time case can be proven in a similar manner. Suppose that the switched linear system is asymptotically stabilizable. Define the function $V : \mathbf{R}^n \mapsto \mathbf{R}_+$ as

$$V(x) = \inf_{\sigma \in \mathcal{S}_{[0,T]}} \int_0^T \big|\phi(t; 0, x, \sigma)\big|\, dt, \tag{4.23}$$

where T will be determined later. Due to the fact that asymptotic stabilizability implies exponential stabilizability [218], the function is well defined, positively homogeneous with degree one, and

$$\int_0^T e^{-\eta t}\, dt |x| \leq V(x) \leq \frac{\beta}{\alpha}|x| \quad \forall x \in \mathbf{R}^n, \tag{4.24}$$

where α and β are as in Definition 4.1, and $\eta = \max\{\|A_1\|, \ldots, \|A_m\|\}$.

Choose T to satisfy the following property: For any given state $x \neq 0$, there is a positive real number ϵ_x such that for any switching path σ with

$$\int_0^T \left|\phi(t; 0, x, \sigma)\right| dt \leq V(x) + \epsilon_x,$$

we have

$$\left|\phi(t; 0, x, \sigma)\right| \leq \frac{|x|}{8} \quad \forall t \in \mathcal{T}_T.$$

This choice is always possible due to the fact that the finiteness of the integration with $T \to +\infty$ implies the convergence of corresponding state trajectory (cf. [220]).

For any fixed switching signal σ, time t, and states x and y, it is clear that

$$\begin{aligned}\int_0^T \left|\phi(t; 0, x+y, \sigma)\right| dt &\leq \int_0^T \left|\phi(t; 0, x, \sigma)\right| dt + \int_0^T \left|\phi(t; 0, y, \sigma)\right| dt \\ &\leq \int_0^T \left|\phi(t; 0, x, \sigma)\right| dt + \int_0^T e^{\eta t}\, dt |y|.\end{aligned}$$

This implies that the function V is globally Lipschitz continuous with Lipschitz constant $\mu = \int_0^T e^{\eta t}\, dt$.

Due to the global Lipschitz continuity, it can be seen that

$$\begin{aligned}\min_{i \in M} \mathcal{D}^+ V(x)|_{A_i x} &= \min_{i \in M} \liminf_{\tau \to 0^+} \frac{V(x + \tau A_i x) - V(x)}{\tau} \\ &= \inf_{\sigma \in \mathcal{S}_{[0,T]}} \liminf_{\tau \to 0^+} \frac{V(\phi(\tau; 0, x, \sigma)) - V(x)}{\tau}.\end{aligned}$$

Finally, let s be a positive time with

$$\left|\phi(t; 0, x, \sigma)\right| \geq \frac{|x|}{2} \quad \forall t \leq s,\ \sigma \in \mathcal{S}_{[0,s)}.$$

For any state $x \neq 0$ and any positive real number $\epsilon \leq \epsilon_x$, there is a switching path σ^x such that

$$\begin{aligned}\inf_{\sigma \in \mathcal{S}_{[0,T]}} V\big(\phi(t; 0, x, \sigma)\big) &= \inf_{\sigma \in \mathcal{S}_{[0,T]}} \inf_{\varrho \in \mathcal{S}_{[0,T]}} \int_0^T \left|\phi\big(\tau; 0, \phi(t; 0, x, \sigma), \varrho\big)\right| d\tau \\ &\leq \int_0^T \left|\phi\big(\tau; 0, x, \sigma^x\big)\right| d\tau - \int_0^t \left|\phi\big(\tau; 0, x, \sigma^x\big)\right| d\tau \\ &\quad + \int_T^{T+t} \left|\phi\big(\tau; 0, x, \sigma^x\big)\right| d\tau \\ &\leq V(x) + \epsilon - \frac{|x|}{4} t \quad \forall t \leq s.\end{aligned}$$

By the arbitrariness of ϵ, this yields

$$\inf_{\sigma\in\mathcal{S}_{[0,T]}}\liminf_{\tau\to 0^+}\frac{V(\phi(\tau;0,x,\sigma))-V(x)}{\tau}\le -\frac{|x|}{4},$$

which shows that the function V satisfies the last item in Definition 4.8. □

Remark 4.13 Note that in the proof of Theorem 4.12, the switched Lyapunov function given in (4.23) may be nonsmooth. While the function can be smoothed by the standard technique, the smoothed function is not necessarily homogeneous any more. Putting Theorems 4.10 and 4.12 together, an asymptotically stabilizable switched linear system admits a switched Lyapunov function that is either smooth or positively homogeneous.

4.3.2 Nonconvexity of Lyapunov Functions

Recall that for guaranteed stability of switched linear systems, asymptotic stability always implies the existence of a convex and homogeneous Lyapunov function. As convexity is a crucial requirement from the computational points of view, it is expected that stabilizability also implies the existence of a convex switched Lyapunov function. This, however, is not true as exhibited by the counterexamples as follows.

For any subset $\mathcal{R}$ of the state space, time $T\ge 0$, and switching signal σ, define the *set of attainable states* to be

$$\mathcal{C}_T(\mathcal{R},\sigma)=\{\phi(t;0,x,\sigma):\, t\in\mathcal{T}_T,\ x\in\mathcal{R}\}.$$

For a positive definite function $V:\mathbf{R}^n\to\mathbf{R}_+$, define the *set of compliant switching signals* w.r.t. V to be

$$\mathcal{S}_V=\{\sigma\in\mathcal{S}:\, V(\phi(t_1;0,x,\sigma))<V(\phi(t_2;0,x,\sigma))\quad \forall t_1>t_2,\ x\neq 0\}.$$

As a preliminary preparation, we introduce the following lemma.

Lemma 4.14 *Suppose that $\mathcal{R}\subset\mathbf{R}^n$ is a compact set and $\mathcal{R}\neq\{0\}$. If switched linear system* (4.4) *admits a convex Lyapunov function V, then, we have*

$$\mathcal{R}\not\subseteq\operatorname{co}\{\mathcal{C}_T(\mathcal{R},\sigma)\}\quad \forall T>0,\ \sigma\in\mathcal{S}_V. \tag{4.25}$$

Proof Suppose that V is a convex switched Lyapunov function and that σ is a switching signal compliant with V. Fix $T>0$. It is clear that

$$\sup_{x\in\mathcal{R},t\in\mathcal{T}_T} V(\phi(t;0,x,\sigma))<\max_{x\in\mathcal{R}}V(x).$$

It follows from the convexity of V that

$$\sup_{x\in \mathrm{co}\{\mathcal{C}_T(\mathcal{R},\sigma)\}} V(x) < \max_{x\in\mathcal{R}} V(x),$$

which immediately leads to relationship (4.25). □

Example 4.15 Consider the planar continuous-time switched linear system given by

$$\dot{x} = A_\sigma x, \quad \sigma \in \{1, 2\} \tag{4.26}$$

$$A_1 = \begin{bmatrix} 1 & 0 \\ 0 & -1 \end{bmatrix}, \qquad A_2 = \begin{bmatrix} \gamma & -1 \\ 1 & \gamma \end{bmatrix},$$

where $\gamma > 0$ is a parameter.

Note that both subsystems are unstable. However, the x_2-axis is a stable invariant subspace for the first subsystem, and the second subsystem rotates at a constant angular speed. Therefore, any initial state away from the x_2-axis can be steered to the axis in a finite time. This clearly means that the switched system is asymptotically stabilizable. However, as proven below, Lemma 4.14 is violated, and the switched system does not admit any convex switched Lyapunov function.

Proposition 4.16 *There exist a time $T > 0$ and a positive real number γ_0 such that*

$$\mathbf{H}_1 \subseteq \mathrm{co}\big\{\mathcal{C}_T(\mathbf{H}_1, \sigma)\big\} \quad \forall \gamma > \gamma_0,\ \sigma \in \mathcal{S}.$$

To prove the proposition, we first consider the convex sector

$$D = \big\{x \in \mathbf{R}^2 \colon x_1 \ge |x_2|\big\}.$$

See Fig. 4.2. Let $V(x) = \frac{1}{2}x^T x$. It is clear that $\dot{V}(x)|_{A_i}$ is nonnegative for any $x \in D$. In addition, for the phase at $x \in D$, $\theta(x) = \arctan(\frac{x_2}{x_1})$, and its derivative can be computed to be

$$\dot{\theta}(x)|_{A_1} = -\frac{2x_1x_2}{x_1^2 + x_2^2}, \qquad \dot{\theta}(x)|_{A_2} = 1.$$

As a result, for the lower part of D which is in the fourth quadrant, the rotation is always counterclockwise. The above analysis means that, for any initial state x at the lower part of D and any switching signal σ that steers the system convergent from x, the state trajectory must intersect the x_1-axis and abandon D by crossing the radial $b = \{x \in \mathbf{R}^2 \colon x_1 = x_2,\ x_1 \ge 0\}$.

Then, let us have a closer look at the possible intersectant states. For this, let $x_0 = [\frac{\sqrt{2}}{2}, -\frac{\sqrt{2}}{2}]^T$, and $\mathcal{S}^{x_0}$ be the set of switching signals that steer the system convergent from x_0. Define

$$x_c = \begin{bmatrix} x_1^* \\ 0 \end{bmatrix}, \quad x_1^* = \inf\big\{x_1 \colon \exists t \in \mathcal{T}_0,\ \sigma \in \mathcal{S}^{x_0} \text{ s.t. } [x_1, 0]^T = \phi(t; 0, x_0, \sigma)\big\},$$

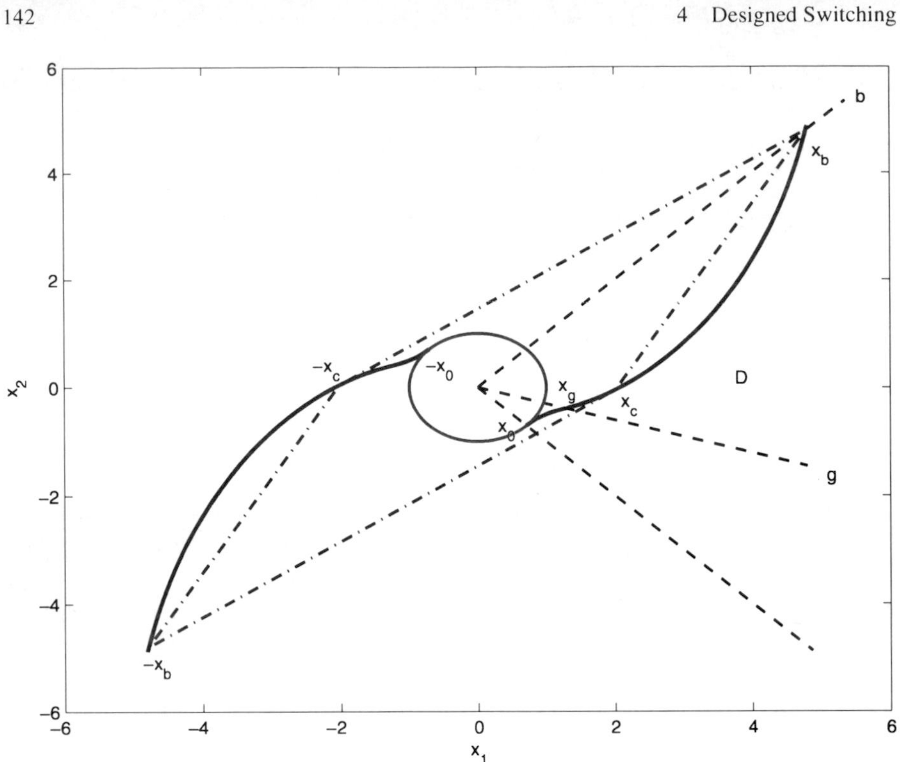

Fig. 4.2 Phase portrait, the convex hull, and the unit circle

which is the state with least norm that is intersectant with x_1-axis, and

$$x_b = \begin{bmatrix} x_2^* \\ x_2^* \end{bmatrix}, \quad x_2^* = \inf\{x_2 : \exists t \in \mathcal{T}_0,\ \sigma \in \mathcal{S}^{x_c},\ \text{s.t. } [x_2, x_2]^T = \phi(t; 0, x_c, \sigma)\},$$

which is the state with least norm that is intersectant with radial b.

Next, we estimate the locations of x_b and x_c. To this end, let g be the radial $g = \{x \in \mathbf{R}^2 : x_1 = -(\gamma + \sqrt{\gamma^2 + 1})x_2, x_1 \geq 0\}$. Rewrite the state equation as

$$\frac{dx_2}{dx_1} = \begin{cases} -\frac{x_2}{x_1}, & \sigma = 1, \\ \frac{x_1 + \gamma x_2}{\gamma x_1 - x_2}, & \sigma = 2. \end{cases}$$

The curve that generates x_c is thus achieved by taking $\sigma = 1$ as long as the state is below radial g and by taking $\sigma = 2$ when the state is on or above g. Denoting by x_g the intersectant state with g, routine calculation yields

$$|x_g| = \sqrt{\frac{\gamma + \sqrt{\gamma^2 + 1}}{2}} \stackrel{\text{def}}{=} \kappa.$$

As V (and the norm) is nondecreasing on D, we have

$$|x_b| \geq |x_c| \geq \kappa.$$

Finally, note that the above argument also holds for $-x_0$ with region $-D = \{-x : x \in D\}$. We thus have $-x_c$ and $-x_b$ accordingly. It can be seen that the unit ball is strictly inside $\mathrm{co}\{x_b, x_c, -x_b, -x_c\}$ if $\kappa \geq \cot\frac{\pi}{8} \simeq 2.6131$, which can be guaranteed when $\gamma \geq 7 \stackrel{\mathrm{def}}{=} \gamma_0$. Note that the above estimate is conservative, and extensive simulation exhibits that $\gamma \geq 1.2$ still works.

To summarize, Proposition 4.16 holds for Example 4.15. It follows from Lemma 4.14 that the switched system does not admit any convex switched Lyapunov function.

Example 4.17 For the planar discrete-time switched linear system

$$x(k+1) = A_{\sigma(k)}x(k), \quad \sigma \in \{1, 2\}, \tag{4.27}$$

$$A_1 = \begin{bmatrix} \mu & 0 \\ 0 & 0 \end{bmatrix}, \qquad A_2 = \nu \begin{bmatrix} \cos(\theta) & -\sin(\theta) \\ \sin(\theta) & \cos(\theta) \end{bmatrix},$$

we assume that $\theta = \frac{5}{8}\pi$ and that ν and μ are sufficiently large positive real numbers.

It is clear that both subsystems are unstable. Notice that, however, the first subsystem steers any state on the x_2-axis to the origin in one step and steers any other state to the x_1-axis in one step. On the other hand, the second subsystem steers any state on the x_1-axis to x_2-axis in four steps. This clearly means that the switched system is asymptotically stabilizable. But it can be proven that Lemma 4.14 is violated, and the switched system does not admit any convex switched Lyapunov function.

Let $\mathcal{R} = \{[x_1, 0]^T : x_1 \in [-1, 1]\}$ and $x_0 = [1, 0]^T$. Note that to make a state x norm contractive in one step, it is necessary that $x \in \Omega = \{[x_1, x_2]^T : x_2^2 \geq (\mu^2 - 1)x_1^2\}$. To steer x_0 or $-x_0$ to region Ω, one needs at least four steps along the second subsystem, while at each step the state norm strictly increases. Applying the first subsystem during the process leads the state back to the x_1-axis with larger norm. These facts clearly indicate that any convergent trajectory starting from x_0 must admit a convex hull that strictly contains $\mathcal{R}$ as an interior. It follows from Lemma 4.14 that switched system (4.27) does not admit any convex switched Lyapunov function.

Remark 4.18 From the above two counterexamples, an asymptotically stabilizable switched linear system does not necessarily admit a convex switched Lyapunov function. As an implication, function sets like piecewise linear and polynomials are not universal as switched Lyapunov function candidates. Recall that, for guaranteed stability, a convex (and homogeneous) Lyapunov function always exists. The subtle difference stems from the following facts. On the one hand, guaranteed stability means stability under arbitrary switching, and a Lyapunov function takes the largest possible energy with respect to all the switching signals; hence the function is convex if the energy function for each switching path is convex, which is indeed the case due to the linearity of the subsystems. On the other hand, stabilizability only implies stability along certain switching signals, and a Lyapunov function takes the least possible energy w.r.t. the switching signals; thus the function might not be convex even when the level set for each subsystem is convex.

4.3.3 Min Quadratic Lyapunov Functions: An Optimization Approach

As convex Lyapunov candidates are not universal for solving the problem of stabilization, we need to seek nonconvex Lyapunov candidates. Recall that any stable linear time-invariant system admits a quadratic Lyapunov function. It is thus natural to extend from quadratic to nonquadratic via proper composite quadratic functions. In the literature, there are various kinds of composite quadratic functions, for example, the maximum (piecewise) quadratic functions

$$V_{\max}^k(x) = \max\{x^T P_1 x, \ldots, x^T P_k\}, \tag{4.28}$$

the convex hull quadratic functions

$$V_c^k(x) = \min_{\substack{\gamma_i \geq 0 \\ \sum \gamma_i = 1}} x^T \left(\sum_i \gamma_i P_i\right)^{-1} x, \tag{4.29}$$

and the minimum quadratic functions

$$V_{\min}^k(x) = \min\{x^T P_1 x, \ldots, x^T P_k x\}, \tag{4.30}$$

where k is a natural number, and $P_1, \ldots, P_k$ are symmetric and positive definite matrices. All these functions are positive definite and homogeneous of degree two, and both the maximum and the convex hull quadratic functions are convex. The minimum quadratic functions, which we will call min functions for simplicity, are nonconvex. This indicates the possibility that the class of min functions is potentially powerful in addressing the problem of stabilization. Fortunately, this is indeed the case, and we are able to prove the following theorem.

Theorem 4.19 *Suppose that the discrete-time switched linear system is exponentially stabilizable. Then, there exist a natural number k and positive definite real matrices $P_1, \ldots, P_k$ such that the min function of* (4.30) *is a switched Lyapunov function of the system.*

To prove the theorem, we need some technical preparations. The discrete-time switched system is given by

$$x(t+1) = A_{\sigma(t)} x(t). \tag{4.31}$$

Let $Q_i, i = 1, \ldots, m$ be positive definite matrices. Define the ith running cost to be

$$L(x, i) = x^T Q_i x.$$

The total cost w.r.t. a switching signal is

$$J(x, \sigma) = \sum_{t=0}^{+\infty} L\big(\phi(t; 0, x, \sigma), \sigma(t)\big), \tag{4.32}$$

which is nonnegative and possibly infinite. The optimal cost is defined to be

$$V^*(x) = \inf_{\sigma \in \mathcal{S}} J(x, \sigma). \tag{4.33}$$

It has been established that the optimal cost is finite for any initial state iff the system is exponentially stabilizable, and in this case the optimal cost is continuous [220]. The problem of switched linear quadratic regulation (SLQR) is to find, if any, a switching law that achieves the minimal cost.

For any mapping $V : \mathbf{R}^n \mapsto \mathbf{R}_+$, define the operator $\zeta[V]$ to be

$$\zeta[V](x) = \inf_{i \in M} \big\{L(x, i) + V(A_i x)\big\}, \tag{4.34}$$

which is called the one-stage cost iteration of the SLQR problem. The composition of the operator could be defined iteratively by

$$\zeta^{k+1}[V](x) = \zeta\big[\zeta^k[V]\big](x), \quad k = 1, 2, \ldots. \tag{4.35}$$

To approach the optimal cost, for a natural number k, we define the k-step cost function by

$$V_k(x) = \inf_{\sigma \in \mathcal{S}_{[0,k-1]}} \sum_{t=0}^{k-1} L\big(\phi(t; 0, x, \sigma), \sigma(t)\big).$$

In particular, let $V_0(x) = 0$ for all $x \in \mathbf{R}^n$.

According to the standard theory of dynamical programming, we have the following result, which can be found in [26].

Lemma 4.20 *We have the following statements*:

(i) $V_k(x) = \zeta^k[V_0](x)$ *for all* $k \in \mathbf{N}_+$ *and* $x \in \mathbf{R}^n$.
(ii) $\lim_{k \to +\infty} V_k(x) = V^*(x)$ *for all* $x \in \mathbf{R}^n$.
(iii) $\zeta[V^*](x) = V^*(x)$ *for all* $x \in \mathbf{R}^n$.

The equality in Item (iii) is the well-known Bellman equation. Furthermore, there is a state-dependent switching law $\sigma(t) = \psi(x(t))$ such that

$$L\big(x, \psi(x)\big) + V^*(A_{\psi(x)} x) = V^*(x) \quad \forall x \in \mathbf{R}^n. \tag{4.36}$$

Suppose that the switched linear system is exponentially stabilizable. Then, it can be seen that

$$qx^T x \le V^*(x) \le \frac{\beta^2}{(1-e^{-\alpha})} x^T x \quad \forall x \in \mathbf{R}^n, \tag{4.37}$$

where α and β are as in Definition 4.1, and

$$q = \min\{\lambda_{\min}(Q_1), \ldots, \lambda_{\min}(Q_m)\}.$$

On the other hand, it follows from (4.36) that there exists a positive real number δ such that

$$\min_{i \in M} V^*(A_i x) - V^*(x) \le -\delta x^T x. \tag{4.38}$$

As a result, the function V^* is a switched Lyapunov function of the switched system. Furthermore, we have the following result.

Lemma 4.21 *Suppose that the switched linear system is exponentially stabilizable. Then, there is a natural number K such that V_k is a switched Lyapunov function of the switched linear system for any $k \ge K$.*

Proof Fix $\mu \in (0, \delta)$, where δ is as in (4.38). Note that, for any $x \in \mathbf{R}^n$, we have

$$0 = V_0(x) \le V_1(x) \le V_2(x) \le \cdots \le V^*(x).$$

It follows from (4.37) that

$$qx^T x \le V_k(x) \le \frac{\beta^2}{(1-e^{-\alpha})} x^T x \quad \forall x \in \mathbf{R}^n,\ k \ge 1. \tag{4.39}$$

By Item (ii) of Lemma 4.20, for any state x with unit norm, there is a natural number j_x such that

$$V^*(x) - V_k(x) \le \mu/2 \quad \forall k \ge j_x.$$

Due to the continuity of V^* and V_k, there is a neighborhood of x, denoted N_x, such that

$$V^*(y) - V_k(y) \le \mu \quad \forall k \ge j_x,\ y \in N_x.$$

As the unit sphere is compact, it follows from the Finite Covering Theorem that there exist a natural number i and states $x_1, \ldots, x_i$ with unit norm such that

$$\bigcup_{s=1}^{i} N_{x_s} = \mathbf{H}_1.$$

Taking $K = \max\{j_{x_1}, \ldots, j_{x_i}\}$, we have

$$V^*(y) - V_k(y) \le \mu \quad \forall y \in \mathbf{H}_1,\ k \ge K. \tag{4.40}$$

Taking advantage of the homogeneity of both V^* and V_k, we have

$$\begin{aligned}
&\min_{i\in M} V_k(A_i x) - V_k(x) \\
&\quad = \min_{i\in M} V_k(A_i x) - \min_{i\in M} V^*(A_i x) + V^*(x) - V_k(x) + \min_{i\in M} V^*(A_i x) - V^*(x) \\
&\quad \le -(\delta - \mu) x^T x \quad \forall k \ge K. \qquad (4.41)
\end{aligned}$$

This, together with inequality (4.39), clearly shows that V_k is a switched Lyapunov function of the switched system when $k \ge K$. □

To connect V^* and V_k with min functions as in (4.30), we define the mapping

$$Z_i(P) = Q_i + A_i^T P A_i,$$

where P is symmetric and positive definite. Furthermore, given a set of real symmetric matrices $\mathcal{Y} = \{P_1, \ldots, P_l\}$, define the *switched Riccati mapping* to be

$$\mathcal{Z}(\mathcal{Y}) = \{Z_i(P_j) : i = 1, \ldots, m,\ j = 1, \ldots, l\}.$$

Finally, define the sequence of matrix sets by

$$\begin{aligned}
\mathcal{Z}_0 &= \{0_{n\times n}\}, \\
\mathcal{Z}_j &= \mathcal{Z}(\mathcal{Z}_{j-1}), \quad j = 1, 2, \ldots.
\end{aligned}$$

Clearly, the elements of $\mathcal{Z}_j$ are positive definite matrices for $j > 0$, and $\mathcal{Z}_j$ consists of up to m^j elements. The following result is a simple yet useful observation that leads to further insight into the SLQR problem.

Lemma 4.22 $V_k(x) = \min\{x^T P x : P \in \mathcal{Z}_k\}$ *for* $k = 0, 1, 2, \ldots$.

Proof We proceed by induction. The equality clearly holds when $k = 0$. Suppose that the relation holds for a general integer k. Then, by Item (i) of Lemma 4.20, we have

$$\begin{aligned}
V_{k+1}(x) = \zeta[V_k](x) &= \inf_{i\in M} \{L(x, i) + V_k(A_i x)\} \\
&= \min_{i\in M} \left(x^T Q_i x + \min\{(A_i x)^T P (A_i x) : P \in \mathcal{Z}_k\}\right) \\
&= \min_{i\in M, P\in\mathcal{Z}_k} x^T \left(Q_i + A_i^T P A_i\right) x = \min_{P\in\mathcal{Z}_{k+1}} x^T P x,
\end{aligned}$$

which completes the proof. □

Proof of Theorem 4.19 Simply combining Lemmas 4.21 and 4.22 leads directly to the conclusion. □

Theorem 4.19 implies that, when the switched linear system is exponentially stabilizable, the system always admits a switched Lyapunov function that is a min function. This exhibits that the class of min functions is universal in characterizing the problem of stabilization. From an invariant set point of view, this indicates that a generally nonconvex level set (of a nonconvex switched Lyapunov function) could be effectively approximated by unions of a finite set of ellipsoids. This also strengthens and generalizes the connection between linear systems and quadratic Lyapunov functions, which is insightful and powerful in system analysis and synthesis.

Suppose that $V_k(x) = \min\{x^T Px : P \in \mathcal{Z}_k\}$ is a switched Lyapunov function. Then, it can be seen that an exponentially stabilizing switching law is

$$\sigma(t) = \arg\min_i \Big\{ \min_{P\in\mathcal{Z}_k} x^T(t)\big(Q_i + A_i^T PA_i\big)x(t)\Big\}. \tag{4.42}$$

When there are two or more indices achieving the minimum, then just take any one. Alternatively, define the state partitions

$$\Omega_i = \Big\{x : \min_{P\in\mathcal{Z}_k} x^T\big(Q_i + A_i^T PA_i\big)x = \min_{j\in M, P\in\mathcal{Z}_k} x^T\big(Q_j + A_j^T PA_j\big)x\Big\}. \tag{4.43}$$

It is clear that Ω_i is a (possibly empty) cone that is not necessarily to be convex. In addition, the union of the partitions cover the whole state space. In terms of the state space partitions, the stabilizing switching law could be given by

$$\sigma(t) = \arg\min_i \big\{x(t) \in \Omega_i\big\}. \tag{4.44}$$

Lemmas 4.21 and 4.22 indicate a constructive approach for computing a switched Lyapunov function. For this, we first need to compute the matrix set $\mathcal{Z}_k$ and then verify whether or not the corresponding min function is strictly decreasing along the unit sphere. Note that the number of elements in $\mathcal{Z}_k$ is up to m^k, which is huge when k is large. To produce a practically implementable computational procedure, we need to further reduce the computational load, as discussed below.

A min function could be equivalently expressed by two or more sets of minimal piecewise quadratic functions. To be more precise, suppose that matrix sets $\{Q_1, \dots, Q_i\}$ and $\{R_1, \dots, R_j\}$ satisfy

$$\min\big\{x^T Q_1 x, \dots, x^T Q_i x\big\} = \min\big\{x^T R_1 x, \dots, x^T R_j x\big\} \quad \forall x \in \mathbf{R}^n.$$

Then, it is clear that the two matrix sets represent the same min function, and the two sets are said to be equivalent. In particular, given a matrix set $\mathcal{Q}$, if a subset $\tilde{\mathcal{Q}} \subset \mathcal{Q}$ is equivalent to $\mathcal{Q}$, then any element in $\mathcal{Q} - \tilde{\mathcal{Q}}$ is redundant, which could be pruned out. A subset with least cardinality that is equivalent to $\mathcal{Q}$ is said to be a minimum equivalent subset. A matrix set could admit more than one minimum equivalent subset. While it is usually hard to find a minimum equivalent subset, we could prune out a redundant element using the following proposition.

Proposition 4.23 *For a matrix set* $\{Q_1, \dots, Q_i\}$, *an element* Q_j *is redundant if there are nonnegative real numbers* α_s *with* $\sum_{s \neq j} \alpha_s = 1$ *and such that*

$$\sum_{s \neq j} \alpha_s Q_s \leq Q_j.$$

Proof Straightforward. □

The verification of the proposition is a convex optimization problem, and it admits numerically efficient algorithms.

Note that matrix set $\mathcal{Z}_k$ is defined in a recursive way. Therefore, the pruning process should be implemented in each iterating step. In this process, it is crucial that useful information could be preserved. This indeed is the case, as shown in the following proposition.

Proposition 4.24 *Suppose that* $\tilde{\mathcal{Z}}_k$ *is an equivalent subset of* $\mathcal{Z}_k$. *Then,* $\mathcal{Z}(\tilde{\mathcal{Z}}_k)$ *is an equivalent subset of* $\mathcal{Z}_{k+1}$.

Proof It is clear that an element of $\mathcal{Z}_{k+1}$ is of the form $Q_i + A_i^T P A_i$, where $P \in \mathcal{Z}_k$ and $i \in M$. For arbitrarily given $x \in \mathbf{R}^n$, let $y = A_i x$. As $\tilde{\mathcal{Z}}_k$ is an equivalent subset of $\mathcal{Z}_k$, there is a matrix $\tilde{P} \in \tilde{\mathcal{Z}}_k$ such that

$$y^T \tilde{P} y \leq y^T P y,$$

which implies that

$$x^T \big(Q_i + A_i^T \tilde{P} A_i\big) x \leq x^T \big(Q_i + A_i^T P A_i\big) x.$$

Since $(Q_i + A_i^T \tilde{P} A_i) \in \mathcal{Z}(\tilde{\mathcal{Z}}_k)$, the conclusion follows due to the arbitrariness of x. □

With the help of the above propositions, we can outline a pruning procedure for removing the redundant elements from sets $\mathcal{Z}_k$ and obtaining the equivalent subsets with smaller cardinalities.

Pruning Procedure for Calculating $\tilde{\mathcal{Z}}_K$

1. Set $k := 0$ and $\mathcal{Y} := \{0\}$.
2. Compute $\hat{\mathcal{Y}} = \mathcal{Z}(\mathcal{Y})$.
3. Prune $\hat{\mathcal{Y}}$ by applying Proposition 4.23 and set the resultant set to be $\tilde{\mathcal{Y}}$.
4. Set $k := k + 1$ and $\mathcal{Y} := \tilde{\mathcal{Y}}$.
5. If $k = K$, set $\tilde{\mathcal{Z}}_K := \mathcal{Y}$ and stop. Otherwise, go to Step 2.

Example 4.25 For the planar discrete-time two-form switched linear system with

$$A_1 = \begin{bmatrix} -1.2299 & 0.9390 \\ 1.6455 & -1.1496 \end{bmatrix}, \qquad A_2 = \begin{bmatrix} 2.6229 & 0.0564 \\ 2.3756 & -0.5879 \end{bmatrix},$$

it can be verified that both subsystems are unstable.

Table 4.1 Cardinalities of $\tilde{\mathcal{Z}}_k$'s

k	1	2	3	4	5	6	7
#	2	4	6	9	10	12	14

To apply the optimization approach, we need to calculate the cost-to-go functions. For this, let $Q_1 = Q_2 = I$. By the pruning procedure, we obtain $\tilde{\mathcal{Z}}_k$ for $k = 1, \dots, 7$. Table 4.1 shows the cardinalities of these sets. It is clear that the pruning algorithm successfully removes many redundant elements. For example, $\mathcal{Z}_8$ contains 128 matrices, while $\tilde{\mathcal{Z}}_8$ contains only 14, which is only one ninth of the former.

Next, we examine the qualification of V_k's as switched Lyapunov functions. For this, we verify the relationship

$$\min_{i \in M, P \in \tilde{\mathcal{Z}}_k} x^T \left(Q_i + A_i^T P A_i\right) x < \min_{P \in \tilde{\mathcal{Z}}_k} x^T P x \quad \forall x \in \mathbf{H}_1$$

for $i = 1, 2, \dots$ and find that the relationship holds for $k = 5$. Therefore, V_k, $k = 5, 6, \dots$, are switched Lyapunov functions for the switched linear system. As a result, the switched system is exponentially stabilizable, and a stabilizing switching

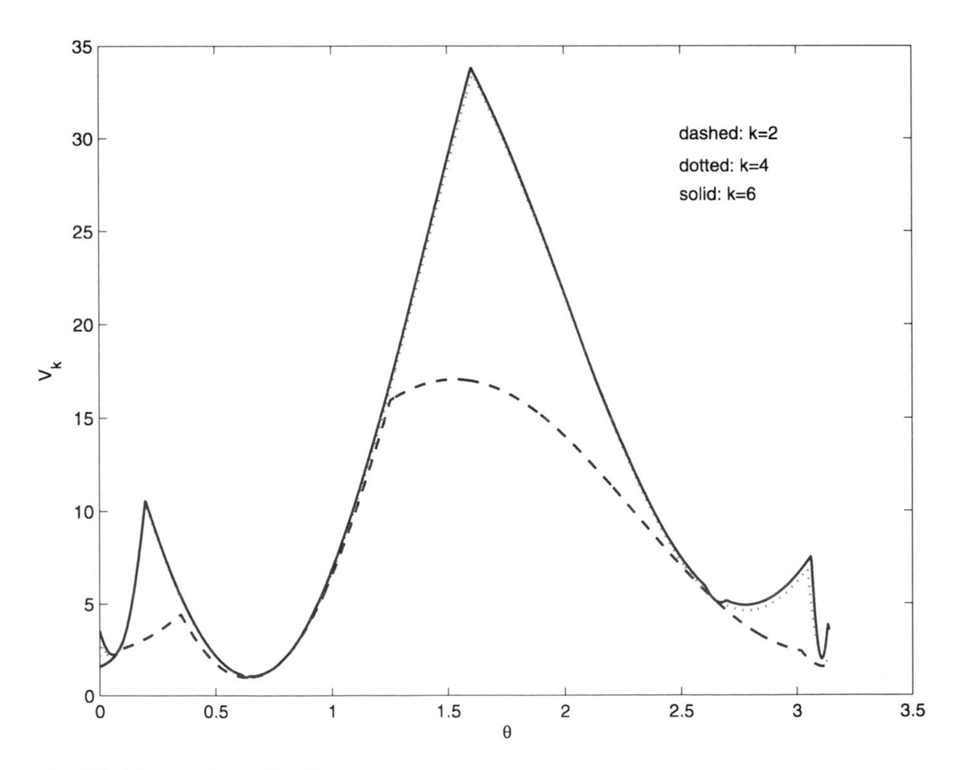

Fig. 4.3 The cost-to-go functions

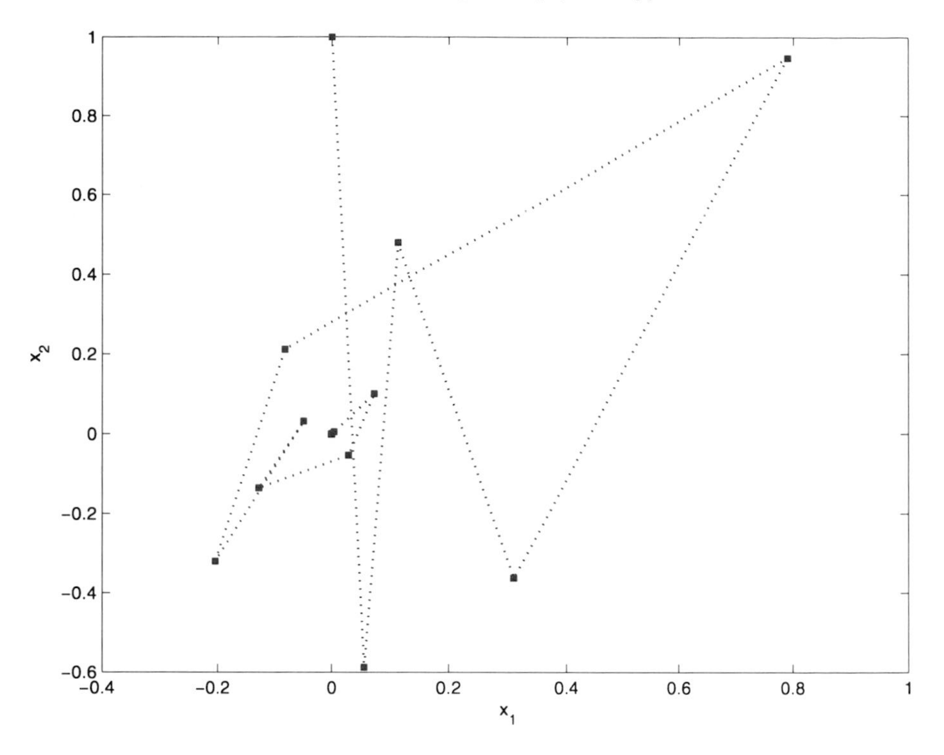

Fig. 4.4 A sample phase portrait

law can be obtained by means of (4.42) or (4.44). Figure 4.3 shows the cost-to-go functions V_k along the unit circle. To be more precise, let θ be an angular in $[0, \pi]$, and $x = [\sin(\theta), \cos(\theta)]^T$. The figure shows V_k along x through θ. It can be seen that, as k becomes larger, $V_k - V_{k-1}$ becomes much smaller, that is, the differences between V_k and V_{k-1} approach zero.

To further present the simulations, we fix V_6 as the switched Lyapunov function. Figures 4.4 and 4.5 depict a sample phase portrait (with $x_0 = [0, 1]^T$) and the level set $\{x : V_6(x) = 1\}$, respectively. It is clear that the state trajectory converges to zero exponentially, and the level set is the union of a set of ellipsoids. Finally, Fig. 4.6 shows the state space partitions, where each partition is a cone.

Finally, we briefly discuss the continuous-time case. While technically more involved, the optimization approach is applicable to continuous-time systems, which yields exactly the same conclusion as in Theorem 4.19 [192]. As a result, the set of min functions is universal in providing switched Lyapunov functions for exponentially stabilizable switched linear systems. To develop a computational procedure for calculating a min function as a switched Lyapunov function, we could first convert the continuous-time switched system into a discrete-time switched system by sampling and then implement the pruning procedure for finding a switched Lyapunov function. We demonstrate this by a numerical example.

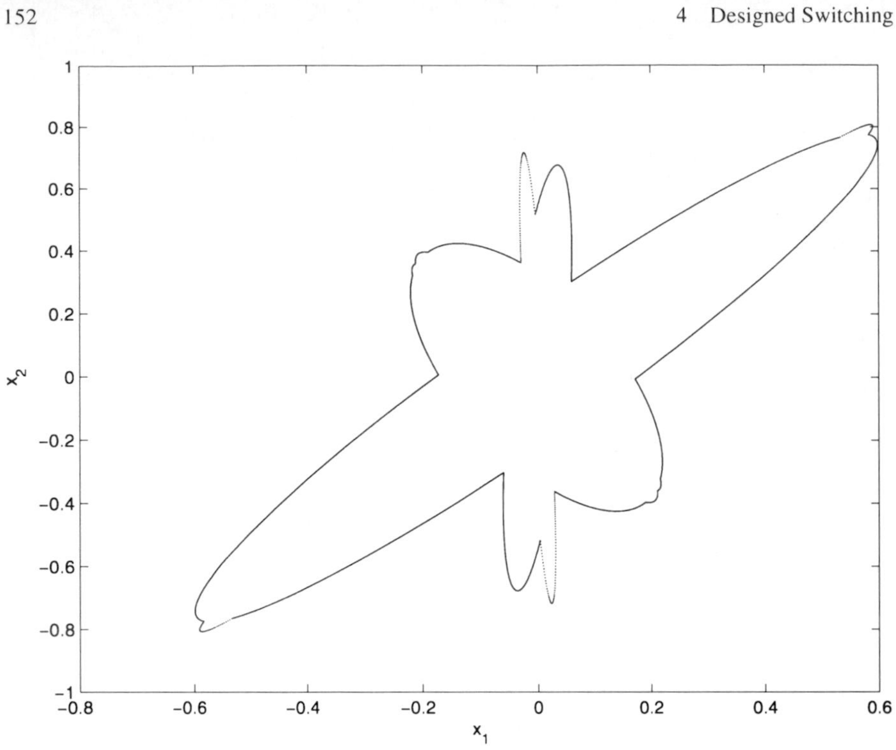

Fig. 4.5 The level set of V_6

Example 4.26 For the continuous-time two-form switched linear system with

$$A_1 = \begin{bmatrix} 0.2341 & -0.9471 & -1.0559 \\ 0.0215 & -0.3744 & 1.4725 \\ -1.0039 & -1.1859 & 0.0557 \end{bmatrix},$$

$$A_2 = \begin{bmatrix} -1.2173 & -1.3493 & 0.1286 \\ -0.0412 & -0.2611 & 0.6565 \\ -1.1283 & 0.9535 & -1.1678 \end{bmatrix},$$

it can be verified that both subsystems are unstable.

Taking the sampling period as $\tau = 0.2$ sec, we convert the continuous-time switched system into the discrete-time switched system

$$z(k+1) = B_{\varrho(k)} z(k), \qquad z(0) = x_0,$$

where $B_i = e^{A_i \tau}$, $i = 1, 2$, and ϱ is the switching signal with $\varrho(k) = \sigma(k\tau)$, $k = 0, 1, \ldots$. When the switching signal of the original system satisfies

$$\sigma(t) = \sigma(k\tau) \quad \forall k = 0, 1, 2, \ldots, \ t \in \big[k\tau, (k+1)\tau\big),$$

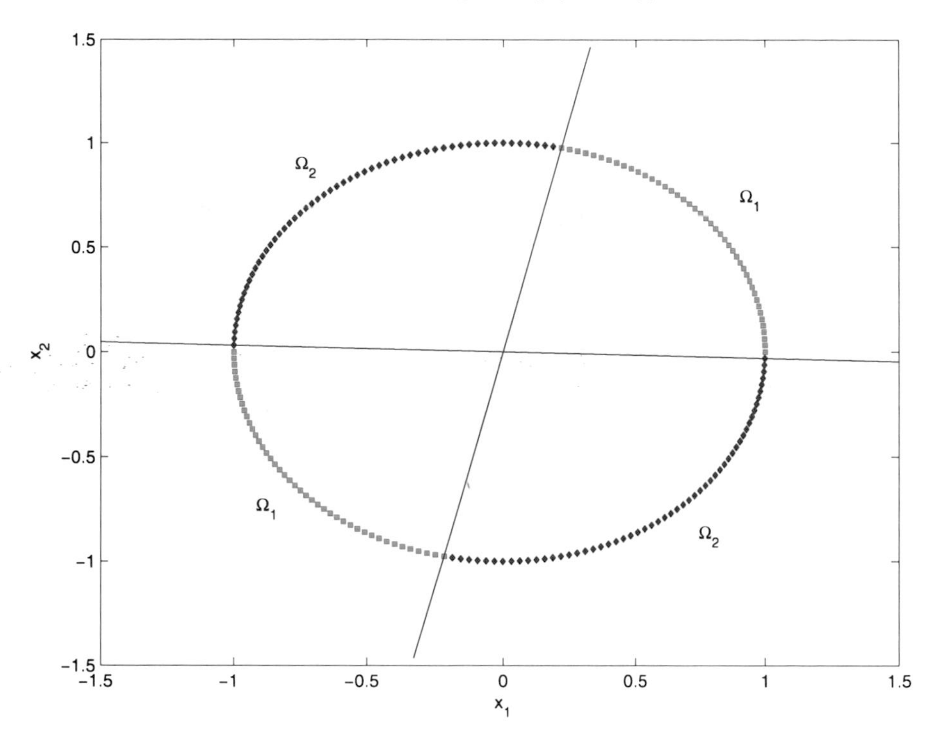

Fig. 4.6 State space partitions

we have $z(k) = x(k\tau)$ for $k = 0, 1, \ldots$. In this case, state $z(\cdot)$ is a sampled trajectory of the original system with the same initial condition.

Let $Q_1 = Q_2 = I$. Applying the pruning procedure to the sampled system, we obtain $\tilde{\mathcal{Z}}_k$ for $k = 1, \ldots, 5$. While $\mathcal{Z}_5$ contains 32 matrices, $\tilde{\mathcal{Z}}_5$ contains 24. Furthermore, it is numerically verified that V_5 is a switched Lyapunov function for the sampled switched system. As a result, the sampled system is exponentially stabilizable, and the original continuous-time system is also exponentially stabilizable. Figure 4.7 depicts the level set $\{x : V_5(x) = 1\}$ on the unit sphere $x = [\cos\theta_1, \sin\theta_1 \cos\theta_2, \sin\theta_1 \sin\theta_2]^T$, $\theta_1, \theta_2 \in [0, 2\pi)$. It can be seen that the level set is the union of a set of ellipsoids and is nonconvex.

Next, fix the initial state $x_0 = [0, 1, -1]^T$. To find a switching signal that steers the original system exponentially stable, we need to calculate a stabilizing switching signal as in (4.42) for the sampled switched system. Then, a stabilizing switching signal for the original continuous-time system is given by

$$\sigma(t) = \varrho(k\tau), \quad t \in \big[k\tau, (k+1)\tau\big).$$

Figures 4.8 and 4.9 depict the switching signal and the corresponding state trajectory, respectively. It is clear that the state trajectory exponentially converges to zero at a satisfactory rate.

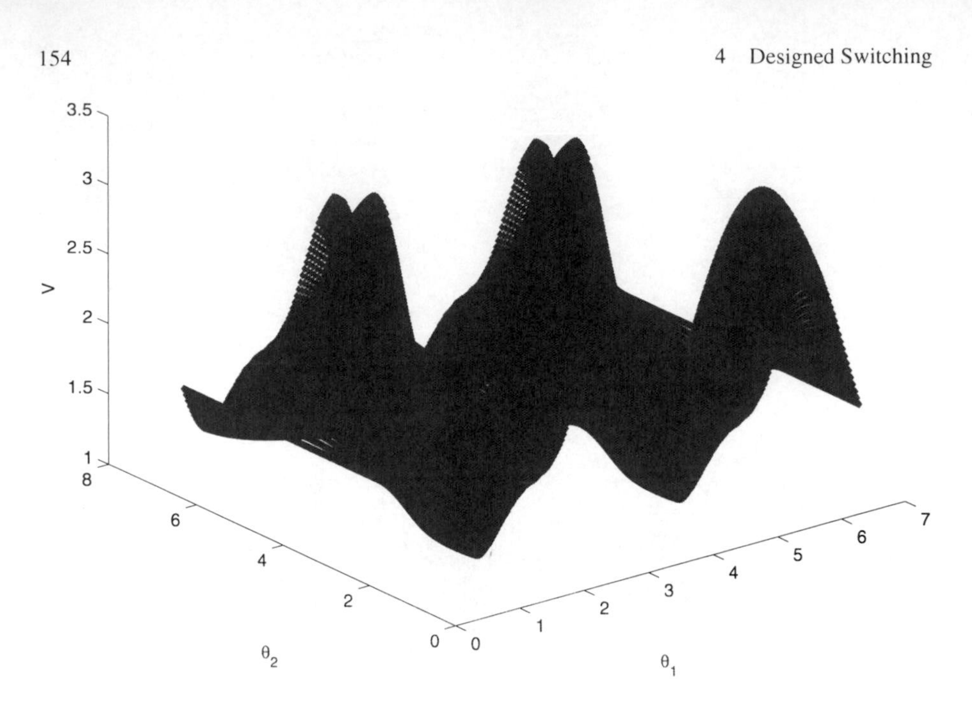

Fig. 4.7 The level set of V_5 on the unit sphere

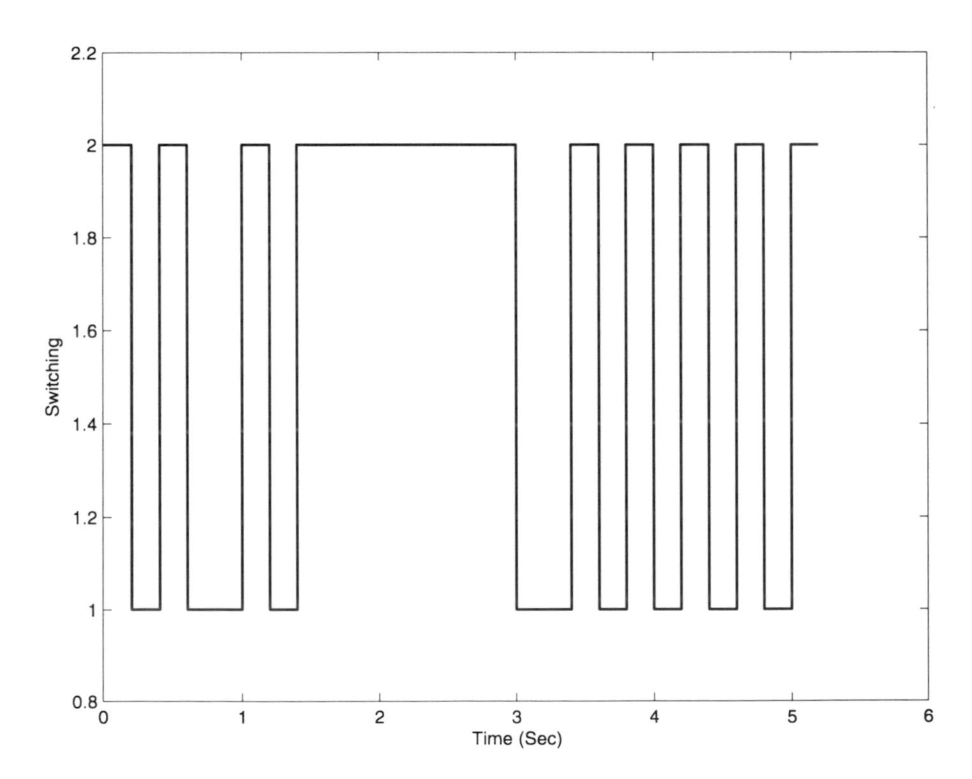

Fig. 4.8 Switching signal

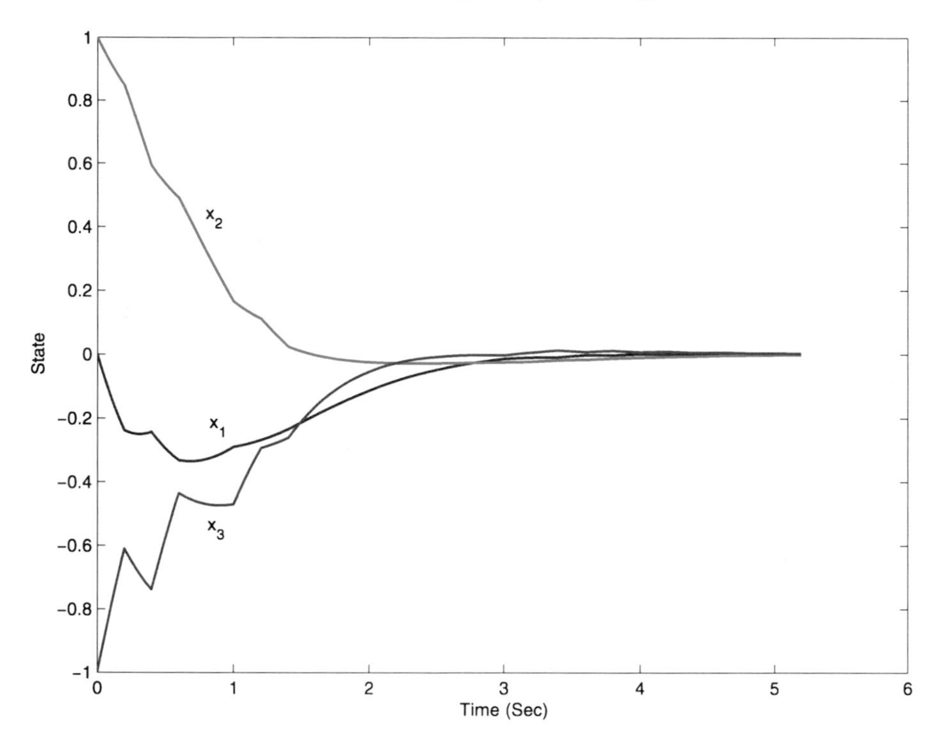

Fig. 4.9 State trajectory

4.3.4 Well-Definedness of State-Feedback Stabilizing Law

A state-feedback switching law can be described by

$$\sigma(t) = \varphi\big(x(t), \sigma(t-)\big)$$

and is said to be *pure-state-feedback* when φ does not explicitly rely on $\sigma(t-)$. Note that, as any asymptotically stabilizable system admits a switched Lyapunov function (cf. Theorem 4.10), a state-feedback stabilizing switching law can be constructed as in (4.11) or (4.13) by means of the Lyapunov function. However, it should be stressed that, while state-feedback switching laws are universal in stabilizing switched systems, they are not always well behaved due to the fact that a well-defined state-feedback switching law for a nominal switched linear system may produce chattering (Zeno) phenomenon for a slightly perturbed system. We show this through a numerical example.

Example 4.27 Consider a perturbed switched linear system given by

$$\dot{x}(t) = A_\sigma x(t) + f_\sigma(t),$$

$$A_1 = \begin{bmatrix} -2 & 0 \\ 0 & 1 \end{bmatrix}, \qquad A_2 = \begin{bmatrix} 1 & 0 \\ 0 & -2 \end{bmatrix},$$

$$f_1(t) = -f_2(t) = \begin{bmatrix} -1 \\ 1 \end{bmatrix} e^{-0.1t}, \tag{4.45}$$

where f_1 and f_2 are perturbations associated to the first and second subsystems, respectively.

It can be easily verified that the function $V(x) = \frac{1}{2}(x_1^2 + x_2^2)$ is a switched Lyapunov function for the nominal system. A stabilizing switching law can be constructed as follows.

First, define the regions

$$\Omega_i = \{x \in \mathbf{R}^2 : \dot{V}|_{A_i} \leq -r_i V(x)\}, \quad i = 1, 2,$$

where r_i, $i = 1, 2$, are nonnegative real numbers. It can be seen that the union of the regions cover the total state space if $r_i < 1$ for $i = 1, 2$. A hysteresis switching law can be defined recursively as

$$\begin{aligned} t_0 &= 0, \\ \sigma(t_0) &= \arg\min_{i=1,2} \{x_0^T Q_i x_0\}, \\ t_{k+1} &= \inf\{t \geq t_k : x(t) \notin \Omega_{\sigma(t_k)}\}, \\ \sigma(t_{k+1}) &= \begin{cases} 1, & \sigma(t_k) = 2, \\ 2, & \sigma(t_k) = 1, \end{cases} \quad k = 0, 1, 2, \ldots. \end{aligned} \tag{4.46}$$

This switching law stabilizes the nominal switched system, and it can be seen that the switching law is well defined for the nominal system.

For the perturbed system, suppose that we use the state-feedback switching law (4.46) with $r_1 = r_2 = 0.4$. Figure 4.10 depicts the state trajectories and the number of switchings starting from $x(0) = [1, -1]^T$. It can be seen that the chattering phenomenon occurs when $t > 11.45$ sec. In fact, as the state converges to the origin, the information of the state is "merged" by the perturbations, that is, $\frac{|f_\sigma(t)|}{|x(t)|} \to +\infty$. Because the state direction and the perturbation direction are always opposite, chattering occurs.

The example clearly exhibits that a well-defined state-feedback switching law might loss its well-definedness if the system is slightly perturbed. This indicates that state-feedback switching laws might not be appropriate for achieving both stability and robustness.

4.4 Stabilization via Mixed-Driven Switching: Aggregation and Calculation

In this section, we focus on stabilizing switching design for the switched linear system

$$x^+(t) = A_{\sigma(t)} x(t), \tag{4.47}$$

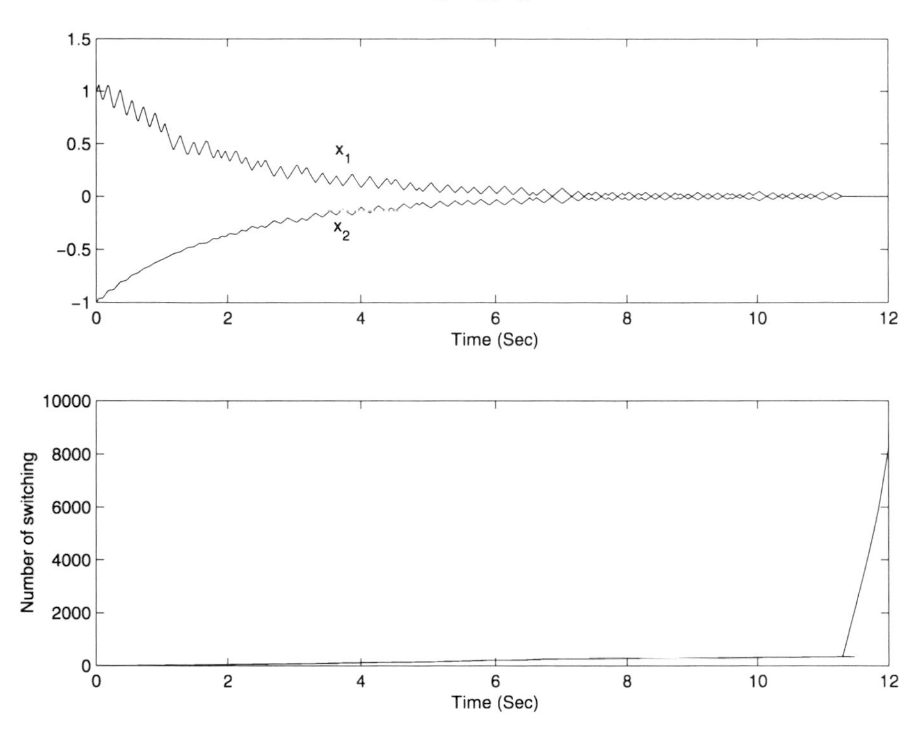

Fig. 4.10 State trajectories and switching numbers of system (4.45)

where $x(t) \in \mathbf{R}^n$, $\sigma(t) \in \{1, \dots, m\}$, and $A_1, \dots, A_m$ are known real constant matrices.

The objective is to find, if possible, a switching law that achieves (i) stability, (ii) robustness against exterior perturbations including structured, unstructured, and switching perturbations, and (iii) well-definedness for the nominal system and the perturbed system.

As discussed in the previous sections, time-driven switching laws are not universal for the problem of stabilization, though its well-definedness is clear as it is independent of the system dynamics. State-feedback switching laws, on the other hand, are universal, but the well-definedness is sensitive to system perturbations. This motivates us to find a new switching law that achieves the merits of both schemes while getting rid of the demerits. The pathwise state-feedback switching law proposed below is exactly such a switching law.

4.4.1 Pathwise State-Feedback Switching

Suppose that the switched system is not consistently stabilizable. This means that any single switching path could not make the total state space contractive. However, it is still possible that a switching path makes a subset of the state space contractive.

For simplicity, we assume that two switching paths, p_1 over $[0, T_1)$ and p_2 over $[0, T_2)$, make regions Ω_1 and Ω_2 contractive, respectively. To be more precise, we have

$$\left|\phi(T_i; 0, x, p_i)\right| < |x| \quad \forall x \in \Omega_i, \ i = 1, 2.$$

Moreover, assume that $\Omega_1 \cup \Omega_2 = \mathbf{R}^n$. We further assume without loss of generality that $\Omega_1 \cap \Omega_2 = \emptyset$, otherwise just redefine $\Omega_2 = \mathbf{R}^n - \Omega_1$. With these in mind, we can construct a stabilizing switching law σ as follows. For any initial state $x_0 = x(t_0)$, let the system operate along switching path p_1 if x belongs to Ω_1, otherwise let the system operate along switching path p_2. Define

$$\begin{aligned} i_0 &= \arg\{x_0 \in \Omega_j\}, \\ t_1 &= t_0 + T_{i_0}, \\ \sigma^{x_0}(t) &= p_{i_0}(t - t_0) \quad \forall t \in [t_0, t_1), \\ x_1 &= \phi(T_{i_0}; 0, x_0, p_{i_0}). \end{aligned}$$

It is clear that $|x_1| \le |x_0|$. Starting from time t_1 at x_1, let the system operate along switching path p_1 if x_1 belongs to Ω_1, otherwise let the system operate along switching path p_2. Define accordingly

$$\begin{aligned} i_1 &= \arg\{x_1 \in \Omega_j\}, \\ t_2 &= t_1 + T_{i_1}, \\ \sigma^{x_0}(t) &= p_{i_1}(t - t_1) \quad \forall t \in [t_1, t_2), \\ x_2 &= \phi(T_{i_1}; 0, x_1, p_{i_1}). \end{aligned}$$

Repeating the process in the same manner, we obtain the switching signal σ^{x_0} over $[0, +\infty)$ such that

$$\begin{aligned} |x_{k+1}| &= \left|\phi\left(t_{k+1}; t_0, x_0, \sigma^{x_0}\right)\right| \\ &\le \left|\phi\left(t_k; t_0, x_0, \sigma^{x_0}\right)\right| = |x_k|, \quad k = 0, 1, \ldots. \end{aligned}$$

Let x_0 vary among the state space, we obtain a switching law that makes the switched system stable.

For any initial state, the above switching law generates a switching signal that concatenates switching paths p_1 and p_2 through the state measurement at the concatenating instants. Therefore, the switching mechanism is mixed time-driven and state-feedback. It is clear that the switching law is well defined if both p_1 and p_2 are well defined, and the well-definedness is independent of the subsystem dynamics. On the other hand, as a state-feedback mechanism is incorporated, the switching law can accommodate the initial state and exotic perturbation information, and it is expected to have some flexibility in achieving both stability and robustness.

We are ready to formally describe the switching law, which will be termed as the *pathwise state-feedback switching law*.

Suppose that k is a natural number, Ω_i, $i = 1, \dots, k$, are regions in $\mathbf{R}^n$ satisfying $\bigcup_{i=1}^k \Omega_i = \mathbf{R}^n$ and $\Omega_i \cap \Omega_j = \emptyset$ for any $i \neq j$, and $\theta_i : [0, s_i] \mapsto M$, $i = 1, \dots, k$, are well-defined switching paths. The pathwise state-feedback switching law via Ω_i and θ_i w.r.t. system (4.47), denoted by $\bigwedge_{i=1}^k \theta_i^{\Omega_i}$, is the concatenation of switching paths $\{\theta_i\}_{i=1}^k$ through $\{\Omega_i\}_{i=1}^k$ as defined below. For any initial state $x(t_0) = x_0$, the generated switching signal σ^{x_0} is recursively defined by

$$\begin{aligned} i_k &= \arg\{x_k \in \Omega_j\}, \\ t_{k+1} &= t_k + s_{i_k}, \\ \sigma^{x_0}(t) &= \theta_{i_k}(t - t_k) \quad \forall t \in [t_k, t_{k+1}), \\ x_{k+1} &= \phi(s_{i_k}; 0, x_k, \theta_{i_k}), \quad k = 0, 1, 2, \dots. \end{aligned} \tag{4.48}$$

It is clear that the pathwise state-feedback switching law is always well defined over $[0, +\infty)$ for any initial state. As each generated switching signal is a concatenation from paths $\{\theta_i\}_{i=1}^k$, the switching law can be equivalently represented by the set of sequences

$$\{(x_0; i_0, i_1, \dots) : x_0 \in \mathbf{R}^n\},$$

where $i_0, i_1, \dots$ are defined as in (4.48). Note that the switching law is independent of the permutation of Ω_i and θ_i, provided that the pairwise relationship keeps unchanged.

Note also that each switching path θ_i corresponds to a state transition matrix $G_i = \Phi(s_i, 0, \theta_i)$ with the property that $\phi(s_i; 0, x, \theta_i) = G_i x$ for all $x \in \Omega_i$.

Lemma 4.28 *Switching law* $\bigwedge_{i=1}^k \theta_i^{\Omega_i}$ *asymptotically stabilizes switched linear system* (4.47) *iff the discrete-time piecewise linear system*

$$z(t+1) = G_i z(t), \qquad z(t) \in \Omega_i, \tag{4.49}$$

is asymptotically stable.

Proof From the definition of $\bigwedge_{i=1}^k \theta_i^{\Omega_i}$ it is clear that

$$x_j = \phi(t_j; 0, x_0, \sigma^{x_0}) = z(j), \quad j = 0, 1, 2, \dots,$$

where t_j and x_j are defined as in (4.48), and $z(j)$ is the state of system (4.49) with $z(0) = x(0)$. As a result, each state trajectory of (4.49) is exactly a sampled state trajectory of the original switched system (4.47) at specified sampled instants. This implies the "only if" part of the lemma. On the other hand, it can be seen that, for any $t \in (t_j, t_{j+1})$, we have

$$|\phi(t; 0, x_0, \sigma^{x_0})| \leq \eta^T |\phi(t_j; 0, x_0, \sigma^{x_0})| = \eta^T |z(j)|,$$

where $T = \max_{i \in M} s_i$, $\eta = \max_{i \in M} \exp(\|A_i\|)$ in continuous time, and $\eta = \max_{i \in M} \|A_i\|$ in discrete time. This, together with the asymptotic stability of system (4.49), guarantees that the original switched system (4.47) is asymptotically stable under switching law $\bigwedge_{i=1}^{k} \theta_i^{\Omega_i}$. □

For clarity, we term the discrete-time switched system (4.49) as the *aggregated system* of the original system (4.47) w.r.t. $\{(\theta_i, \Omega_i)\}_{i=1}^{k}$. It is clear that each pathwise state-feedback switching law corresponds to an aggregated system, and the switching law stabilizes the original system iff the aggregated system is asymptotically stable.

Lemma 4.28 indicates a way to find a stabilizing switching law for the switched linear system. Indeed, if we can find a number of switching paths θ_i and a set of state partitions Ω_i that make the aggregated system asymptotically stable, then the original system is stabilized by the pathwise state-feedback switching law $\bigwedge_{i=1}^{k} \theta_i^{\Omega_i}$. For this, we need to find the switching paths and the state partitions at the same time, which may be not an easy task. To reduce the complexity, observe that, if the switching paths θ_i are properly designed in the following sense, then the corresponding state partitions can then be determined accordingly.

Lemma 4.29 *Suppose that V is a continuous positive definite function defined on $\mathbf{R}^n$ and that θ_i, $i = 1, \ldots, k$, are switching paths defined over $[0, s_i)$, respectively. Then, switched system (4.47) is asymptotically stabilizable if*

$$\min_{i=1}^{k} V\big(\phi(s_i; 0, x, \theta_i)\big) < V(x) \qquad \forall x \in \mathbf{R}^n,\ x \neq 0. \tag{4.50}$$

The proof of the lemma is immediate as condition (4.50) implies that the aggregated system

$$z(t+1) = G_i z(t), \qquad z(t) \in \Omega_i,$$

with $G_i = \Phi(s_i, 0, \theta_i)$ is asymptotically stable. In this case, let

$$\Omega_1 = \Big\{x :\ V\big(\phi(s_1; 0, x, \theta_1)\big) = \min_{i=1}^{k} V\big(\phi(s_i; 0, x, \theta_i)\big)\Big\} \tag{4.51}$$

and

$$\Omega_j = \Big\{x :\ V\big(\phi(s_j; 0, x, \theta_j)\big) = \min_{i=1}^{k} V\big(\phi(s_i; 0, x, \theta_i)\big)\Big\} - \bigcup_{l=1}^{j-1} \Omega_l \tag{4.52}$$

for $j = 2, \ldots, k$. It can be seen that the pathwise state-feedback switching law $\bigwedge_{i=1}^{k} \theta_i^{\Omega_i}$ asymptotically stabilizes the original system.

Lemma 4.29 enables us to design the switching paths instead of designing both the switching paths and the corresponding state-space partitions, as indicated by

Lemma 4.28. Note that, if V is 0-symmetric and positively homogeneous of degree one, i.e.,

$$V(\lambda x) = |\lambda| V(x) \quad \forall x \in \mathbf{R}^n,\ \lambda \in \mathbf{R},$$

then the state-space partitions Ω_i as defined in (4.51) and (4.52) are 0-symmetric cones, i.e.,

$$x \in \Omega_i \quad \Longleftrightarrow \quad \lambda x \in \Omega_i \quad \forall x,\ i, \lambda \neq 0.$$

Therefore, the resultant pathwise state-feedback switching law $\bigwedge_{i=1}^{k} \theta_i^{\Omega_i}$ for the nominal switched linear system is radially invariant in the sense $\sigma^x = \sigma^{\lambda x}$ for any $x \in \mathbf{R}^n$ and $\lambda \neq 0$. This further implies that the state trajectories are radially linear, i.e.,

$$\phi\left(t; 0, \lambda x, \bigwedge_{i=1}^{k} \theta_i^{\Omega_i}\right) = \lambda \phi\left(t; 0, x, \bigwedge_{i=1}^{k} \theta_i^{\Omega_i}\right) \quad \forall x \in \mathbf{R}^n,\ \lambda \in \mathbf{R}. \tag{4.53}$$

The above discussion encourages us to look for an origin-symmetric and positively homogeneous function serving as the Lyapunov function in Lemma 4.29. For this, note that vector norms are natural candidates. A question that arises is whether the existence of such Lyapunov functions is universal or not. The following theorem presents a confirmative answer.

Theorem 4.30 *Let* $|\cdot|$ *be any vector norm in* $\mathbf{R}^n$, *and* μ *be any real number with* $0 < \mu < 1$. *Suppose that switched linear system* (4.47) *is switched attractive. Then, there exist a natural number* k, *positive times* $s_i > 0$ *for* $i = 1, \ldots, k$, *and well-defined switching paths* θ_i *over* $[0, s_i)$ *for* $i = 1, \ldots, k$ *such that*

$$\min_{i=1}^{k} \left|\phi(s_i; 0, x, \theta_i)\right| \leq \mu |x| \quad \forall x \in \mathbf{R}^n. \tag{4.54}$$

Proof Note that relationship (4.54) implies (4.50) with $V(x) = |x|$. As the norm is positively homogeneous of degree one, (4.54) holds if it holds for the unit sphere, i.e.,

$$\min_{i=1}^{k} \left|\phi(s_i; 0, x, \theta_i)\right| \leq \mu \quad \forall x \in \mathbf{H}_1. \tag{4.55}$$

Fix $x \in \mathbf{H}_1$. It follows from the switched attractivity that there exist a positive time t^x and a well-defined switching path $\theta^x_{[0, t^x]}$ such that

$$\left|\phi(t^x; 0, x, \theta^x)\right| \leq \frac{\mu}{2}. \tag{4.56}$$

In terms of the transition matrix, it follows from (4.56) that

$$\left|\Phi(t^x, 0, \theta^x)\, x\right| \leq \frac{\mu}{2}.$$

As a result, there is an open neighborhood $\mathcal{N}_x$ of x such that

$$\left|\Phi\left(t^x, 0, \theta^x\right) y\right| \le \mu \quad \forall y \in \mathcal{N}_x. \tag{4.57}$$

Letting x vary along the unit sphere, it is obvious that

$$\bigcup_{x \in \mathbf{H}_1} \mathcal{N}_x \supseteq \mathbf{H}_1.$$

As the unit sphere is a compact set in $\mathbf{R}^n$, by the Finite Covering Theorem, there exist a natural number k and a set of states $x_1, \ldots, x_k$ on the unit sphere such that

$$\bigcup_{i=1}^{k} \mathcal{N}_{x_i} \supseteq \mathbf{H}_1.$$

Accordingly, we can partition the unit sphere into k regions $\mathcal{M}_1, \ldots, \mathcal{M}_k$ such that

(a) $\bigcup_{i=1}^{k} \mathcal{M}_i = \mathbf{H}_1$, and $\mathcal{M}_i \cap \mathcal{M}_j = \emptyset$ for $i \neq j$ and
(b) $x_i \in \mathcal{M}_i$ for any $i = 1, \ldots, k$, and

$$\left|\Phi\left(t^{x_i}, 0, \theta^{x_i}\right) y\right| \le \mu \quad \forall y \in \mathcal{M}_i. \tag{4.58}$$

Let $s_i = t^{x_i}$ and $\theta^i = \theta^{x_i}$. It is clear that relationship (4.58) implies (4.55), and the theorem follows. □

Corollary 4.31 *For the switched linear system, the following statements are equivalent*:

(1) *The system is switched attractive.*
(2) *The system is asymptotically stabilizable.*
(3) *The system is exponentially stabilizable.*

Proof Suppose that the system is switched attractive. It follows from Theorem 4.30 that, there exist switching paths $\theta^i_{[0,s_i)}$, $i = 1, \ldots, k$, such that the aggregated system (4.49) is exponentially stable. In fact, let $z_0, z_1, \ldots$ be a state trajectory of the aggregated system. It can be seen that

$$|z_j| \le \mu^j |z_0|, \quad j = 0, 1, \ldots.$$

Back to the original system, it is clear that

$$\phi\left(t_j; 0, z_0, \bigwedge_{i=1}^{k} \theta_i^{\Omega_i}\right) = z_j \quad \forall j = 0, 1, 2, \ldots, \tag{4.59}$$

where Ω_i and t_i are given in (4.51)–(4.52) and (4.48), respectively.

Let $h=\max\{s_1,\dots,s_k\}$, $\alpha=-\frac{\ln\mu}{h}$, $\eta=\max\{\|A_1\|,\dots,\|A_m\|\}$, and $\beta=e^{h\eta}$ in continuous time and $\beta=\eta^h$ in discrete time. It can be seen that

$$\begin{aligned}\left|\phi\left(t;0,x_0,\bigwedge_{i=1}^{k}\theta_i^{\Omega_i}\right)\right| &\le \beta\left|\phi\left(t_{j+1};0,x_0,\bigwedge_{i=1}^{k}\theta_i^{\Omega_i}\right)\right|\\ &=\beta|z_{j+1}|\le\beta\mu^{j+1}|x_0|\le\beta e^{-\alpha(j+1)h}|x_0|\\ &\le\beta e^{-\alpha t}|x_0|\quad\forall t\in[t_j,t_{j+1}),\ j=0,1,2,\dots,\end{aligned}$$

which clearly exhibits that the original system is exponentially stable under the pathwise state-feedback switching law $\bigwedge_{i=1}^{k}\theta_i^{\Omega_i}$. This concludes the proof. □

Theorem 4.30 reveals a couple of important facts, which are listed below as remarks.

Remark 4.32 The most important implication of the theorem is that the class of pathwise state-feedback switching laws is universal for the purpose of stabilizing design of switched linear systems. Indeed, suppose that switching paths $\theta^i_{[0,s_i)}$ satisfy relationship (4.54). Let the state-space partitions Ω_i be

$$\begin{aligned}\Omega_1&=\left\{x:\ \left|\phi(s_1;0,x,\theta_1)\right|=\min_{i=1}^{k}\left|\phi(s_i;0,x,\theta_i)\right|\right\},\\ \Omega_j&=\left\{x:\ \left|\phi(s_j;0,x,\theta_j)\right|=\min_{i=1}^{k}\left|\phi(s_i;0,x,\theta_i)\right|\right\}-\bigcup_{l=1}^{j-1}\Omega_l,\\ &\quad j=2,\dots,k.\end{aligned}\tag{4.60}$$

Then, the pathwise state-feedback switching law $\bigwedge_{i=1}^{k}\theta_i^{\Omega_i}$ exponentially stabilizes the switched linear system. As a result, the aggregated system (4.49) is exponentially stable.

Remark 4.33 It should be noticed that number k is independent of the number of the subsystems, m. As can be seen in the proof of Theorem 4.30, it relies on the norm and the particular choice of the contractive switching paths for the state on the unit sphere. Due to the nonconstructiveness of the Finite Covering Theorem, we do not have a general upper bound of number k, though it is indeed a finite number. The same comment applies to the lengths s_i of the switching paths θ^i. Nevertheless, for any consistently stabilizable system, it is always possible to find a single switching path that makes the unit sphere contractive, and thus the system admits a linear time-invariant system as its aggregated system. Here we present an example to show how k varies w.r.t. the norm.

Example 4.34 For the continuous-time switched linear system

$$\dot{x}(t)=A_{\sigma(t)}x(t),\quad \sigma(t)\in\{1,2\},$$

$$A_1 = \begin{bmatrix} 1 & -\epsilon \\ \epsilon & 1 \end{bmatrix}, \qquad A_2 = \begin{bmatrix} 1 & 0 \\ 0 & -\epsilon \end{bmatrix},$$

where ϵ is a positive real number, it can be verified that the system is switched attractive and hence is exponentially stabilizable.

For the standard Euclidean norm, it can be seen that the contractive cone for the second subsystem is contained in the region

$$\Omega_1 \stackrel{\text{def}}{=} \left\{x : \sqrt{1-\epsilon^2}|x_1| < \epsilon|x_2|\right\}.$$

Note that, for any initial state outside the cone, the stabilizing switching law that achieves the largest possible convergence rate is to activate the first subsystem until the state reaches region Ω_1. To steer the state $x_0 = \left[\begin{smallmatrix} 1 \\ 0 \end{smallmatrix}\right]$ to region Ω_1, it needs at least time $h \geq \frac{\pi/2 - \arcsin(\epsilon)}{\epsilon}$, which tends to infinity as ϵ approaches 0. At the same time, any states x_1 and x_2 on the unit sphere can be steered to region Ω_1 by a single switching path only if they can be simultaneously steered to the region by activating the first subsystem. This means that to steer the whole unit sphere into region Ω_1, we need at least $N = \lceil \frac{\pi}{2\arcsin(\epsilon)} \rceil - 1$ switching paths, where $\lceil a \rceil$ denotes the least natural number greater than or equal to a. It is clear that N approaches infinity as ϵ approaches 0.

On the other hand, if we take the norm as $|x| = \sqrt{\epsilon^5 x_1^2 + x_2^2}$, it can be verified that the region $\Omega_2 \stackrel{\text{def}}{=} \{x : 2\epsilon^2|x_1| \leq |x_2\}$ is contractive under the switching path $\hat{2}_{[0,s_1)}$, where $s_1 \stackrel{\text{def}}{=} \frac{1}{4}\min\{1, \epsilon\}$. For any state outside the region, it can be verified that the switching path $\hat{2}_{[0,s_1)} \wedge \hat{1}_{[0,s_2)}$ makes the state contractive, where $\wedge$ denotes the concatenation of switching paths, and $s_2 \stackrel{\text{def}}{=} 2\arctan(2\epsilon^2)$. As a result, $k = 2$, and s_i, $i = 1, 2$, are small for small ϵ.

Finally, we present the following definition which is motivated from Theorem 4.30.

Definition 4.35 Suppose that $|\cdot|$ is a given vector norm. A pathwise state-feedback switching law $\bigwedge_{i=1}^{k} \theta_i^{\Omega_i}$ is said to be *piecewise contractive* w.r.t. the switched linear system and the norm if

$$\left|\phi(s_i; 0, x, \theta_i)\right| < |x| \quad \forall x \in \Omega_i,\ x \neq 0,\ i = 1, \ldots, k.$$

4.4.2 Computational Algorithms

Theorem 4.30 establishes that pathwise state-feedback switching laws are universal in achieving exponential stability of switched linear systems. To solve the stabilization problem, we need to develop a constructive procedure for computing a pathwise state-feedback switching law that stabilizes the switched linear system. In this subsection, we present several algorithms for calculating a stabilizing switching law.

The following procedure summarizes the main steps toward the construction of a stabilizing switching law.

Conceptual Procedure for Finding a Stabilizing Switching Law

1. Fix a vector norm $|\cdot|$ and a real number $\mu \in (0,1)$. For each state x in the unit sphere, find a time t^x and a switching path $\theta^x_{[0,t^x]}$ such that

$$\left|\phi\left(t^x, 0, x, \theta^x\right)\right| \leq \mu.$$

2. By the Finite Covering Theorem, there exist a finite natural number k, a state set $\{x_1,\ldots,x_k\}$, and a partition of state space $\{\Omega_1,\ldots,\Omega_k\}$ such that

$$\bigcup_{i=1}^{k} \Omega_k = \mathbf{R}^n, \qquad \Omega_i \cap \Omega_j = \emptyset \ \forall i \neq j,$$

and

$$\left|\phi\left(t^{x_i}, 0, x, \theta^{x_i}\right)\right| \leq \mu|x_i| \quad \forall x \in \Omega_i,\ i = 1,\ldots,k.$$

Denote $\theta_i = \theta^{x_i}$ for $i = 1,\ldots,k$.

3. The state-feedback pathwise switching law $\bigwedge_{i=1}^{k} \theta_i^{\Omega_i}$ exponentially stabilizes the switched system.

It can be seen from the procedure that, while Step 3 is implementable online by means of a timer and a state-measure device, Steps 1 and 2 are generally not constructive. To tackle this problem, we propose an alternative procedure to find k, θ_i, and Ω_i simultaneously.

First, we show that, for any asymptotically stabilizable continuous-time switched linear system, it is always possible to find a positive sampling period such that the sampled system is also asymptotically stabilizable as a discrete-time switched system. Note that a pathwise state-feedback switching law for the sampled system can be converted into a pathwise state-feedback switching law for the original continuous-time system in a clear manner. As a result, the design procedure of a stabilizing switching law for both continuous-time and discrete-time systems can be presented in a unified manner.

Lemma 4.36 *For an asymptotically stabilizable continuous-time switched linear system*

$$\dot{x}(t) = A_{\sigma(t)}x(t), \qquad \sigma(t) \in M,$$

there is a positive real number ρ such that for any sampling period $T < \rho$, the sampled switched system

$$y(t+1) = B_{\varrho(t)}y(t), \qquad \varrho(t) \in M, \qquad B_i = e^{A_i T} \tag{4.61}$$

is also asymptotically stabilizable.

Proof Suppose that $\bigwedge_{i=1}^{k}\theta_i^{\Omega_i}$ is a stabilizing switching law for the continuous-time system with

$$\left|\phi(s_i;t_0,x,\theta_i)\right|\le\mu|x|\quad\forall x\in\Omega_i,\ i=1,\dots,k,$$

where $\mu<1$. Rewrite the inequality as

$$\left|e^{A_{j^i_{l_i-1}}(t^i_{l_i}-t^i_{l_i-1})}\cdots e^{A_{j^i_0}(t^i_1-t^i_0)}x\right|\le\mu|x|\quad\forall x\in\Omega_i,\ i=1,\dots,k,\tag{4.62}$$

where $t_0=t^i_0<t^i_1<\cdots<t^i_{l_i-1}$ and $j^i_0,j^i_1,\dots,j^i_{l_i-1}$ are the switching time/index sequences of θ_i over $[0,s_i)$, respectively, and $t^i_{l_i}=s_i$. Define the functions $\xi_i:\mathbf{R}^{l_i}\times\Omega_i\mapsto\mathbf{R}$ for $i=1,\dots,k$ by

$$\xi_i\left(\tau^i_1,\dots,\tau^i_{l_i},x\right)=\left|e^{A_{j^i_{l_i-1}}\tau^i_s}e^{A_{j^i_{l_i-2}}\tau^i_{l_i-1}}\cdots e^{A_{j^i_0}\tau^i_1}x\right|.$$

Fix $\bar\mu\in(\mu,1)$. It follows from (4.62) and the continuity of the functions that there exists a positive real number ρ such that

$$\xi_i\left(\tau^i_1,\dots,\tau^i_{l_i},x\right)\le\bar\mu|x|\quad\forall x\in\Omega_i,\ i=1,\dots,k,$$

whenever $\tau^i_j-(t^i_j-t^i_{j-1})\in[0,\rho]$, $j=1,\dots,l_i$. As a result, when the sampling rate T is less than or equal to ρ, then there are natural numbers $\nu^i_1,\dots,\nu^i_{l_i}$ such that

$$\left|\left(\exp(A_{j^i_{l_i-1}}T)\right)^{\nu^i_{l_i}}\cdots\left(\exp(A_{j^i_0}T)\right)^{\nu^i_1}x\right|\le\bar\mu|x|\quad\forall x\in\Omega_i,\ i=1,\dots,k.$$

This means that, under the sampling rate T, the sampled-data system (4.61) is switched attractive and is asymptotically stabilizable. □

Lemma 4.36 enables us to focus on the discrete-time switched linear system

$$x(t+1)=A_{\sigma(t)}x(t),\quad\sigma(t)\in\{1,\dots,m\}.\tag{4.63}$$

The system can be represented by the m-tuple $(A_1,\dots,A_m)$.

We can arrange an ordered set of all the possible switching paths and their corresponding state transition matrices. For example, let

$$\begin{aligned}&\vartheta_1=(1),\quad\dots,\quad\vartheta_m=(m),\quad\vartheta_{m+1}=(1,1),\quad\dots,\quad\vartheta_{2m}=(1,m),\\&\vartheta_{2m+1}=(2,1),\quad\dots,\quad\vartheta_{3m}=(2,m),\quad\dots,\quad\vartheta_{m^2+1}=(m,1),\quad\dots,\\&\vartheta_{m^2+m}=(m,m),\quad\vartheta_{m^2+m+1}=(1,1,1),\quad\dots,\end{aligned}$$

and the corresponding state transition matrices be

$$C_1=A_1,\quad\dots,\quad C_m=A_m,\quad C_{m+1}=A_1^2,\quad\dots,\quad C_{2m}=A_mA_1,$$

$$C_{2m+1} = A_1 A_2, \quad \ldots, \quad C_{3m} = A_m A_2, \quad \ldots, \quad C_{m^2+1} = A_1 A_m, \quad \ldots,$$
$$C_{m^2+m} = A_m^2, \quad C_{m^2+m+1} = A_1^3, \quad \ldots.$$

The next algorithm presents a schematic procedure for finding the requested parameters k, θ_i, and Ω_i, $i = 1, \ldots, k$.

Schematic Algorithm for Computing the Parameters

Initialization.

(0) Set $k := 0$, *ContractiveRegion* $:= \emptyset$, $j := 0$. Fix μ in $(0, 1)$.

Recursion.

(1) Set $j := j + 1$.
(2) Set $C := C_j$ and compute the singular values of matrix C. If all the singular values are greater than or equal to one, then go back to Step (1). Otherwise, compute the set $\Omega = \{x \in \mathbf{R}^n : |Cx| \le \mu|x|\}$.
(3) If $\Omega \subseteq$ *ContractiveRegion*, then go back to Step (1). Otherwise, set

$$k := k + 1, \quad \Omega_k := \Omega, \quad \theta_k := \vartheta_j, \quad \text{and}$$
$$\textit{ContractiveRegion} := \textit{ContractiveRegion} \cup \Omega.$$

(4) If *ContractiveRegion* $= \mathbf{R}^n$, then go to Step (5). Otherwise, go back to Step (1).

Reduction.

(5) For $i = 1, \ldots, k$, verify if $\Omega_i \subseteq \bigcup_{l \ne i} \Omega_l$. If yes, set $\Omega_l := \Omega_{l+1}$ and $\theta_l := \theta_{l+1}$ for $l = i, \ldots, k-1$ and set $k := k - 1$. Continue the process until the partition is irreducible.
(6) Finally, set $\Omega_i := \Omega_i - \bigcup_{l=1}^{i-1} \Omega_l$ for $i = 2, \ldots, k$.

The termination condition *ContractiveRegion* $= \mathbf{R}^n$ means that, for each initial state x, there exists a region Ω_i such that $x \in \Omega_i$ and the switching path θ_i that makes the region (and the state) contractive. Upon termination of the algorithm, Ω_i and θ_i for $i = 1, \ldots, k$ give the state partitions and the corresponding switching paths, respectively.

Note that the algorithm terminates only when *ContractiveRegion* $= \mathbf{R}^n$. That is to say, if the switched linear system is not switched attractive, then the algorithm will not terminate in a finite time. However, if the system is switched attractive, the algorithm does terminate in a finite number of steps, although we do not have an upper bound of the step number. Other termination conditions may apply, for example, if all the singular values of matrices $A_1, \ldots, A_m$ are greater than or equal to one, then the system is by no means to be switched attractive [213], and we can terminate the algorithm in m steps ($j \le m$).

In the schematic algorithm, the major computational loads include the determination of the regions Ω_i and the verification of $\bigcup_{i=1}^{l} \Omega_i = \mathbf{R}^n$. For this, we briefly discuss the computation of the region $\mathcal{X}_C = \{x \in \mathbf{R}^n : |Cx| \le \mu|x|\}$ for a matrix $C \in \mathbf{R}^{n\times n}$ and the verification of the relationship

$$\bigcup_{j=1}^{k} \mathcal{X}_{C_j} = \mathbf{R}^n \tag{4.64}$$

for a matrix set $\{C_1, \dots, C_k\}$.

Note that the region $\mathcal{X}_C$ is norm-dependent, that is, different norms may correspond to different regions $\mathcal{X}_C$ for a matrix C. Note also that, for the ℓ_2-norm, the region is generally nonpolyhedric in nature, that is, it is not a union of a finite set of polyhedra. In this case, it is very hard to verify whether or not $\bigcup_{j=1}^{k} \mathcal{X}_{C_j} = \mathbf{R}^n$ for a matrix set $\{C_1, \dots, C_k\}$. On the other hand, for the ℓ_1-norm, the region is indeed polyhedric. Accordingly, we use the ℓ_1-norm.

Given a matrix $C = (c_{i,j})_{n\times n}$, it is clear that

$$\mathcal{X}_C = \left\{ x \in \mathbf{R}^n : \sum_{i=1}^{n} \left| \sum_{j=1}^{n} c_{i,j} x_j \right| \le \mu \sum_{j=1}^{n} |x_j| \right\}, \tag{4.65}$$

which can be routinely converted into the union of a finite number of regions in the form

$$\{x : E_1 x + F_1 \le 0\}, \tag{4.66}$$

where E_1 and F_1 are a matrix and a column vector with compatible dimensions, respectively. It is clear that any region in form (4.66) is a convex polyhedron.

Next, we propose a reduced-order method which enables us to compute the region (4.65) and to verify relation (4.64) in a unifying scheme.

Take the half unit sphere

$$\mathbf{H}_1^{x_n \ge 0} = \left\{ x = [x_1, \dots, x_n]^T : x_n \ge 0,\ |x|_1 = 1 \right\}.$$

This half-sphere can be projected onto the $(n-1)$-dimensional unit ball by the map

$$\mathcal{P}_n : \mathbf{H}_1^{x_n \ge 0} \mapsto \mathbf{B}_1^{n-1}, \quad \mathcal{P}_n x = [x_1, \dots, x_{n-1}]^T.$$

The inverse map is

$$\mathcal{P}_n^{-1} y = \left[y_1, \dots, y_{n-1}, 1 - |y|_1 \right]^T, \quad y \in \mathbf{B}_1^{n-1}.$$

Let

$$\mathcal{Y}_C = \left\{ y \in \mathbf{B}_1^{n-1} : \left| C\mathcal{P}_n^{-1} y \right|_1 \le \mu \left| \mathcal{P}_n^{-1} y \right|_1 \right\},$$

which is also a finite union of regions in form (4.66). It can be seen that

$$\mathcal{X}_C = \left\{ r\mathcal{P}_n^{-1} y : r \in \mathbf{R},\ y \in \mathcal{Y}_C \right\}.$$

Moreover, given matrices $C_1, \dots, C_k$, a necessary and sufficient condition for $\bigcup \mathcal{X}_{C_i} = \mathbf{R}^n$ is

$$\bigcup \mathcal{Y}_{C_i} = \mathbf{B}_1^{n-1}. \tag{4.67}$$

When $n = 2$, the set $\mathcal{Y}_C$ is the union of several (possibly empty) pieces of intervals, and the verification of (4.67) is simply of dimension one. Similarly, when $n = 3$, the verification of (4.67) is to check whether the regions cover the planar unit ball, which can be conducted conveniently by phase plane portraits. For higher-dimensional systems, due to the polyhedric structure of $\mathcal{Y}_{C_i}$ and $\mathbf{B}_1^{n-1}$, there are commercial numerical softwares available (for example, MATLAB GBT Toolbox [252]) for computing the regions $\mathcal{Y}_{C_i}$ and for verifying relation (4.67).

Integrating the conceptual procedure, the schematic algorithm, and Lemma 4.36, we obtain the following constructive procedure for stabilizing switching design of the switched linear system.

Computational Algorithm for Stabilizing Switching Design

1. Given n, m, and $A_i \in \mathbf{R}^{n\times n}$ for $i \in M = \{1, \dots, m\}$. Set *flag* $:= 0$ for continuous time and *flag* $:= 1$ for discrete time.
2. If *flag* $= 1$, then set $B_{i-1} := A_i$ for $i \in M$, and $\tau := 1$. Otherwise, choose a positive real number τ (sampling rate), and let $B_{i-1} := \exp(A_i \tau)$ for $i \in M$.
3. Set $k := 0$, *FeasibleSet* $:= \emptyset$, *UnionOfRegions* $:= \emptyset$.
4. Express k in base m, i.e., $k = \sum_{i=1}^{s} j_i m^{i-1}$ with $j_i \in M$ and $j_s \neq 0$. Set *path* $:= [j_1, \dots, j_s]$ and $C := B_{j_s} \cdots B_{j_1}$.
5. Compute $\mathcal{Y}_C$.
6. If $\mathcal{Y}_C \not\subset$ *UnionOfRegions*, set
   FeasibleSet $:=$ *FeasibleSet* $+ \{(C, \mathit{path}, \mathcal{Y}_C)\}$, and
   UnionOfRegions $:=$ *UnionOfRegions* $\cup \mathcal{Y}_C$.
7. If *UnionOfRegions* $\neq \mathbf{B}_1^{n-1}$, set $k := k + 1$ and go to Step 4).
8. Remove the reducible triplet elements of *FeasibleSet* whose $\mathcal{Y}_C$'s are subsets of the union of the other $\mathcal{Y}_C$'s (cf. the *reduction* part of the Schematic Algorithm).
9. Set k to be the number of elements of *FeasibleSet*. For $i = 1, \dots, k$, set G_i be the C in *FeasibleSet*(i), path_i be the *path* in *FeasibleSet*(i), and Λ_i be the $\mathcal{Y}_C$ in *FeasibleSet*(i).
10. Set $\Lambda_i := \Lambda_i - \bigcup_{s=1}^{i-1} \Lambda_s$ for $i = 2, \dots, k$.
11. Input an initial state x_0 and a terminal time t_f.
12. Set *State* $:= \emptyset$, *IterativeState* $:= x_0$, *IterativeTime* $:= 0$.
13. Set $y := \mathcal{P}_n(\mathit{sgn}(\mathit{IterativeState}(n)) * \mathit{IterativeState}/|\mathit{IterativeState}|)$, find $i \in \{1, \dots, k\}$ such that $y \in \Lambda_i$, and set $j := 1$.
14. Set $\mathit{index} := \mathit{path}_i(j) + 1$.
15. Compute the solution $[t, x]$ of the differential/difference equation

$$x^+(t) = A_{\text{index}} x(t)$$

   over time interval $[\mathit{IterativeTime}, \mathit{IterativeTime} + \tau]$ and set
   $\mathit{IterativeTime} := \mathit{IterativeTime} + \tau$ and
   $\mathit{State} := \mathit{State} \cup [t, x]$.

16. If $j < length(path_i)$, set $j := j + 1$ and go to Step 14).
17. If *IterativeTime* $< t_f$, set *IterativeState* := *State*(*IterativeTime*) and go to Step 13).
18. Plot the state trajectory *State*.

4.5 Stabilization via Mixed-Driven Switching: Robustness Analysis

Suppose that switched linear system (4.47) is asymptotically stabilizable and that the pathwise state-feedback switching law $\bigwedge_{i=1}^{k} \theta_{i\,s_i}^{\Omega_i}$ exponentially stabilizes the system with convergence rate α. That is,

$$\left|\phi\left(t; 0, x, \bigwedge_{i=1}^{k} \theta_{i\,s_i}^{\Omega_i}\right)\right| \leq \beta e^{-\alpha t}|x| \quad \forall x \in \mathbf{R}^n,\ t \in \mathcal{T}_0 \tag{4.68}$$

for some $\beta > 0$.

Let us consider the situation that the switched linear system undergoes exotic disturbances or perturbations. The perturbations can enter either structurally or unstructurally and can enter into either the subsystems or the switching signals. A general description of the perturbed system can thus be given as

$$x^+(t) = (A_{\bar{\sigma}(t)} + \tilde{A}_{\bar{\sigma}(t)})x(t) + f_{\bar{\sigma}(t)}(t) + f_0(t),$$

where $\bar{\sigma}(t)$ is the (structural) perturbation of the nominal switching signal, $\tilde{A}_i$ and f_i are the structural and unstructural perturbations on the ith subsystem, respectively, and $f_0(t)$ is the unstructural perturbation of the nominal system. As f_0 can be absorbed by f_i in an obvious manner, we can assume that this part is null and rewrite the perturbed system as

$$x^+(t) = (A_{\bar{\sigma}(t)} + \tilde{A}_{\bar{\sigma}(t)})x(t) + f_{\bar{\sigma}(t)}(t). \tag{4.69}$$

The objective of this section is to analyze the robustness of the nominal switched linear system w.r.t. the perturbations. For this, note that the structural and unstructural perturbations for the subsystems can be handled by the conventional scheme of robust analysis in a normed space. The perturbation analysis for switching signal, however, is a new topic that needs more attention.

The study of the robustness against switching signal perturbations is well motivated for the reasons below. First, in practice we cannot implement a switching signal precisely. For example, time delay is unavoidable in many practical situations, and exact online measure of the state variable is usually impossible. Second, the switching device may mismanipulate in certain cases. For instance, the system should activate the ith subsystem, but it activates the jth instead. Third, component (subsystem) failures lead to displacement of switching paths, and the ability of fault tolerance is always an important issue for practical implementation. Finally, from the design viewpoint, we prefer to choose a switching law which still works (for the stability purpose) under small perturbations.

In the next subsection, we present an approach to characterize the distance between two switching signals generated by a pathwise state-feedback switching law, which paves the way for the robust analysis conducted later.

4.5.1 Distance Between Switching Signals

To implement a pathwise state-feedback switching law $\bigwedge_{i=1}^{k} \theta_i^{\Omega_i}$, it involves a repeated process of first determining an index j according to the online state measurement, and then coordinating the switching along the switching path θ_j. In this process, perturbations might enter either in index determining or in executing the switching law. The distance between the nominal switching signal and perturbed switching signal should be the summation of the variations of both types in some sense. For this, we first define the distance between two switching paths and then extend it to the case of pathwise state-feedback switching laws.

For a right-continuous switching path $p_{[t_0,t_f)}$, let $t_0 < t_1 < \cdots < t_s$ be its switching time sequence. Its switching index sequence is $p(t_0), p(t_1), \ldots, p(t_s)$. For convenience, we refer to the sequence

$$SS_p = \{(t_0, p(t_0)), (t_1, p(t_1)), \ldots, (t_s, p(t_s)), \ldots\}$$

as its switching sequence and the sequence

$$DS_p = \{(p(t_0), t_1 - t_0), (p(t_1), t_2 - t_1), \ldots, (p(t_{s-1}), t_s - t_{s-1}), \ldots\}$$

as its duration sequence. It is clear that p, SS_p, and DS_p are equivalent in the sense that one can determine the others and vice versa. As a result, a variation of a switching path means a variation of the switching/duration sequence, which can also be seen as a variation of the transition chain defined later. This simple observation provides a basic insight into the way for characterizing the distance between two switching paths.

For a switching duration sequence $(i_0, h_0), (i_1, h_1), \ldots$, the transition chain is defined as the sequence of state transition matrices at the switching instants, i.e.,

$$\varphi(i_0, h_0)|\varphi(i_1, h_1)\varphi(i_0, h_0)|\varphi(i_2, h_2)\varphi(i_1, h_1)\varphi(i_0, h_0)|\cdots, \tag{4.70}$$

where $\varphi(i, h) = e^{A_i h}$ in continuous time, and $\varphi(i, h) = A_i^h$ in discrete time. The transition chain clearly exhibits the state transition process along the switching path.

Before formulating the distance between two switching paths, we present a couple of motivating examples.

Example 4.37 Consider a continuous-time switching signal p_1 whose switching sequence is $(0, 1), (1, 2), (2, 1), (3, 2), \ldots$.

Suppose that there is a small time delay ϵ at the first switching time. It is obvious that the delayed switching signal p_2 is with switching sequence $(0, 1)$, $(1+\epsilon, 2)$, $(2+\epsilon, 1)$, $(3+\epsilon, 2)$, ... The corresponding transition chains for the nominal and perturbed switching signals are

$$e^{A_1}|e^{A_2}e^{A_1}|e^{A_1}e^{A_2}e^{A_1}|\cdots$$

and

$$e^{A_1(1+\epsilon)}|e^{A_2}e^{A_1(1+\epsilon)}|e^{A_1}e^{A_2}e^{A_1(1+\epsilon)}|\cdots,$$

respectively. It can be seen that the two chains possess the same convergence property, that is, the perturbed matrix chain is convergent iff the nominal matrix chain is convergent. However, it is clear that

$$p_1(t) \neq p_2(t) \quad \forall t \in \Theta,$$

where $\Theta \stackrel{\text{def}}{=} \bigcup_{k=1}^{+\infty}[k, k+\epsilon)$. That is, the two switching signals are not equal at time intervals of an infinite length no matter how small ϵ is. This excludes the reasonableness of formulating the distance between p_1 and p_2 by $\text{meas}\{t : p_1(t) \neq p_2(t)\}$, though this seems to be the most straightforward way. Another observation from this example is that, comparing with the switching sequence, the switching duration sequence is more suitable for describing the transition chain of the switched system.

Example 4.38 For the following three switching paths

(i) $DS_{p_1} = \{(i_0, h_0), (i_1, h_1), (i_2, h_2), (i_3, h_3), \ldots\}$
(ii) $DS_{p_2} = \{(i_1, h_0), (i_0, h_1), (i_2, h_2), (i_3, h_3), \ldots\}$ and
(iii) $DS_{p_3} = \{(i_0, h_1), (i_1, h_0), (i_2, h_2), (i_3, h_3), \ldots\}$

which two are closer?

The transition chains of the paths are

$$\varphi(i_0, h_0) \mid \varphi(i_1, h_1)\varphi(i_0, h_0) \mid \varphi(i_2, h_2)\varphi(i_1, h_1)\varphi(i_0, h_0) \mid \cdots,$$

$$\varphi(i_1, h_0) \mid \varphi(i_0, h_1)\varphi(i_1, h_0) \mid \varphi(i_2, h_2)\varphi(i_0, h_1)\varphi(i_1, h_0) \mid \cdots,$$

$$\varphi(i_0, h_1) \mid \varphi(i_1, h_0)\varphi(i_0, h_1) \mid \varphi(i_2, h_2)\varphi(i_1, h_0)\varphi(i_0, h_1) \mid \cdots,$$

respectively.

Roughly speaking, though the switching time sequences totally coincide for the first and second switching paths, the index sequences are different for the first two periods, so the difference in time measure is $h_0 + h_1$. Similarly, the difference between the first and third is $2|h_1 - h_0|$, and between the second and third it is $2\min\{h_0, h_1\}$. In this sense, if $h_1 \geq 2h_0$ or $h_0 \geq 2h_1$, then the distance between the second and third is shortest, otherwise the distance between the first and the third is shortest. In both cases, the distance between the first and the second paths is the longest.

Example 4.39 Consider two switching paths θ_1 and θ_2 in discrete time, which are represented by sequences

$$\begin{aligned}\theta_1 &= 3\ 1\ 2\ 4\ 2\ 1\ 3\ 1\ 1\ 1\ 3\ 1\ 1\ 2\ 2\ 2\ 1\ 2\ 2\ 1,\\ \theta_2 &= 3\ 1\ 2\ 4\ 3\ 1\ 3\ 1\ 1\ 3\ 3\ 1\ 1\ 2\ 2\ 1\ 2\ 2\ 1,\end{aligned} \tag{4.71}$$

which are of lengths 20 and 19, respectively. To compare the sequences, two intuitive ways are aligning the sequence by adding and by removing certain elements to/from the sequences. By adding new elements, we have

$$\begin{aligned}\bar{\theta}_1 &= 3\ 1\ 2\ 4\ 2\ \bar{3}\ 1\ 3\ 1\ 1\ 1\ \bar{3}\ 3\ 1\ 1\ 2\ 2\ 2\ 1\ 2\ 2\ 1,\\ \bar{\theta}_2 &= 3\ 1\ 2\ 4\ \bar{2}\ 3\ 1\ 3\ 1\ 1\ \bar{1}\ 3\ 3\ 1\ 1\ 2\ \bar{2}\ 2\ 1\ 2\ 2\ 1,\end{aligned} \tag{4.72}$$

where the numbers with a bar are newly added. It is clear that five new elements are incorporated, and $\bar{\theta}_1$ is identical to $\bar{\theta}_2$. Similarly, by removing elements from the sequences, we have

$$\begin{aligned}&\quad\quad\ \ \uparrow 2\quad\ \ \uparrow 1\quad\ \ \uparrow 2\\ \tilde{\theta}_1 &= 3\ 1\ 2\ 4\ 1\ 3\ 1\ 1\ 3\ 1\ 1\ 2\ 2\ 1\ 2\ 2\ 1,\\ \tilde{\theta}_2 &= 3\ 1\ 2\ 4\ 1\ 3\ 1\ 1\ 3\ 1\ 1\ 2\ 2\ 1\ 2\ 2\ 1,\\ &\quad\quad\ \ \downarrow 3\quad\ \ \downarrow 3\end{aligned}$$

where five elements are removed, and $\tilde{\theta}_1$ is identical to $\tilde{\theta}_2$. Note also that the cardinality of $\bar{\theta}_1$ is 22, the cardinality of $\tilde{\theta}_1$ is 17, and the difference is five. As a result, it is intuitively proper to define the absolute distance between θ_1 and θ_2 to be five.

The above examples motivate us to define the distance between two switching paths through the time variation while ignoring the dynamic difference between the system matrices, that is, we assume, for simplicity, that the distance between any A_i and A_j is normalized.

Let Δ be the set of intervals of the form $[a, b)$ with $0 \leq a < b$ and of unions of such intervals. Given $\pi \in \Delta$, we can define the map $\psi_\pi : (\mathbf{R}_+ - \pi) \mapsto \mathbf{R}_+$ by

$$\psi_\pi(t) = t - \text{meas}\{s \leq t : s \in \pi\}, \quad t \in \mathbf{R}_+ - \pi.$$

Given two switching paths $p_{1[t_1,t_2)}$ and $p_{2[t_3,t_4)}$, p_1 is said to be a *child-path* of p_2, or p_2 is said to be a *parent-path* of p_1, denoted by $p_1 \preceq p_2$, if there exist $\pi \in \Delta$ and a time transition $\delta \in \mathbf{R}$ such that

$$\begin{aligned}[t_1, t_2) &= \bigcup_{t \in ([t_3,t_4)-\pi)} \{\psi_\pi(t) - \delta\},\\ p_2(t) &= p_1\big(\psi_\pi(t) - \delta\big) \quad \forall t \in [t_3, t_4) - \pi.\end{aligned} \tag{4.73}$$

Correspondingly, let $\Delta_{p_1}^{p_2}$ be the set of π that satisfies (4.73). It should be stressed that the set $\Delta_{p_1}^{p_2}$ may contain more than one element.

For two switching paths $p_1 \preceq p_2$, define the distance to be

$$|p_2 - p_1| = \inf_{\pi \in \Delta_{p_1}^{p_2}} \operatorname{meas} \pi. \tag{4.74}$$

It is clear that $|p_2 - p_1| = 0$ iff p_1 is a pure time transition of p_2, that is, $t_4 - t_3 = t_2 - t_1$ and $p_1(t_1 + s) = p_2(t_3 + s)$ for all $s \in [0, t_2 - t_1)$. In this case, we denote $p_2 = p_1^{\mapsto t_3 - t_1}$.

Given switching paths p_1, p_2, and p_3, p_3 is said to be a *common parent-path* of p_1 and p_2, denoted $p_3 \in CP(p_1, p_2)$, if $p_1 \preceq p_3$ and $p_2 \preceq p_3$. It can be seen that, for any two switching paths, there must exist a common parent-path. Indeed, suppose that

$$DS_{p_j} = \{(i_0^j, h_0^j), (i_1^j, h_1^j), (i_2^j, h_2^j), \dots\}, \quad j = 1, 2.$$

Then, the switching path

$$DS_{p_3} = \{(i_0^1, h_0^1), (i_0^2, h_0^2), (i_1^1, h_1^1), (i_1^2, h_1^2), (i_2^1, h_2^1), (i_2^2, h_2^2), \dots\}$$

is a common parent-path of p_1 and p_2.

Definition 4.40 For any switching paths p_1 and p_2, the distance between them is defined as

$$d(p_1, p_2) = \inf_{p_3 \in CP(p_1, p_2)} \bigl(|p_3 - p_1| + |p_3 - p_2|\bigr). \tag{4.75}$$

Remark 4.41 It is interesting to note that the switching distance is closely related to sequence alignment, which is one of the basic tasks of bioinformatics. A primary usage of alignment is to attempt to identify regions of sequences that share a common evolutionary origin. That is, the purpose of sequence alignment is to identify the similarity of two or more biological sequences. It is clear that the smaller distance, the more similarity. Taking mRNA sequences as an example, they are taking values from alphabet $\{A, C, G, T\}$. In fact, if we relabel A, C, G, and T with 1, 2, 3, and 4, respectively, then the sequences (4.71) are exactly segments of mRNA sequences as shown in [191, Fig. 2B]. While numerous computer algorithms were developed for sequence alignment, the distance defined here captures the major feature of many algorithms. The reader is referred to [181, 185] for recent development of sequence alignment.

Proposition 4.42 *The distance between switching paths possesses the following properties*:

(1) (*positive definiteness*) $d(p_1, p_2) \geq 0$, *and* $d(p_1, p_2) = 0$ *iff* $p_1 = p_2^{\mapsto s}$ *for some* $s \in \mathbf{R}$

(2) (*symmetricalness*) $d(p_1, p_2) = d(p_2, p_1)$ *and*
(3) (*triangular inequality*) $d(p_1, p_2) \le d(p_1, p_3) + d(p_2, p_3)$

Proof Properties (1) and (2) straightforwardly follow from the definition. To prove (3), let ε be an arbitrarily small positive real number. By definition, there is a common parent-path p_4 of paths p_1 and p_3 such that

$$d(p_1, p_3) \ge |p_4 - p_1| + |p_4 - p_3| - \varepsilon.$$

Similarly, there is a common parent-path p_5 of paths p_2 and p_3 such that

$$d(p_2, p_3) \ge |p_5 - p_2| + |p_5 - p_3| - \varepsilon.$$

On the other hand, we can find a common parent-path p_6 of paths p_4 and p_5 such that

$$|p_6 - p_4| \le |p_5 - p_3| \quad \text{and} \quad |p_6 - p_5| \le |p_4 - p_3|.$$

As p_6 is also a common parent-path of p_1 and p_2, we have

$$\begin{aligned} d(p_1, p_2) &\le |p_6 - p_1| + |p_6 - p_2| \\ &\le |p_6 - p_4| + |p_4 - p_1| + |p_6 - p_5| + |p_5 - p_2| \\ &\le |p_5 - p_3| + |p_4 - p_1| + |p_4 - p_3| + |p_5 - p_2| \\ &\le d(p_1, p_3) + d(p_2, p_3) + 2\varepsilon. \end{aligned}$$

Due to the arbitrariness of ε, the triangular inequality holds. □

It follows from the proposition that the set of switching paths forms a metric space.

Next, we formulate the distance between two switching signals generated by a pathwise state-feedback switching law. Fix a pathwise state-feedback switching law $\bigwedge_{i=1}^{k} \theta_i^{\Omega_i}$. For any given switched system, it generates a switching signal σ^x for any initial state x that is the concatenation of the switching paths θ_i, $i = 1, \dots, k$. This switching signal can be described in a recursive way, as described below.

(1) *Initiation*. From the initial state x_0, choose the active subsystem index $j \in \{1, \dots, k\}$ such that $x_0 \in \Omega_j$, and let the system evolve along switching path θ_j for the duration of s_j. At the end of the process, the state is $x_1 = \phi(s_j; 0, x_0, \theta_j)$, which is termed as the *current relay state at* $t_1 = s_j$.
(2) *Recursion*. From the current relay state x_ν at t_ν, choose the active subsystem index $l \in \{1, \dots, k\}$ such that $x_\nu \in \Omega_l$, and let the system evolve along switching path θ_l for the duration of s_l. At the end of the process, the state is $x_{\nu+1} = \phi(s_l; 0, x_\nu, \theta_l)$, which is the current relay state at $t_{\nu+1} = t_\nu + s_l$.

Suppose that we are to implement the (nominal) switching signal in the above-mentioned recursive way. Then, perturbations may enter in the following possible manners:

(1) When implementing a switching path θ_i for a duration of s_i, mismanipulations may occur in certain cases. For instance, a time delay in the operation or component (subsystem) failures lead to displacement of switching paths, etc.
(2) At the concatenating instants t_j, the next piece of switching path is wrongly selected due to the imprecise measure of the current initial state.

In either case, due to the interaction between the continuous state and the switching signal, for the same initial state, any small perturbation may lead the perturbed switching signal to totally deviated from the nominal switching signal. This indicates that the distance between the perturbed switching signal and the nominal switching signal should be formulated based on the following rules:

(1) The distance should be counted locally (piece by piece) rather than globally.
(2) Any mismanipulation should be counted only once in the sense that the future effluence should not be taken into account to the distance.
(3) Once a mismanipulation occurs, the next current relay state should be taken as the nominal initial state for future comparison between the switching signals.

In the light of the rules, we rewrite the nominal switching signal σ_n as

$$\sum_{i=0}^{+\infty}(x_i, j_i, \theta_{j_i}, s_{j_i}): \quad x_{i+1} = \phi(s_{j_i}; 0, x_i, \theta_{j_i}) \in \Omega_{j_{i+1}}. \tag{4.76}$$

Suppose that the perturbed switching signal σ_p can be expressed in a similar way as

$$\sum_{i=0}^{+\infty}(y_i, \kappa_i, p_i, \tau_i): \quad y_{i+1} = \phi(\tau_i; 0, y_i, p_i), \quad y_0 = x_0, \tag{4.77}$$

where κ_i is the actually selected (perturbed) index of the switching path for the current initial state y_i, and $p_{i[0,\tau_i)}$ is the actual (perturbed) implementation of $\theta_{j_{\kappa_i}}$. It can be seen that a deviation occurs when either $y_i \notin \Omega_{\kappa_i}$ or $p_i \neq \theta_{j_{\kappa_i}}$.

Definition 4.43 For a pathwise state-feedback switching law $\bigwedge_{i=1}^{k} \theta_i^{\Omega_i}$, let σ_n^x and σ_p^x be the nominal switching signal and perturbed switching signal with initial state x, respectively.

(1) For a natural number N, the N-distance between σ_n^x and σ_p^x is

$$D_x^N(\sigma_p, \sigma_n) = \sum_{i=0}^{N-1} d(p_i, \theta_{l_i}), \tag{4.78}$$

where l_i is the index that $y_i \in \Omega_{l_i}$ for $i = 0, \dots, N-1$, and $d(\cdot, \cdot)$ is the distance between two switching paths as defined in (4.75).

(2) The relative distance between the nominal switching signal σ_n^x and the perturbed switching signal σ_p^x is defined as

$$RD_x(\sigma_p, \sigma_n) = \limsup_{N \to +\infty} \frac{1}{N} D_x^N(\sigma_p, \sigma_n). \tag{4.79}$$

(3) The supremal relative distance between the nominal switching signal σ_n and the perturbed switching signal σ_p is defined as

$$SRD(\sigma_p, \sigma_n) = \limsup_{x \in \mathbf{R}^n} RD_x(\sigma_p, \sigma_n). \tag{4.80}$$

Remark 4.44 The N-distance measures the absolute distance between the nominal switching and the perturbed switching over the first N-concatenating periods. It is the summation of the distances between the nominal and perturbed switching signals. The relative distance, on the other hand, measures the average distance in time over an infinite horizon. It should be stressed that the relative distance is more subtle than the (absolute) N-distance in characterizing the distance between two switching signals. In fact, for any two switching signals defined on an infinite horizon, the absolute ∞-distance must be infinite if the relative distance is positive, but the reverse is not necessarily true.

4.5.2 Robustness Analysis

In this subsection, we establish that a well-designed pathwise state-feedback switching law is robust against various types of perturbations including structural perturbations, unstructural perturbations, and switching perturbations.

Suppose that the pathwise state-feedback switching law $\sigma = \bigwedge_{i=1}^k \theta_i^{\Omega_i}$ is piecewise contractive for the switched linear system

$$x^+(t) = A_{\sigma(t)} x(t), \tag{4.81}$$

where $A_1, \dots, A_m$ are known real matrices. Define

$$\rho_i \stackrel{\text{def}}{=} \sup_{x \in \Omega_i \cap \mathbf{H}_1} \left|\phi(s_i; 0, x, \theta_i)\right| < 1, \quad i = 1, \dots, k. \tag{4.82}$$

According to the analysis in Sect. 4.4.1, we have

$$\left|\phi\left(t; 0, x, \bigwedge_{i=1}^k \theta_i^{\Omega_i}\right)\right| \le \beta e^{-\alpha t}|x| \quad \forall x \in \mathbf{R}^n,\ t \in \mathcal{T}_0,$$

where α and β are positive real numbers that can be explicitly estimated.

The first property of the switching law is its robustness with respect to structural perturbations.

Theorem 4.45 *There is a positive real number ε such that the switching law exponentially stabilizes any switched linear system*

$$x^+(t) = \bar{A}_{\sigma(t)} x(t) \tag{4.83}$$

with $\max_{i=1}^m \|A_i - \bar{A}_i\| < \varepsilon$.

Proof The main idea is to prove that the switching law is still piecewise contractive w.r.t. the perturbed system when ε is sufficiently small. To see this, we examine the state transition matrices of the nominal and perturbed systems, respectively. For switching path θ_i, denote by $t_0, t_1, \dots, t_k$ its switching time sequence. Let $\Phi(s_i, 0, \theta_i)$ and $\bar{\Phi}(s_i, 0, \theta_i)$ be the state transition matrices of the nominal system and the perturbed system, respectively. It is obvious that

$$\Phi(s_i, 0, \theta_i) = e^{A_{\sigma(t_k)}(s_i - t_k)} \dots e^{A_{\sigma(t_0)}(t_1 - t_0)},$$

$$\bar{\Phi}(s_i, 0, \theta_i) = e^{\bar{A}_{\sigma(t_k)}(s_i - t_k)} \dots e^{\bar{A}_{\sigma(t_0)}(t_1 - t_0)}$$

in continuous time and

$$\Phi(s_i, 0, \theta_i) = A_{\sigma(s_i - 1)} \cdots A_{\sigma(0)}, \qquad \bar{\Phi}(s_i, 0, \theta_i) = \bar{A}_{\sigma(s_i - 1)} \cdots \bar{A}_{\sigma(0)}$$

in discrete time. Define the error

$$\tilde{\Phi} = \bar{\Phi}(s_i, 0, \theta_i) - \Phi(s_i, 0, \theta_i).$$

It is clear that $\tilde{\Phi}$ is a continuous function of the entries of matrices $\bar{A}_i$ for $i = 1, \dots, m$. Note that $\tilde{\Phi} = 0$ when $\bar{A}_i = A_i$ for $i = 1, \dots, m$. It follows from (4.82) that

$$\sup_{x \in \Omega_i \cap \mathbf{H}_1} \left| \bar{\Phi}(s_i, 0, \theta_i) x \right| < 1, \quad i = 1, \dots, m,$$

when $\max_{i=1}^m \|A_i - \bar{A}_i\|$ is sufficiently small. This clearly leads to the conclusion. □

Remark 4.46 The robustness against structural perturbations is sometimes termed as *structural stability*. The property is important in that the system cannot work properly if it is not structurally stable. An interesting open problem is how to estimate the margin ε explicitly.

Next, we turn to the robustness w.r.t. unstructural perturbations. We formulate the problem by means of the framework of input-to-state stability (ISS).

For the perturbed system

$$x^+(t) = A_{\sigma(t)} x(t) + w(t), \tag{4.84}$$

where $w(t) \in \mathbf{R}^n$ is the unstructural system perturbation, we take the perturbation as an uncontrolled input.

Definition 4.47 For a switching law σ, system (4.84) is said to be *input-to-state stable* (ISS) w.r.t. the switching law, if there exist real-valued functions $\alpha \in \mathcal{K}_\infty$ and $\beta \in \mathcal{KL}$ such that

$$\bigl|\phi(t; x_0, \sigma, w)\bigr| \le \max\bigl\{\beta(x_0, t), \alpha\bigl(|w^t|_\infty\bigr)\bigr\}, \quad \forall x_0,\ w, t \in \mathcal{T}_0, \tag{4.85}$$

where $|w^t|_\infty = \sup_{s\in[t_0,t)} |w(s)|$, and $\phi(\cdot; x_0, \sigma, w)$ denotes the solution of (4.84) with initial condition $x(0) = x_0$.

This notion degenerates into the standard ISS for nonswitched systems [211].

The second property of the switching law is that it steers the switched system ISS.

Theorem 4.48 *The switched system is input-to-state stable under the switching law* $\bigwedge_{i=1}^k \theta_i^{\Omega_i}$.

Proof For the perturbed system (4.84), its aggregated system is

$$z_{l+1} = G_i z_l + \nu_l, \quad z_l \in \Omega_i, \tag{4.86}$$

where

$$\nu_l = \int_0^{s_i} \Phi(s_i, \tau, \theta_i) w(t_l + \tau)\, d\tau, \quad l \in \mathbf{N}$$

in continuous time, and

$$\nu_l = \sum_{j=0}^{s_i - 1} \Phi(s_i, j, \theta_i) w(t_l + j), \quad l \in \mathbf{N}$$

in discrete time, and $0, t_1, t_2, \ldots$ is the switching time sequence. This implies that, for any nonnegative integer l, we have

$$\omega^l \overset{\text{def}}{=} \max\bigl\{|\nu_0|, |\nu_1|, \ldots, |\nu_l|\bigr\} \le \upsilon \sup\bigl\{|w(t)| : t \in [t_0, t_l]\bigr\}, \tag{4.87}$$

where

$$\upsilon = \max\bigl(e^{\|A_1\| s}, \ldots, e^{\|A_m\| s}\bigr)$$

in continuous time, and

$$\upsilon = \max\bigl(\|A_1\|^s, \ldots, \|A_m\|^s\bigr)$$

in discrete time. Here $s \overset{\text{def}}{=} \max\{s_1, \ldots, s_k\}$.

In view of (4.82), let $\mu = \max\{\rho_1, \ldots, \rho_k\}$, and let $\bar{\mu}$ be a real number in $(\mu, 1)$. Fix a natural number j. For any $l \le j$, if $|z_l| \le \frac{\omega^j}{\bar{\mu} - \mu}$, then we have

$$|z_{l+1}| = |G_i z_l + \nu_l| \le \mu |z_l| + \omega^j \le \frac{\omega^j}{\bar{\mu} - \mu}.$$

This means that the closed ball $\mathbf{B}_{\frac{\omega^j}{\bar{\mu}-\mu}}$ is invariant in that the successor of any state within the ball is still kept in the ball. On the other hand, if $|z_l| > \frac{\omega^j}{\bar{\mu}-\mu}$, then we have

$$|z_{l+1}| \le \mu|z_l| + \omega^j \le \bar{\mu}|z_l|.$$

The above facts clearly lead to

$$|z_l| \le \max\left\{\bar{\mu}^l|x_0|, \frac{\omega^j}{\bar{\mu}-\mu}\right\} \quad \forall l = 1, \dots, j. \tag{4.88}$$

It is also clear that

$$\left|\phi(t; x_0, \sigma, w)\right| \le \upsilon|z_l| \quad \forall t \in (t_l, t_{l+1}),\ l = 0, 1, 2, \dots. \tag{4.89}$$

Combining (4.87), (4.88), and (4.89) yields

$$\left|\phi(t; x_0, \sigma, w)\right| \le \left\{\upsilon\bar{\mu}e^{-\alpha t}, \frac{\upsilon^2 w^t}{\bar{\mu}-\mu}\right\} \quad \forall t \in \mathcal{T}_0,$$

where $\alpha = \frac{-\ln\bar{\mu}}{\max(s_1,\dots,s_k)}$. This clearly exhibits that the original system is input-to-state stable under the switching law. □

The next property is the robustness of the switching law against switching perturbations. In terms of the distance between switching signals defined in Definition 4.43, we have the following result.

Theorem 4.49 *There is a positive real number γ such that the switching law $\bigwedge_{i=1}^k \theta_i^{\Omega_i}$ exponentially stabilizes any switched linear system*

$$x^+(t) = A_{\bar{\sigma}(t)}x(t) \tag{4.90}$$

with SRD$(\bar{\sigma}, \bigwedge_{i=1}^k \theta_i^{\Omega_i}) < \gamma$.

To proceed with the proof of the theorem, we need the following technical lemma.

Lemma 4.50 *Suppose that p_1 and p_2 are switching paths defined over intervals $[0, \tau_1)$ and $[0, \tau_2)$, respectively, and x is an arbitrarily given state. If $|\phi(\tau_1; 0, x, p_1)| \le \psi|x|$ and $d(p_1, p_2) = \zeta$, then, we have*

$$\left|\phi(\tau_2; 0, x, p_2)\right| \le \left(\eta^{\tau_1+\zeta+1}\zeta + \psi\right)|x|, \tag{4.91}$$

where $\eta \stackrel{\text{def}}{=} \max_{i=1}^m e^{\|A_i\|}$ in continuous time, and $\eta \stackrel{\text{def}}{=} \max(1, \max_{i=1}^m \|A_i\|)$ in discrete time.

Proof Here we only prove for the continuous time, and the discrete-time case can be proceeded in a similar manner. Let ε be an arbitrarily given positive real number. By the definition of the distance between two switching paths, there is a common parent path p over $[0, \tau)$ of p_1 and p_2 such that

$$|p - p_1| + |p - p_2| \le \zeta + \varepsilon.$$

Denote $\zeta_1 = |p - p_1|$ and $\zeta_2 = |p - p_2|$. Suppose, for instance, that

$$\Phi(\tau_1, 0, p_1) = e^{A_2 h_2} e^{A_1 h_1}, \qquad \Phi(\tau, 0, p) = e^{A_2 h_2} e^{A_4 h_4} e^{A_1 h_1} e^{A_3 h_3}.$$

Other cases can be treated in exactly the same way. Simple computation yields

$$\begin{aligned}
\left|\phi(\tau; 0, x, p)\right| &\le \left|\phi(\tau; 0, x, p) - \phi(\tau_1; 0, x, p_1)\right| + \left|\phi(\tau_1; 0, x, p_1)\right| \\
&\le \left|e^{A_2 h_2} e^{A_4 h_4} e^{A_1 h_1} e^{A_3 h_3} x - e^{A_2 h_2} e^{A_1 h_1} x\right| + \psi|x| \\
&\le \left|e^{A_2 h_2}\left(e^{A_4 h_4} - I\right) e^{A_1 h_1} e^{A_3 h_3} x\right| \\
&\quad + \left|e^{A_2 h_2} e^{A_1 h_1}\left(e^{A_3 h_3} - I\right) x\right| + \psi|x| \\
&\le \left(\eta^{\tau_1 + \zeta_1 + 1} \zeta_1 + \psi\right)|x|,
\end{aligned}$$

where the relationships $\tau_1 = h_1 + h_2$, $\zeta_1 = h_3 + h_4$ and $\|e^{At} - I\| \le \|A\| e^{\|A\| t} |t|$ have been used. Further calculation leads to

$$\begin{aligned}
\left|\phi(\tau_2; 0, x, p_2)\right| &\le \left|\phi(\tau_2; 0, x, p_2) - \phi(\tau; 0, x, p)\right| + \left|\phi(\tau; 0, x, p)\right| \\
&\le \eta^{\tau_2 + \zeta_2 + 1} \zeta_2 |x| + \left|\phi(\tau; 0, x, p)\right| \\
&\le \left(\eta^{\tau_2 + \zeta_2 + 1} \zeta_2 + \eta^{\tau_1 + \zeta_1 + 1} \zeta_1 + \psi\right)|x| \\
&\le \left(\eta^{\tau_1 + \zeta + \varepsilon + 1}(\zeta + \varepsilon) + \psi\right)|x|,
\end{aligned}$$

where the relationship $\tau_2 \le \tau_1 + \zeta_1$ was utilized. This completes the proof due to the arbitrariness of ε. □

Proof of Theorem 4.49 Let x_0 be any given but fixed state, ϵ be any given positive real number, and σ^{x_0} and $\bar{\sigma}^{x_0}$ be the nominal switching signal and the perturbed switching signal generated by the switching law $\bigwedge_{i=1}^{k} \theta_i^{\Omega_i}$ w.r.t. initial state $x(0) = x_0$ for the nominal system and the perturbed system, respectively. Recall that $SRD(\bar{\sigma}, \bigwedge_{i=1}^{k} \theta_i^{\Omega_i}) < \gamma$ means that $RD_{x_0}(\bar{\sigma}^{x_0}, \sigma^{x_0}) < \gamma$, which further means that the N-distance between the perturbed switching and the nominal switching is upper bounded by $N(\gamma + \epsilon)$ for sufficiently large N.

Rewrite the nominal switching law as in (4.76):

$$\sigma^{x_0} = \sum_{i=0}^{+\infty} (x_i, j_i, \theta_{j_i}, s_{j_i}) : \quad x_{i+1} = \phi(s_{j_i}; 0, x_i, \theta_{j_i}) \in \Omega_{j_{i+1}}.$$

Similarly, rewrite the perturbed switching law as in (4.77):

$$\bar{\sigma}^{x_0} = \sum_{i=0}^{+\infty} (y_i, \kappa_i, p_i, \tau_i): \quad y_{i+1} = \phi(\tau_i; 0, y_i, p_i), \ y_0 = x_0.$$

We are to prove that the state sequence $y_0, y_1, \ldots$ is exponentially convergent, which implies the exponential stability of the perturbed system. For this, we only examine the continuous time, and the discrete-time case can be proven in a similar manner.

In view of (4.82), let $\mu = \max\{\rho_1, \ldots, \rho_k\}$, and let $\bar{\mu}$ be a real number in $(\mu, 1)$. Fix an arbitrarily given natural number i. Let $\tau \stackrel{\text{def}}{=} \max\{s_1, \ldots, s_k\}$, and let $\vartheta = \eta^{-(\tau+2)}(\bar{\mu} - \mu)$. It follows from Lemma 4.50 that

$$|y_i| \le \bar{\mu}|y_{i-1}| \quad \text{if } d(p_i, \theta_{j_i}) \le \vartheta. \tag{4.92}$$

Applying Lemma 4.50 again gives

$$|y_i| \le \left(\eta^{\tau+d_i+1} + \frac{\mu}{\vartheta}\right) d_i |y_{i-1}| \le \eta^{\lambda+d_i} d_i |y_{i-1}| \quad \text{if } d(y_i, \theta_{j_i}) > \vartheta, \tag{4.93}$$

where $\lambda = \tau + 2 + \max\{0, (\ln\mu - \ln\vartheta)/\ln\eta\}$, and $d_i \stackrel{\text{def}}{=} d(y_i, \theta_{j_i})$. Now choose $\gamma = \epsilon = \varpi\vartheta$ where ϖ is a positive real number to be determined later. Let l be a (sufficiently large) natural number such that, for any $N \ge l$, the N-distance between the perturbed switching and the nominal switching is upper bounded by $N(\gamma + \epsilon)$. Fix $N \ge l$ and define

$$N_1 = \#\{i \le N : |y_i| > \bar{\mu}|y_{i-1}|\}, \qquad N_2 = N - N_1,$$

where # denotes the cardinality of a set. By the definition of N-distance, we have

$$N_1 \le \lceil N\varpi \rceil, \qquad N_2 \ge \lfloor N(1-\varpi) \rfloor,$$

where $\lceil a \rceil$ ($\lfloor a \rfloor$) denotes the smallest (largest) integer equal to or greater (less) than a. Based on the above facts, routine calculation gives

$$\begin{aligned} |y_N| &\le |y_0|\bar{\mu}^{N_2} \prod_{d_i > \vartheta} \eta^{\lambda+d_i} d_i \le e^{-\nu(N(1-\varpi)-1)} e^{2N\varpi(1+e^{\lambda-1})} |x_0| \\ &\le e^{-(N-1)\nu} e^{N(\nu+2+2e^{\lambda-1})\varpi} |x_0|, \end{aligned} \tag{4.94}$$

where $\nu \stackrel{\text{def}}{=} -\ln\bar{\mu}$, and the fact that $\max_x (\frac{a}{x})^x = e^{\frac{a}{e}}$ was used.

Finally, let $\varpi = \frac{\nu}{2(\nu+2+2e^{\lambda-1})}$, which is clearly independent of N. Then, it can be seen from inequality (4.94) that the sequence $y_0, y_1, \ldots$ is exponentially convergent with rate $\nu/2$. This completes the proof. □

Remark 4.51 Theorem 4.49 reveals that the pathwise state-feedback switching law is robust against perturbations from the switching law itself, which means that the switching law is fault tolerant—a very important property from the practical point of view. Moreover, it can be seen from the proof that the allowable robustness margin γ can be explicitly estimated.

Finally, by combining Theorems 4.45, 4.48, and 4.49, we can prove the following comprehensive robustness property of the switched system with the stabilizing pathwise state-feedback switching law.

Theorem 4.52 *Suppose that the switching law $\bigwedge_{i=1}^{k}\theta_i^{\Omega_i}$ exponentially stabilizes the nominal system*

$$x^+(t) = A_{\sigma(t)}x(t).$$

Then, there are positive real numbers ϵ_1 and ϵ_2 such that the switching law makes the perturbed system

$$x^+(t) = \bar{A}_{\bar{\sigma}(t)}x(t) + w(t)$$

input-to-state stable if

$$\max_{k=1}^{m} \|\bar{A}_k - A_k\| \le \epsilon_1, \qquad SRD\left(\bar{\sigma}, \bigwedge_{i=1}^{k}\theta_i^{\Omega_i}\right) \le \epsilon_2. \tag{4.95}$$

4.5.3 Examples and Simulations

Example 4.53 Let us examine the discrete-time switched linear system

$$x(t+1) = A_{\sigma(t)}x(t), \quad x(t) \in \mathbf{R}^2, \ \sigma(t) \in \{1, 2, 3\} \tag{4.96}$$

with

$$A_1 = \begin{bmatrix} -0.7113 & 0.5333 \\ 1.8498 & 0.0968 \end{bmatrix}, \qquad A_2 = \begin{bmatrix} -0.0378 & 0.4588 \\ 2.4130 & 0.4437 \end{bmatrix},$$

and

$$A_3 = \begin{bmatrix} -0.7714 & 0.2266 \\ -0.8239 & -1.4026 \end{bmatrix}.$$

Simple computation exhibits that

$$\lambda^1(A_i)\lambda^2(A_i) > 1, \quad i = 1, 2, 3,$$

where $\{\lambda^1(A), \lambda^2(A)\}$ is the spectrum of matrix $A \in \mathbf{R}^{2\times 2}$. It follows from Lemma 4.3 that the switched linear system is not consistently stabilizable. That is, the system cannot be made stable via any single switching signal.

To verify the stabilizability of the switched system, we conduct the computational algorithm in Sect. 4.4.2, which yields

$$\bigcup \mathcal{Y}_{G_i} = \mathbf{B}_1^1 = [-1, 1],$$

where

$$G_1 = A_1, \qquad G_2 = A_2, \qquad G_3 = A_3, \qquad G_4 = A_1^2 A_3, \qquad G_5 = A_2 A_1 A_3,$$

and the corresponding switching paths are

$$\theta_1 = (1), \qquad \theta_2 = (2), \qquad \theta_3 = (3), \qquad \theta_4 = (3, 1, 1), \qquad \theta_5 = (3, 1, 2).$$

As a result, the algorithm terminates with $k = 5$, which exhibits that the switched system is exponentially stabilizable. To further calculate a stabilizing pathwise state-feedback switching law, we choose

$$\begin{aligned}
\Lambda_1 &= [-0.0995, 0.4540) \subset \mathcal{Y}_{G_1}, \\
\Lambda_2 &= [-0.3895, -0.0995) \subset \mathcal{Y}_{G_2}, \\
\Lambda_3 &= [-0.7100, -0.3895) \subset \mathcal{Y}_{G_3}, \\
\Lambda_4 &= [-1.0000, -0.9625) \cup [0.4540, 1.0000] \subset \mathcal{Y}_{G_4}, \\
\Lambda_5 &= [-0.9625, -0.7100) \subset \mathcal{Y}_{G_5},
\end{aligned}$$

which generate the state space partitions

$$\begin{aligned}
\Omega_1 &= \{r\mathcal{P}_2^{-1}y : r \in \mathbf{R},\ y \in \Lambda_1\}, \\
\Omega_i &= \{r\mathcal{P}_2^{-1}y : r \neq 0,\ y \in \Lambda_i\}, \quad i = 2, \dots, k.
\end{aligned}$$

Figure 4.11 shows the $\mathcal{Y}_{G_i}$'s, the Λ_i's, and Ω_i's.

With the state partitions, the aggregated system is

$$z_{j+1} = G_i z_j, \quad z_j \in \Omega_i$$

with $z_0 = x(0)$. It is clear that the aggregated system is exponentially stable, and the pathwise state-feedback switching law $\bigwedge_{i=1}^k \theta_i^{\Omega_i}$ exponentially stabilizes the original switched system.

Take the initial state to be

$$x(0) = [1.1908, -1.2025]^T.$$

In what follows, we present simulations for the nominal systems and its perturbed systems, respectively.

Firstly, Fig. 4.12 shows sample state phase portraits for the original system and the aggregated system. It can be seen that the aggregated state trajectory is always

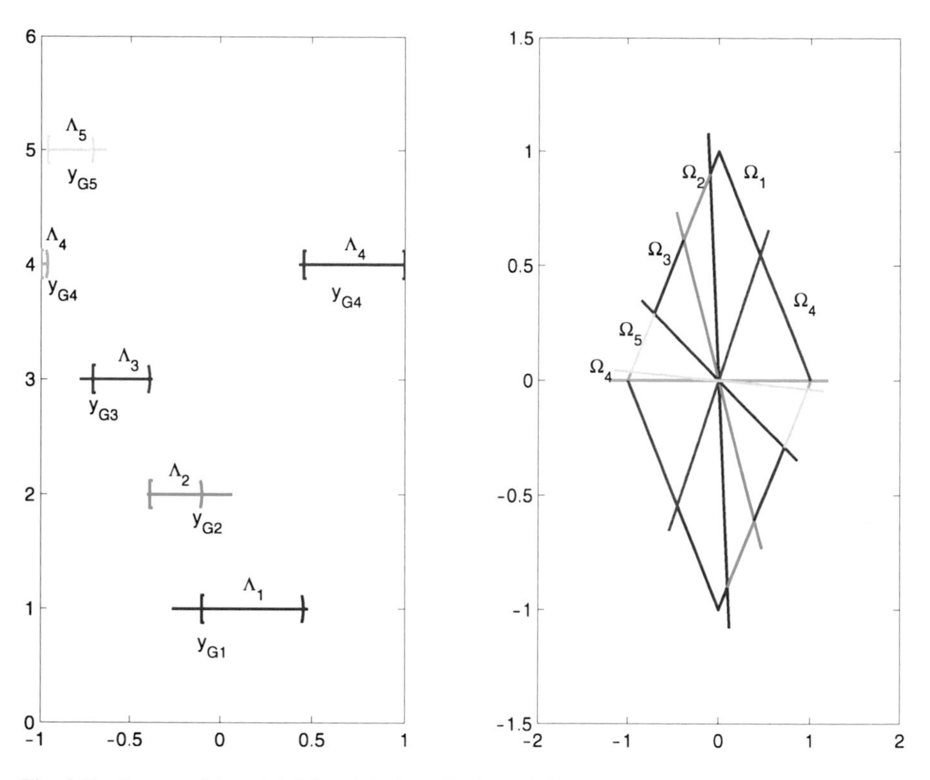

Fig. 4.11 State partitions (*right*) and their projections (*left*)

contractive relative to the norm $|\cdot|_1$, while the original state trajectory is not necessarily contractive at the nonaggregated instants.

Secondly, to illustrate the robustness against structural perturbations, we assume that the matrices A_i are perturbed by the matrices $\tilde{A}_i$, where

$$\tilde{A}_1 = \begin{bmatrix} 0.2117 & -0.3501 \\ 0.1243 & 0.1395 \end{bmatrix}, \qquad \tilde{A}_2 = \begin{bmatrix} 0.1623 & 0.2620 \\ 0.1273 & 0.0654 \end{bmatrix},$$

and

$$\tilde{A}_3 = \begin{bmatrix} -0.1346 & -0.4898 \\ -0.0299 & 0.0947 \end{bmatrix}.$$

Figure 4.13 depicts the state phase portrait for the perturbed system

$$x(t+1) = (A_{\sigma(t)} + \tilde{A}_{\sigma(t)})x(t).$$

It is clear that the perturbed system is also exponentially convergent, though the convergence rate is lower than that of the nominal system.

Thirdly, to illustrate the robustness of the switching law with respect to switching perturbations, we assume that the nominal switching signal is perturbed in the following way: (i) a delay occurs at each time implementing the switching path

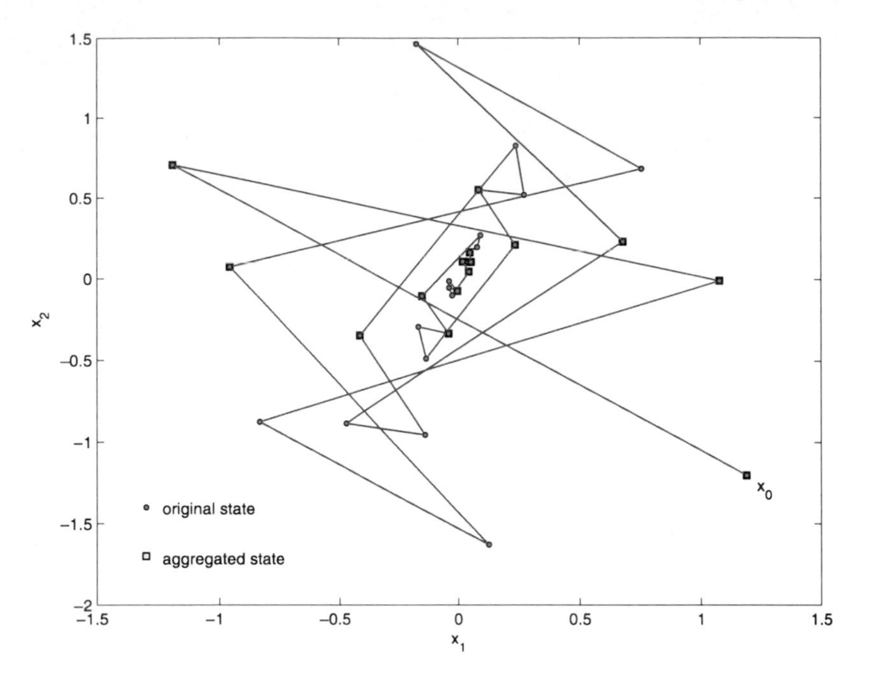

Fig. 4.12 State phase portraits of the original system and the aggregated system

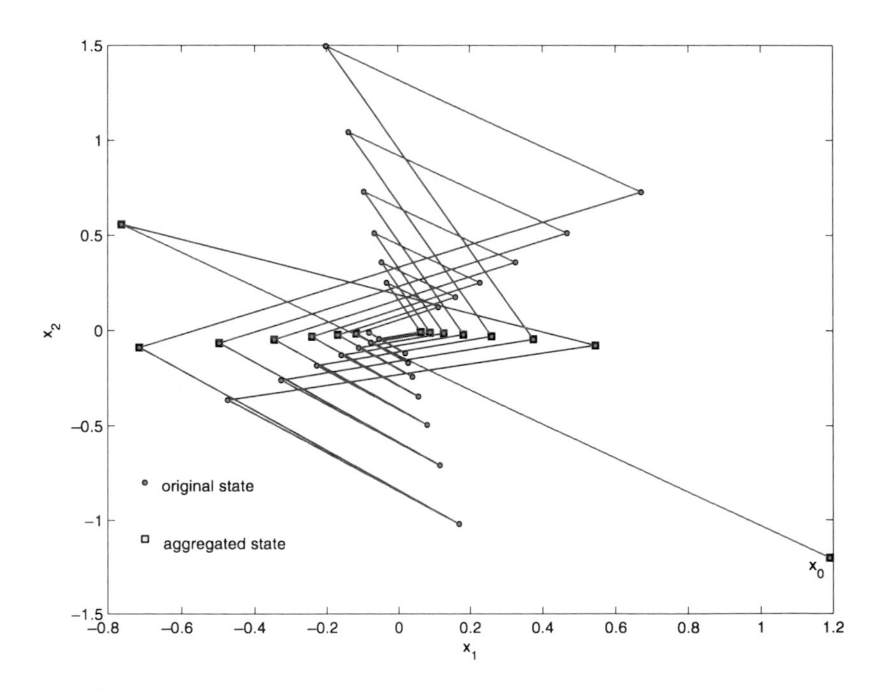

Fig. 4.13 State phase portraits of the structurally perturbed system

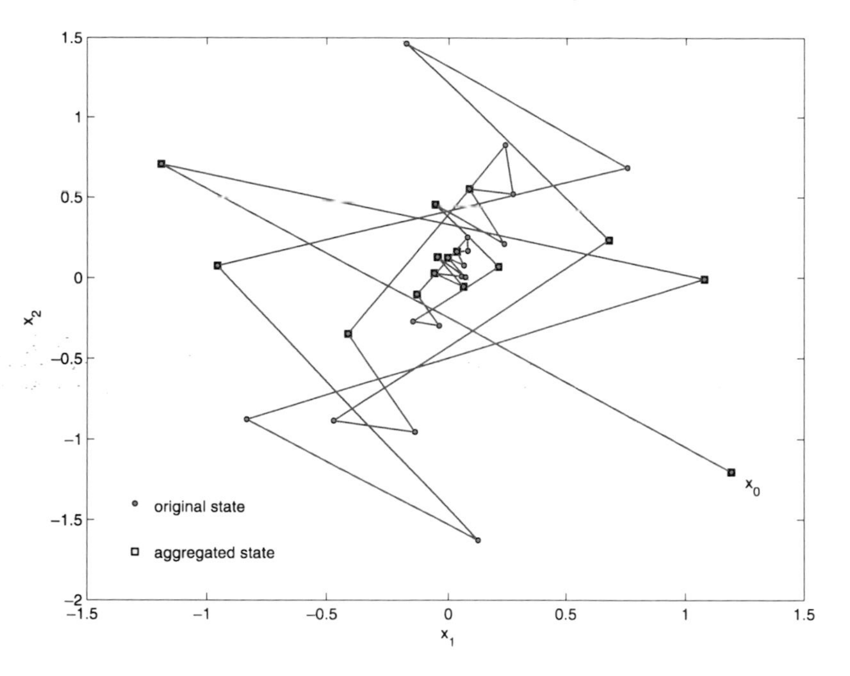

Fig. 4.14 State phase portraits with perturbed switching

$\theta_1 = (1)$, that is, the perturbed switching path is $\hat{\theta}_1 = (1, 1)$; and (ii) the state measurement is inexact in that the measured state is $\bar{x} = [1.1x_1, x_2]^T$ for any state $x = [x_1, x_2]^T$, that is, there is a ten-percent overdue for the first state variable. Figure 4.14 shows the state phase portrait for the perturbed system. It is clear that the state is still convergent. A routine calculation gives that the relative distance in the simulated horizon is 0.0526. This means that over five percent length of the switching signal is wrongly implemented.

Fourthly, to illustrate the input-to-state stability, let us examine the perturbed system

$$x(t+1) = A_{\sigma(t)}x(t) + w(t)$$

with $w(t) = \mathrm{sgn}(\tan(t))$, and $\mathrm{sgn}(\cdot)$ denotes the signum function. The state phase portrait for the perturbed system is shown in Fig. 4.15. It can be seen that the system is input-to-state stable with a reasonable bound.

Finally, by putting all the above perturbations together, we have the perturbed system with three types of perturbations. Figure 4.16 depicts its state phase portrait, which is still input-to-state stable with a bigger overshoot.

Example 4.54 For the continuous-time switched linear system

$$\dot{x}(t) = A_{\sigma(t)}x(t), \quad x(t) \in \mathbf{R}^3, \ \sigma(t) \in \{1, 2\}, \tag{4.97}$$

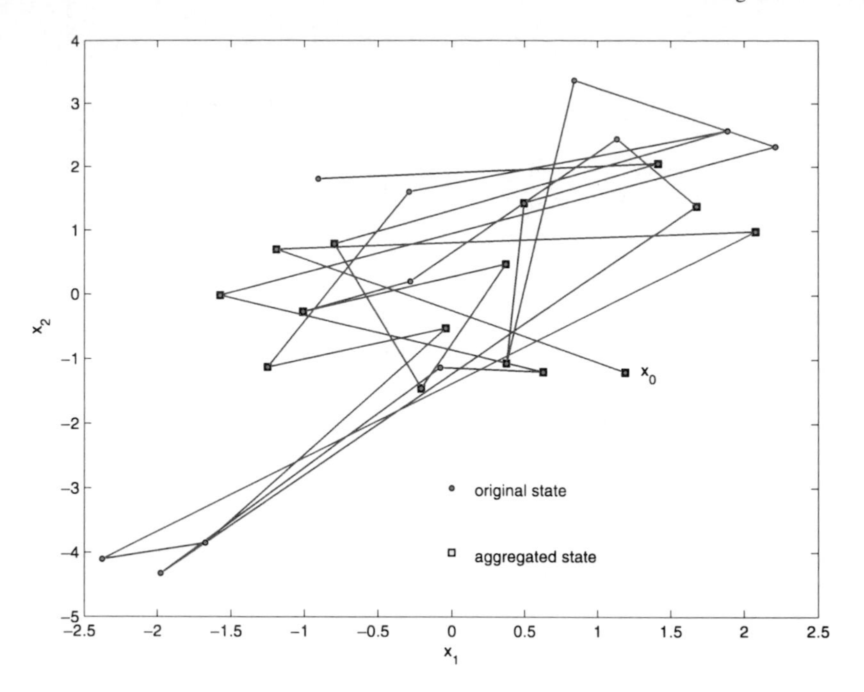

Fig. 4.15 State phase portrait illustrating the ISS property

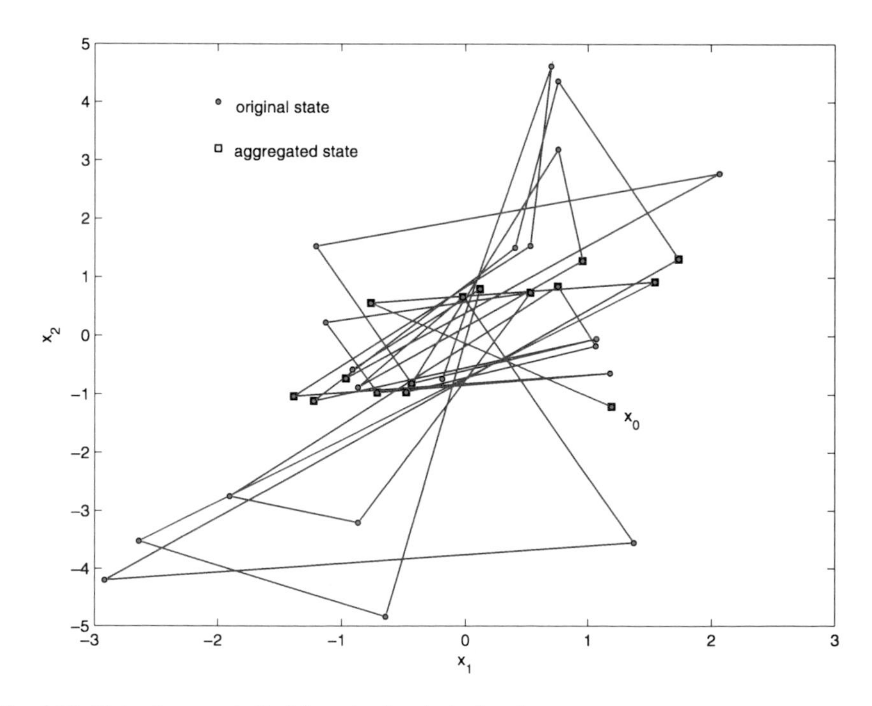

Fig. 4.16 State phase portrait of the mixed perturbed system

with

$$A_1 = \begin{bmatrix} -0.4203 & 1.4913 & 0.3085 \\ -1.4281 & 1.2367 & -1.1178 \\ -0.2091 & 0.7129 & -0.7467 \end{bmatrix},$$

$$A_2 = \begin{bmatrix} 0.1967 & -0.5520 & 0.9880 \\ 0.5359 & -0.7562 & -1.0518 \\ -1.0481 & 0.7114 & 0.8080 \end{bmatrix},$$

simple computation exhibits that

$$\sum_{j=1}^{3} \lambda^j(A_i) > 0, \quad i = 1, 2.$$

By Lemma 4.3, the switched linear system is not consistently stabilizable. In particular, the matrices A_1 and A_2 do not admit any stable convex combination, and the system is not quadratically stabilizable [80].

Let the sampling period be $\tau = 0.38$, and $B_i = e^{A_i \tau}$ for $i = 1, 2$. For the sampled-data switched system

$$x\big((j+1)\tau\big) = B_{\sigma(j\tau)} x(j\tau), \quad j \in \mathbf{N},$$

we apply the computational algorithm in Sect. 4.4.2. This gives

$$\bigcup \mathcal{Y}_{G_i} = \mathbf{B}_1^2,$$

where

$$\begin{aligned} &G_1 = B_1 B_2^2, \quad G_2 = B_2 B_1^2, \quad G_3 = B_1^2 B_2 B_1, \quad G_4 = B_1 B_2^2 B_1, \\ &G_5 = B_2^3 B_1, \quad G_6 = B_1^3 B_2 B_1, \quad G_7 = B_1^2 B_2 B_1 B_2, \\ &G_8 = B_1 B_2 B_1^2 B_2, \quad G_9 = B_2^2 B_1 B_2 B_1, \quad G_{10} = B_2^3 B_1^2. \end{aligned}$$

As a result, the algorithm terminates with $k = 10$, which exhibits that the switched system is exponentially stabilizable. The corresponding switching paths can be formulated routinely as, for example,

$$\theta_1(t) = \begin{cases} 2 & t \in [0, 2\tau), \\ 1 & t \in [2\tau, 3\tau), \end{cases}$$

$$\theta_2(t) = \begin{cases} 1 & t \in [0, 2\tau), \\ 2 & t \in [2\tau, 3\tau), \end{cases}$$

and so forth. To determine the state partitions, let

$$\Lambda_1 = \mathcal{Y}_{G_1},$$

$$\Lambda_j = \mathcal{Y}_{G_j} - \bigcup_{i=1}^{j-1} \Lambda_i, \quad j = 2, \ldots, k,$$

which generate the state space partitions

$$\Omega_1 = \{r\mathcal{P}_3^{-1}y : r \in \mathbf{R},\ y \in \Lambda_1\},$$
$$\Omega_i = \{r\mathcal{P}_3^{-1}y : r \neq 0,\ y \in \Lambda_i\}, \quad i = 2, \ldots, k.$$

Hence, the aggregated system is

$$z_{j+1} = G_i z_j, \quad z_j \in \Omega_i,$$

with $z_0 = x(0)$.

From the above preparation, we are ready to implement the pathwise state-feedback switching law $\bigwedge_{i=1}^k \theta_i^{\Omega_i}$. Take the initial state to be

$$x(0) = [-0.6918, 0.8580, 1.2540]^T.$$

Figure 4.17 shows the original state and the aggregated state of the closed-loop system. It can be seen that the systems converge exponentially with satisfactory rate.

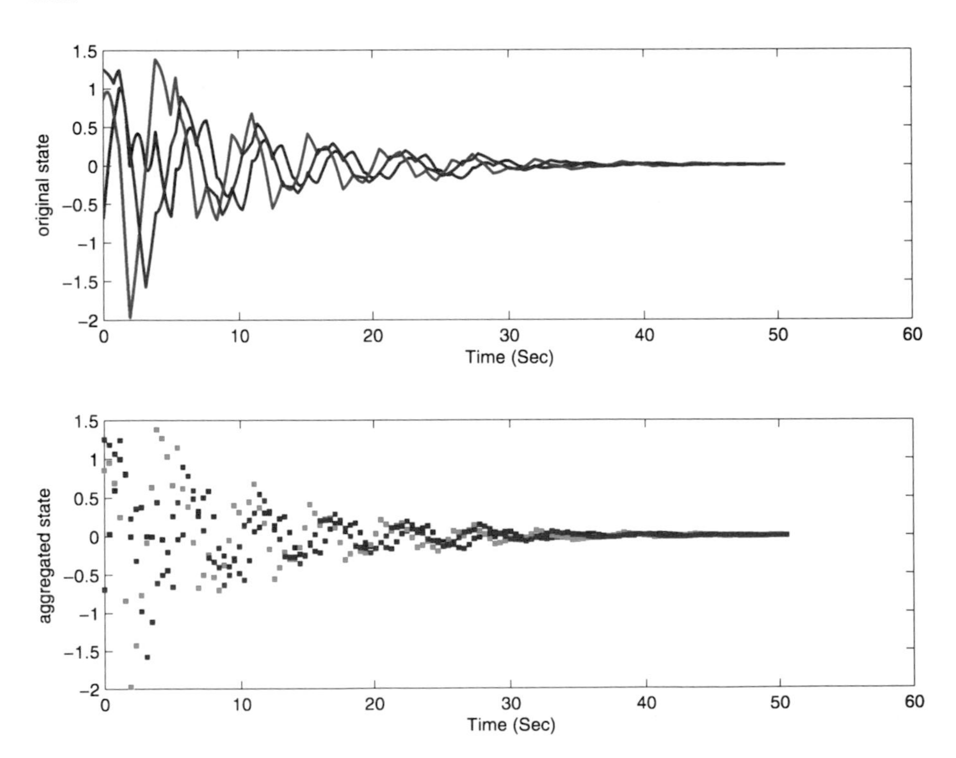

Fig. 4.17 Original (*upper*) and aggregated (*lower*) state trajectories

The next steps are to verify the robustness of the switching law w.r.t. various types of perturbations. For this, we take the perturbed matrices to be

$$\tilde{A}_1 = \begin{bmatrix} 0.3915 & -0.0680 & 0.2380 \\ 0.1009 & -0.2280 & -0.2232 \\ 0.3729 & -0.0422 & 0.1271 \end{bmatrix},$$

$$\tilde{A}_2 = \begin{bmatrix} -0.1203 & 0.0172 & 0.0924 \\ 0.1102 & -0.4009 & -0.0642 \\ -0.2200 & -0.0986 & 0.2473 \end{bmatrix},$$

and the unstructural perturbation to be

$$w(t) = \cos\big(\sin(t)\big), \quad t \in \mathcal{T}_0.$$

As for the switching signal, we assume that the nominal switching signal is perturbed in the following way: (i) a delay of $d = 0.1$ sec occurs at each duration of the second subsystem in a sample period. That is, the sampling period for the second subsystem is $\tau + d$ instead of τ; and (ii) the state measurement is inexact in that the measured state is $\bar{x} = [x_1, x_2, 0.9x_3]^T$ for any state $x = [x_1, x_2, x_3]^T$, that is, there is a ten-percent error in measuring the third state variable. Let $\bar{\sigma}$ denote the perturbed switching signal for the initial state x_0. Figures 4.18, 4.20, and 4.19 show the state motions for the perturbed systems

$$\dot{x}(t) = (A_{\sigma(t)} + \tilde{A}_{\sigma(t)})x(t),$$

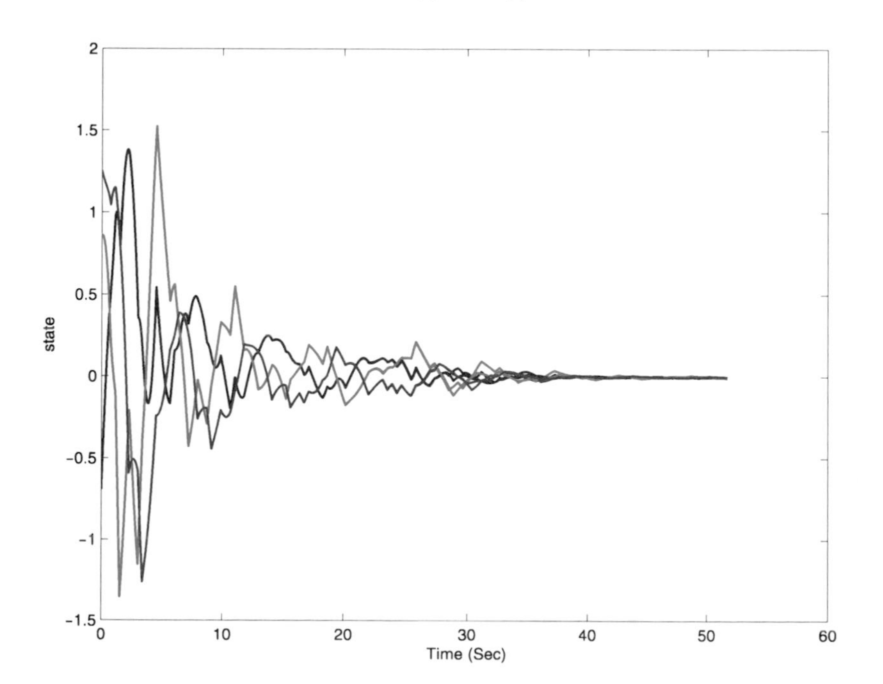

Fig. 4.18 State trajectory of the structurally perturbed system

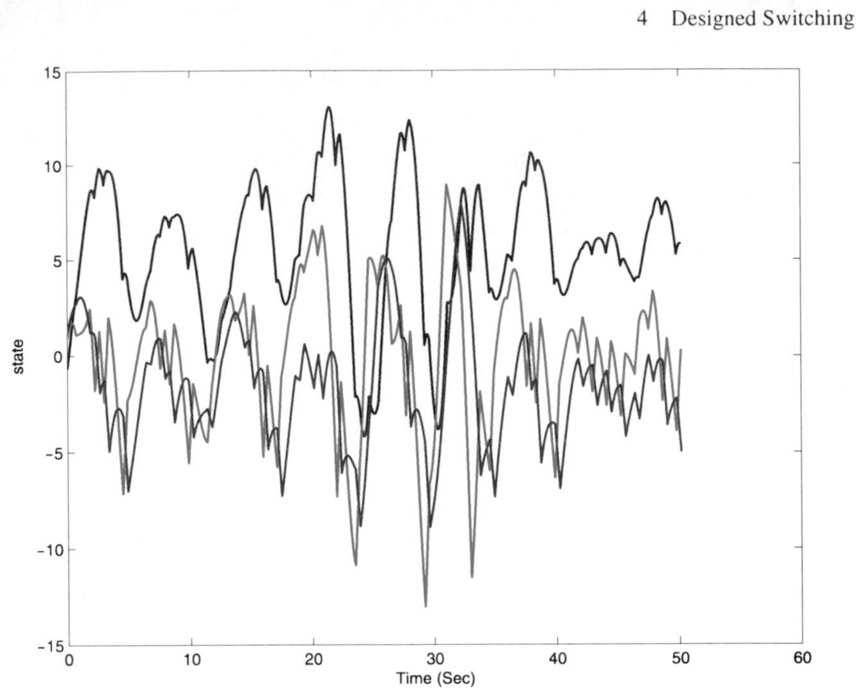

Fig. 4.19 ISS stability of the unstructurally perturbed system

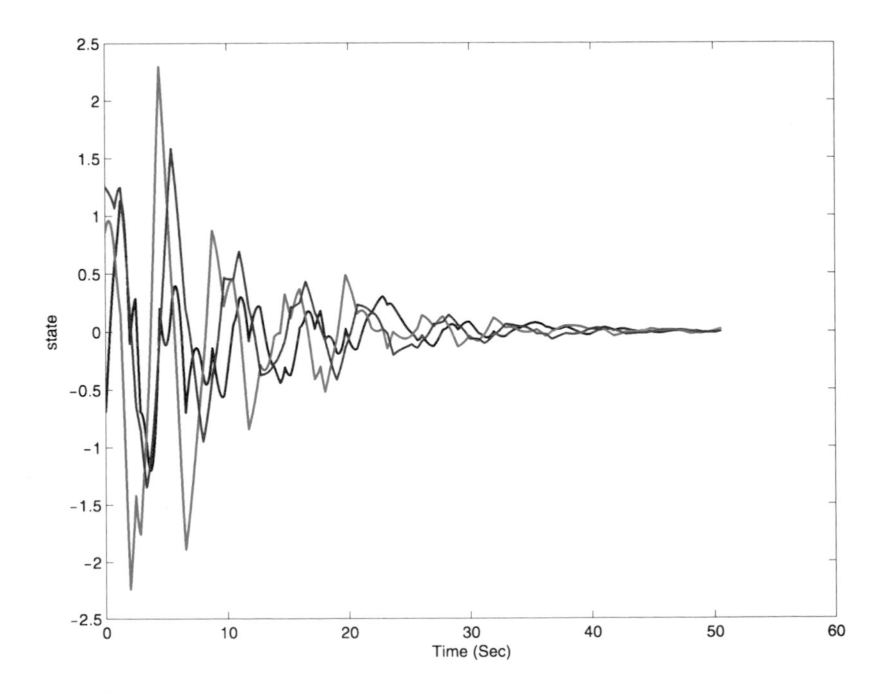

Fig. 4.20 State trajectory with perturbed switching signal

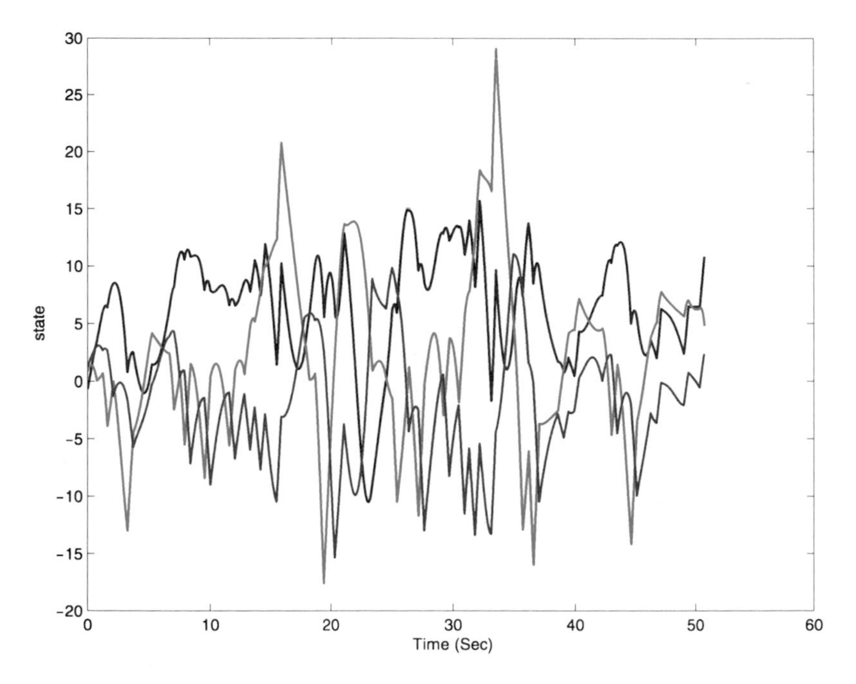

Fig. 4.21 State trajectory of the mixed-perturbed system

$$\dot{x}(t) = A_{\sigma(t)}x(t) + w(t),$$

and

$$\dot{x}(t) = A_{\bar{\sigma}(t)}x(t),$$

respectively. It is clear that the state trajectories are still convergent and bounded, respectively.

Finally, by putting all the above perturbations together, we have the perturbed system with three types of perturbations given by

$$\dot{x}(t) = (A_{\bar{\sigma}(t)} + \tilde{A}_{\bar{\sigma}(t)})x(t) + w(t).$$

Figure 4.21 depicts the state trajectories, which is still input-to-state stable with a bigger overshoot.

4.6 Notes and References

When the switching signal is a design variable for the switched system, the primary issues for stability and robustness include (1) establishing stabilizability criteria, or equivalently, identifying the class of switched systems which could be made stable by means of suitable switching laws; and (2) stabilizing switching design, or equivalently, developing computational algorithms for finding switching laws that achieve stability and robustness. For both issues, many works have been published in the literature. The stabilizing switching design problem received a lot of attention in the literature. In particular, it was found that, when there is a stable linear convex combination of the subsystems, the switched systems are quadratically stabilizable [15, 258, 259]. It was also proved that the converse is also true for switched linear systems with two subsystems [80]. For the class of switched linear systems that admit stable convex combinations, quite a few stabilizing switching design procedures have been reported, among which the most notable are the periodic switching design based on the average approach [250, 257, 258, 267, 277], the state-feedback switching design based on the Lyapunov method [87, 217, 259], the passivity/dissipativity approach [281, 282], and the combined switching design to reduce the switching frequency [219, 263, 264]. However, it is well recognized that the existence of a stable convex combination is too conservative for quadratic stabilizability, not to mention the general stabilizability which is not necessarily quadratic or even convex [35, 36]. For the robust switching design problem, which is to find switching laws that steer the switched systems stable and robust in the presence of perturbations and disturbances, there were also some delightful developments. In particular, it has been revealed that the class of consistently stabilizable switched linear systems is also robust against small structural perturbations [213, 216], Lyapunov-like techniques were also utilized to tackle the problem [276], and a comprehensive treatment can be found in [150, 151].

This chapter presented the state-of-the-art development on stabilizing switching design. We examined the problems of stabilization by means of time-driven switching, state-feedback switching, and mixed-driven switching, respectively. In particular, we show that asymptotic stabilizability is equivalent to the existence of a smooth switched Lyapunov function, and furthermore, the set of min functions is proven to be universal in providing switched Lyapunov functions for switched linear systems. By proposing the pathwise state-feedback switching law, the stabilizing switching design problem and the robust switching design problem were addressed in a unified and rigorous framework for both continuous-time and discrete-time switched systems.

Section 4.2 presented several properties of consistent stabilizability, which is stable under a time-driven switching signal. The material was mostly taken from [216]. While the periodic switching is not able to generally address the problem of stabilization, recent progress indicates that eventually periodic switching signals are capable of fully characterizing the problem [16].

In Sect. 4.3, the switched Lyapunov function in Definition 4.8 is an autonomous function of the state, and the word "switched" here reflects the fact that it needs only to decrease along the most descendent switching signal. In the literature, the concept of switched Lyapunov functions appeared in [60], which is switching-signal-dependent and is with a different meaning. The converse Lyapunov theorem for switched nonlinear system, Theorem 4.10, could be seen as a special case of the converse Lyapunov theorems established in [58, 198] and [129] for more general class of dynamical systems. The converse Lyapunov theorem for switched linear systems, Theorem 4.12, could be found in [222]. The two counterexamples, Examples 4.15 and 4.17, are owned to Blanchini and Savorgnan [35, 36]. The composite quadratic Lyapunov functions have been investigated in [114, 115]. Theorem 4.19 and the supporting material were adopted from [279, 280] for discrete-time systems and from [192] for continuous-time systems.

The main content of Sects. 4.4 and 4.5 was adopted from [225]. Example 4.27 was taken from [217]. Corollary 4.31 can be found in [218]. Section 4.5.1 and Theorem 4.49 were adopted from [231, 235].

Chapter 5
Connections and Implications

A notable and attractive feature of a switched dynamical system is its wide connections with many other known types of dynamical systems. As a result, stability of switched systems has clear implications in stability and robustness analysis for many well-known system frameworks. In particular, the guaranteed stability under arbitrary switching can be seen as robustness against the switching signal, which is an aggregation of the robustness for a nominal plant with structural/unstructural linear/nonlinear time-invariant/time-variant uncertainties or disturbances. The autonomous stability of piecewise linear systems is closely related to stability analysis of highly nonlinear systems by means of piecewise linear approximation. The stabilizing switching design methodology provides hybrid control approaches for controlling and optimizing highly nonlinear systems with possibly unknown parameters.

In this chapter, we investigate several typical problems in systems and control that are closely related with stability of switched linear systems. The problems include the absolute stability of Lur'e systems, stability of T–S fuzzy systems, consensus of multiagent systems, supervisory adaptive control, and stabilization of controllable systems. By establishing their connections with switching-oriented analysis and design, the ideas and methods developed in the previous chapters can be applied to the problems either directly or indirectly.

5.1 Absolute Stability for Planar Lur'e Systems

Lur'e systems are closed-loop systems consisting of linear plants and sector-bounded uncertain output feedbacks. Absolute stability is a problem that looks for the largest possible sector bounds within which the Lur'e system is globally asymptotically stable with respect to all the sector-bounded uncertainties. As a classic topic of interest, the problem provides a most originating and stimulating source of motivation for the development of the nonlinear control theory. In Sect. 2.5, we discussed the connection between the absolute stability of a Lur'e system and the guaranteed stability of a switched linear system. In this section, by taking advantage

Z. Sun, S.S. Ge, *Stability Theory of Switched Dynamical Systems*,
Communications and Control Engineering,
DOI 10.1007/978-0-85729-256-8_5,

of the connection, we present an algebraic criterion for the guaranteed stability of planar switched systems and apply the criterion to the problem of absolute stability.

5.1.1 Guaranteed Stability in the Plane

Consider the planar continuous-time two-form switched linear system given by

$$\dot{x} = A_\sigma x, \tag{5.1}$$

where $x \in \mathbf{R}^2$, $\sigma \in \{1,2\}$, and A_1 and A_2 are stable matrices in $\mathbf{R}^{2\times 2}$.

For a matrix $A \in \mathbf{R}^{2\times 2}$, let $\det A$ and $\operatorname{tr}(A)$ be the determinant and the trace, respectively. Furthermore, introduce the notation

$$
\begin{aligned}
\operatorname{disc}(A) &= \operatorname{tr}(A)^2 - 4\det A,\\
\Gamma(A_1,A_2) &= \frac{1}{2}\big(\operatorname{tr}(A_1)\operatorname{tr}(A_2) - \operatorname{tr}(A_1A_2)\big),\\
\tau_i &= \begin{cases} \frac{\operatorname{tr}(A_i)}{\sqrt{|\operatorname{disc}(A_i)|}} & \text{if } \operatorname{disc}(A_i)\neq 0,\\ \frac{\operatorname{tr}(A_i)}{\sqrt{|\operatorname{disc}(A_j)|}} & \text{if } \operatorname{disc}(A_i)=0 \ \& \ \operatorname{disc}(A_j)\neq 0,\\ \frac{\operatorname{tr}(A_i)}{2} & \text{if } \operatorname{disc}(A_1)=\operatorname{disc}(A_2)=0,\end{cases}\\
\psi &= \frac{2\tau_1\tau_2}{\operatorname{tr}(A_1)\operatorname{tr}(A_2)}\left(\operatorname{tr}(A_1A_2) - \frac{1}{2}\operatorname{tr}(A_1)\operatorname{tr}(A_2)\right),\\
\Delta &= 4\big(\Gamma(A_1,A_2)^2 - \Gamma(A_1,A_1)\Gamma(A_2,A_2)\big),\\
\eta_i &= \begin{cases} \frac{\pi}{2} - \arctan\frac{\operatorname{tr}(A_1)\operatorname{tr}(A_2)(\psi\tau_i+\tau_j)}{2\tau_1\tau_2\sqrt{\Delta}} & \text{if } \operatorname{disc}(A_i)<0,\\ \operatorname{arctanh}\frac{2\tau_1\tau_2\sqrt{\Delta}}{\operatorname{tr}(A_1)\operatorname{tr}(A_2)(\psi\tau_i-\tau_j)} & \text{if } \operatorname{disc}(A_i)>0,\\ \frac{2\sqrt{\Delta}}{(\operatorname{tr}(A_1A_2)-\operatorname{tr}(A_1)\operatorname{tr}(A_2)/2)\tau_i} & \text{if } \operatorname{disc}(A_i)=0,\end{cases}\\
\varpi &= \frac{2\Gamma(A_1,A_2)+\sqrt{\Delta}}{2\sqrt{\det A_1\det A_2}}e^{\tau_1\eta_1+\tau_2\eta_2},
\end{aligned}
$$

where $i = 1,2$ and $j \in \{1,2\} - \{i\}$. Note that all the above constants are invariant w.r.t. coordinate transformations. In particular, $\operatorname{disc}(A)$ is the discriminant of matrix A, which is defined as the discriminant of its characteristic polynomial. For matrix $A \in \mathbf{R}^{2\times 2}$, $\operatorname{disc}(A) \geq 0$ ($\operatorname{disc}(A) < 0$) means that the matrix admits real (complex) eigenvalues. Observe that $\tau_i < 0$, $i = 1,2$, due to the fact that both A_i are Hurwitz. As a result, we have $\operatorname{sgn}\Gamma(A_1,A_2) = \operatorname{sgn}(\tau_1\tau_2 - \psi)$. Finally, observe that $\Gamma(A,A) = \det A$ and $\operatorname{sgn}\psi = \operatorname{sgn}(2\operatorname{tr}(A_1A_2) - \operatorname{tr}(A_1)\operatorname{tr}(A_2))$.

Theorem 5.1 *For planar switched linear system* (5.1), *we have the following statements*:

(1) *If* $\Gamma(A_1, A_2) < -\sqrt{\det A_1 \det A_2}$, *then the system is unstable.*
(2) *If* $\Gamma(A_1, A_2) = -\sqrt{\det A_1 \det A_2}$, *then the system is marginally stable.*
(3) *If* $\Gamma(A_1, A_2) > -\sqrt{\det A_1 \det A_2}$ *and* $\operatorname{tr}(A_1 A_2) > -2\sqrt{\det A_1 \det A_2}$, *then the system is quadratically stable.*
(4) *Otherwise, the system is stable, marginally stable, and unstable if* $\varpi < 1$, $\varpi = 1$, *and* $\varpi > 1$, *respectively.*

To prove the theorem, we need some preparations.

Lemma 5.2 *If switched linear system* (5.1) *is* (*marginally*) *stable, then, for any positive real numbers* κ_1 *and* κ_2, *the rescaled switched system*

$$\dot{x} = \bar{A}_\sigma x, \qquad \bar{A}_i = \kappa_i A_i, \quad i = 1, 2,$$

is also (*marginally*) *stable.*

Proof As the system is (marginally) stable, it admits a common (weak) Lyapunov function that decreases along any subsystems. It can be seen that the function is also a common (weak) Lyapunov function for the rescaled switched system. As a result, the rescaled system is (marginally) stable. □

Based on the lemma, a normal form for the switched system can be described as follows.

Lemma 5.3 *Suppose that the commutator* $[A_1, A_2]$ *is nonsingular. Then, up to a linear coordinate transformation, exchanging of* A_1 *and* A_2, *and a rescaling according to Lemma* 5.2, A_1 *admits the normal form*

$$A_1 = \begin{bmatrix} \tau_1 & 1 \\ \operatorname{sgn}(\operatorname{disc}(A_1)) & \tau_1 \end{bmatrix}, \tag{5.2}$$

and

(1) *when* $\det[A_1, A_2] < 0$, *there exists a real number* F *with* $|F| > 1$ *such that* $F + \frac{\operatorname{sgn}(\operatorname{disc}(A_1)\operatorname{disc}(A_2))}{F} = 2\psi$, *and*

$$A_2 = \begin{bmatrix} \tau_2 & \operatorname{sgn}(\operatorname{disc}(A_2))/F \\ F & \tau_2 \end{bmatrix} \tag{5.3}$$

(2) *when* $\det[A_1, A_2] > 0$, *we have* $\psi \in (-1, 1)$, *and*

$$A_2 = \begin{bmatrix} \tau_2 + \sqrt{1-\psi^2} & \psi \\ \psi & \tau_2 - \sqrt{1-\psi^2} \end{bmatrix} \tag{5.4}$$

Proof We proceed with the case where $\det[A_1, A_2] < 0$, and the other case can be proven in a similar manner.

Suppose first that both $\operatorname{disc}(A_1)$ and $\operatorname{disc}(A_2)$ are negative. Let $\alpha_i \pm \sqrt{-1}\beta_i$ be the eigenvalues of A_i for $i = 1, 2$, where $\alpha_i < 0$ and $\beta_i > 0$, $i = 1, 2$. By a coordinate change, we can put A_1 and A_2 into normal forms as

$$A_1 = \begin{bmatrix} \alpha_1 & \beta_1 \\ -\beta_1 & \alpha_1 \end{bmatrix}, \qquad A_2 = \begin{bmatrix} \alpha_2 & -\beta_2/F \\ \beta_2 F & \alpha_2 \end{bmatrix},$$

where F is a real number. This corresponds to putting A_1 into the normal form and then rotating the coordinates so that the integral curves of A_2 are elliptical spirals with axes along the x_1 and x_2 directions. Simple calculation gives

$$[A_1, A_2] = \beta_1\beta_2(F - 1/F)\begin{bmatrix} 1 & 0 \\ 0 & -1 \end{bmatrix}.$$

It is clear that $\det[A_1, A_2] < 0$ implies that $F - 1/F > 0$. On the other hand, simple computation yields $F + 1/F = 2\psi$. It is clear that we can choose F with $|F| > 1$. Rescaling A_i by $\frac{A_i}{\beta_i}$, $i = 1, 2$, we obtain the normal forms as in (5.2) and (5.3).

Next, suppose that $\operatorname{disc}(A_1)\operatorname{disc}(A_2) = 0$. Up to exchanging A_1 and A_2, we assume that $\operatorname{disc}(A_1) = 0$. By a suitable coordinate change, we have

$$A_1 = \begin{bmatrix} \alpha_1 & 1 \\ 0 & \alpha_1 \end{bmatrix}, \qquad A_2 = \begin{bmatrix} a & b \\ c & d \end{bmatrix}.$$

It is clear that $\det[A_1, A_2] = -c^2$, which implies that $c \neq 0$. Taking the linear transformation $T = \left[\begin{smallmatrix} 1 & \frac{a-d}{2c} \\ 0 & 1 \end{smallmatrix}\right]$, we have

$$T^{-1}A_1T = A_1, \qquad T^{-1}A_2T = \begin{bmatrix} \frac{a+d}{2} & \frac{\operatorname{disc}(A_2)}{4c} \\ c & \frac{a+d}{2} \end{bmatrix}.$$

If $\operatorname{disc}(A_2) = 0$, then $2\psi = c$ and $\frac{a+d}{2} = \tau_2$, and hence the transformed matrices are exactly of the normal forms. Otherwise, by further implementing the linear transformation

$$\bar{T} = \begin{bmatrix} \frac{\sqrt{2}}{|\operatorname{disc}(A_2)|^{1/4}} & 0 \\ 0 & \frac{|\operatorname{disc}(A_2)|^{1/4}}{\sqrt{2}} \end{bmatrix}$$

and properly re-scaling, we can also obtain the normal forms.

Finally, suppose that $\max\{\operatorname{disc}(A_1), \operatorname{disc}(A_2)\} > 0$ and $\operatorname{disc}(A_1)\operatorname{disc}(A_2) \neq 0$. Up to exchanging A_1 and A_2, we assume that $\operatorname{disc}(A_1) > 0$. By a suitable coordinate change, we have

$$A_1 = \begin{bmatrix} \alpha_1 & 0 \\ 0 & \alpha_2 \end{bmatrix}, \qquad A_2 = \begin{bmatrix} a & b \\ c & d \end{bmatrix},$$

where $\alpha_1 < \alpha_2$. Simple computation gives $\det[A_1, A_2] = bc(\alpha_1 - \alpha_2)^2$, which implies that $bc < 0$. By implementing the linear transformation

$$T = \begin{bmatrix} -\sqrt{-b/c} & \sqrt{-b/c} \\ 1 & 1 \end{bmatrix},$$

we have

$$\frac{2}{\sqrt{\operatorname{disc}(A_1)}} T^{-1} A_1 T = \begin{bmatrix} \tau_1 & 1 \\ 1 & \tau_1 \end{bmatrix}$$

and

$$\frac{2}{\sqrt{|\operatorname{disc}(A_2)|}} T^{-1} A_2 T = \begin{bmatrix} \tau_2 & \operatorname{sgn}(\operatorname{disc}(A_2))/F \\ F & \tau_2 \end{bmatrix},$$

where F satisfies $F + \operatorname{sgn}(\operatorname{disc}(A_2))/F = 2\psi$. Note that this is exactly of the normal form. The proof is completed. □

Remark 5.4 The lemma presents normal forms for the case that $[A_1, A_2]$ is nonsingular. When the commutator $[A_1, A_2]$ is singular, it is not hard to see that A_1 and A_2 are simultaneously triangularizable. As a result, the switched system is quadratically stable when both A_1 and A_2 are stable (cf. Sect. 2.3.5).

Lemma 5.5 *Switched linear system* (5.1) *is quadratically stable iff for any* $\omega \in [0, 1]$, *we have*

$$\det(\omega A_1 + (1-\omega) A_2) > 0, \qquad \det(\omega A_1 + (1-\omega) A_2^{-1}) > 0. \tag{5.5}$$

Proof Suppose first that the switched linear system is quadratically stable, that is, there is $P > 0$ such that $A_i^T P + P A_i < 0$ for $i = 1, 2$. It follows that $A_i^{-T} P + P A_i^{-1} < 0$ [154]. As a result, we have

$$\begin{aligned} &(\omega A_1 + (1-\omega) A_2)^T P + P(\omega A_1 + (1-\omega) A_2) < 0, \\ &(\omega A_1 + (1-\omega) A_2^{-1})^T P + P(\omega A_1 + (1-\omega) A_2^{-1}) < 0 \end{aligned}$$

for any $\omega \in [0, 1]$, which implies (5.5).

When the switched system is not quadratically stable, then either $A_1 A_2$ or $A_1 A_2^{-1}$ admits (at least) one negative real eigenvalue [206]. We proceed with the case where $A_1 A_2^{-1}$ admits a negative real eigenvalue, and the other case can be addressed in a similar way. It can be seen that there is a positive real number μ such that

$$\det(\mu I_2 + A_1 A_2^{-1}) = 0,$$

which implies that

$$\det(\mu A_2 + A_1) = 0.$$

This violates the first inequality of (5.5). □

Proof of Theorem 5.1 First, assume that $\Gamma(A_1, A_2) \le -\sqrt{\det A_1 \det A_2}$. Taking $\varphi(\omega) = \det(\omega A_1 + (1-\omega) A_2)$ as a function of ω in $[0, 1]$, we have

$$\varphi(\omega) = \omega^2 \det A_1 + 2\omega(1-\omega)\Gamma(A_1, A_2) + (1-\omega)^2 \det A_2.$$

Straightforward calculation demonstrates that the minimum of the function is achieved at

$$\omega_0 = \frac{\det A_2 - \Gamma(A_1, A_2)}{\det A_1 + \det A_2 - 2\Gamma(A_1, A_2)} \in (0, 1)$$

with

$$\varphi(\omega_0) = \frac{-\Delta}{4(\det A_1 + \det A_2 - \Gamma(A_1, A_2))} \leq 0.$$

In particular, when $\Gamma(A_1, A_2) < -\sqrt{\det A_1 \det A_2}$, we have $\varphi(\omega_0) < 0$, which implies that the matrix $\omega_0 A_1 + (1 - \omega_0)A_2$ admits a positive real eigenvalue, and the switched system admits an unstable convex combination. As a result, the switched system is unstable. On the other hand, when $\Delta = 0$, we have $\varphi(\omega_0) = 0$. Similar analysis leads to the conclusion that the switched system is not (asymptotically) stable. However, it is marginally stable since it admits the following common weak quadratic Lyapunov function:

$$V(x) = x_1^2 + \frac{(\operatorname{sgn}(\operatorname{disc}(A_1)) \operatorname{sgn}(\operatorname{disc}(A_2)) - F^2)^2}{4F^2(\tau_1 F - \tau_2 \operatorname{sgn}(\operatorname{disc}(A_1)))^2} x_2^2.$$

This proves Statements (1) and (2).

Next, assume that

$$\Gamma(A_1, A_2) > -\sqrt{\det A_1 \det A_2}, \qquad \operatorname{tr}(A_1 A_2) > -2\sqrt{\det A_1 \det A_2}.$$

It is clear that $\Gamma(A_1, A_2) > -\sqrt{\det A_1 \det A_2}$ is equivalent to $\psi(\omega) > 0$ for any $\omega \in [0, 1]$. Similar analysis shows that $\operatorname{tr}(A_1 A_2) > -2\sqrt{\det A_1 \det A_2}$ is equivalent to $\det(\omega A_1 + (1 - \omega)A_2^{-1}) > 0$ for any $\omega \in [0, 1]$. It follows from Lemma 5.5 that the switched system is quadratically stable. This proves Statement (3).

Finally, we proceed with the case where $\Gamma(A_1, A_2) > -\sqrt{\det A_1 \det A_2}$ and $\operatorname{tr}(A_1 A_2) \leq -2\sqrt{\det A_1 \det A_2}$. Note that

$$\operatorname{tr}(A_1 A_2) \leq -2\sqrt{\det A_1 \det A_2},$$

which implies that

$$\Gamma(A_1, A_2) = \frac{1}{2}\big(\operatorname{tr}(A_1)\operatorname{tr}(A_2) - \operatorname{tr}(A_1 A_2)\big) > \sqrt{\det A_1 \det A_2}.$$

Therefore, we only need to consider the case where

$$\Gamma(A_1, A_2) > \sqrt{\det A_1 \det A_2}, \qquad \operatorname{tr}(A_1 A_2) \leq -2\sqrt{\det A_1 \det A_2}.$$

From the normal form we compute that

$$\operatorname{tr}(A_1 A_2) = 2\psi + 2\tau_1\tau_2,$$

which implies that $\psi < 0$ and further $F < -1$. Denote $Q(x) = \det[A_1 x, A_2 x]$ for $x \in \mathbf{R}^2$, and it is clear that the discriminant of Q is exactly Δ. Define $\Omega = \{x \in \mathbf{R}^2 :$

$Q(x) = 0\}$. It follows from the fact $\Delta > 0$ that $\Omega = D_1 \cup D_2$, where D_1 and D_2 are (noncoinciding) lines passing through the origin. For an $x \in \Omega - \{0\}$, we say that Ω is direct (inverse, respectively) at x if $A_2 x = \mu A_1 x$ for some $\mu > 0$ ($\mu < 0$, respectively). It is clear that there are real constants μ_1 and μ_2 such that

$$A_2 A_1^{-1} A_1 x = A_2 x = \mu_i A_1 x \quad \forall x \in D_i,\ i = 1, 2.$$

It follows that both μ_1 and μ_2 are eigenvalues of matrix $A_2 A_1^{-1}$. As a result, $\mu_1 \mu_2 = \det(A_2 A_1^{-1}) = \det A_2 / \det A_1 > 0$. On the other hand, it follows from the fact $A_1^{-1} = (2\tau_1 I_2 - A_1)/\det A_1$ that

$$\begin{aligned} \mu_1 + \mu_2 &= \operatorname{tr}(A_2 A_1^{-1}) = \operatorname{tr}(2\tau_1 A_2 - A_2 A_1)/\det A_1 \\ &= \Gamma(A_1, A_2)/\det A_1 > 0, \end{aligned}$$

which implies that both μ_1 and μ_2 are positive, i.e., Ω is always direct. Furthermore, we can establish that

$$\det[A_1 x, x] > 0, \qquad \det[A_2 x, x] > 0 \quad \forall x \in \Omega - \{0\},$$

which implies that both A_1 and A_2 rotate clockwise at Ω.

It is clear that D_1 and D_2 divide the state space into four cones, and within a cone the sign of $Q(x)$ keeps unchanged. Note that, from an initial state, the most stabilizing/destabilizing phase portrait (along all possible switching signals) is achieved when switches occur at Ω. To be more precise, from a nonzero initial state x_0, the most destabilizing state trajectory, denoted by $\phi^*(t, x_0)$, is that $\dot{\phi}^*(t, x_0) = A_i \phi^*(t, x_0)$ forms the smaller angle with the exiting radial direction than the other at any nonswitching time t. Note that the most destabilizing trajectory rotates clockwise in the phase plane, and the switched system is stable iff the trajectory is convergent (to the origin). Now take an initial state $x_0 \in D_1$, and let t^* be the least (positive) time that $x_1 = \phi^*(t^*, x_0) \in D_1$. Denote the ratio $R = \frac{|x_1|}{|x_0|}$. Then the switched system is stable (marginal stable, unstable) iff $R < 1$ ($R = 1$, $R > 1$, respectively). Tedious but straightforward analysis based on the normal forms shows that R is exactly ϖ, and the conclusion follows. □

5.1.2 Application to Absolute Stability of Planar Lur'e Systems

A typical Lur'e system consists of a linear plant with an output feedback whose gain is sector-bounded. Suppose that the planar linear plant is both controllable and stable and is described by

$$\begin{aligned} \dot{x} &= Ax + bu = \begin{bmatrix} 0 & 1 \\ -\alpha_1 & -\alpha_2 \end{bmatrix} \begin{bmatrix} x_1 \\ x_2 \end{bmatrix} + \begin{bmatrix} 0 \\ 1 \end{bmatrix} u, \\ y &= cx = [c_1 \quad c_2] \begin{bmatrix} x_1 \\ x_2 \end{bmatrix}, \end{aligned} \tag{5.6}$$

where α_1 and α_2 are positive real numbers. For a positive real number k, the system is said to be $[0, k)$-*absolutely stable* if the system is globally asymptotically stable under any (continuous) time-varying nonlinear output feedback law $u = \kappa(y, t)$ with $0 \le \kappa(y, t)y < ky^2$. The problem of absolute stability is to determine the largest possible number k^* such that the system is $[0, k)$-absolutely stable for any $k < k^*$. If the system is $[0, k)$-absolutely stable for any k, then let $k^* = +\infty$.

Note that the absolute stability implies that, for any $k \in [0, k^*)$, the switched system with subsystems $A_1 = A$ and $A_2 = A + kbc$ is guaranteed asymptotically stable. The converse is also true. Therefore, the stability criterion presented in the previous subsection applies. To this end, observe that

$$A + bkc = \begin{bmatrix} 0 & 1 \\ -\alpha_1 + kc_1 & -\alpha_2 + kc_2 \end{bmatrix},$$

which is also of the companion form. Let $k_1^* = \sup_k\{kc_1 \le \alpha_1, kc_2 \le \alpha_2\}$. It is clear that $k^* \le k_1^*$. Another observation is that, for any $k \le k_1^*$, we have $\Gamma(A, A + kbc) \ge 0$. According to Theorem 5.1, if $k^* < +\infty$, it should satisfy the relationships

$$\operatorname{tr}\big(A(A + k^*bc)\big) \le -2\sqrt{\det A \det(A + k^*bc)}, \quad \varpi = 1. \tag{5.7}$$

Take the equation

$$\operatorname{tr}\big(A(A + kbc)\big)^2 = 4 \det A \det(A + kbc), \tag{5.8}$$

where k is a variable to be determined. A solution of the equation is a root of a second-order polynomial, and it admits an analytical expression. Denote the two solutions to be k_1^2 and k_2^2. If the numbers are real, then there is at most one k_i^2 satisfying

$$k_i^2 \in (0, k_1^*], \quad \operatorname{tr}\big(A\big(A + k_i^2 bc\big)\big) < 0.$$

In this case, let $k_2^* = k_i^2$. Otherwise, let $k_2^* = k_1^*$. It follows from Theorem 5.1 that the Lur'e system is $[0, k_2^*)$-absolutely stable. As a result, $k^* \ge k_2^*$. In the case where $k_1^* = k_2^*$, we have $k^* = k_1^*$, and the problem is solved.

When $k_1^* > k_2^*$, we are to determine the k^* such that the corresponding ϖ is equal to one. As ϖ relies on k in a highly nonlinear manner, we propose a computational searching procedure for finding k^*. As an initiation, substitute k_1^* into a very large number (in the computational sense) when $k_1^* = +\infty$ and verify whether or not k_i^*, $i \in \{1, 2\}$, corresponds to unit ϖ. If so, set $k^* := k_i^*$, and the procedure is terminated. Otherwise, set $k_i := k_i^*$, $i = 1, 2$, and go to the recursive steps described in the following.

Set $k := \sqrt{k_1 k_2}$ and compute the corresponding ϖ. If $\varpi = 1$, then set $k^* := k$ and terminate. Else if $\varpi < 1$, then set $k_1 := k$. Otherwise, set $k_2 := k$. Repeat the process.

Note that in the above procedure, $\varpi = 1$ should be understood to be $\varpi \in (1 - \epsilon, \frac{1}{1-\epsilon})$, where $\epsilon > 0$ is a small number fixed in advance.

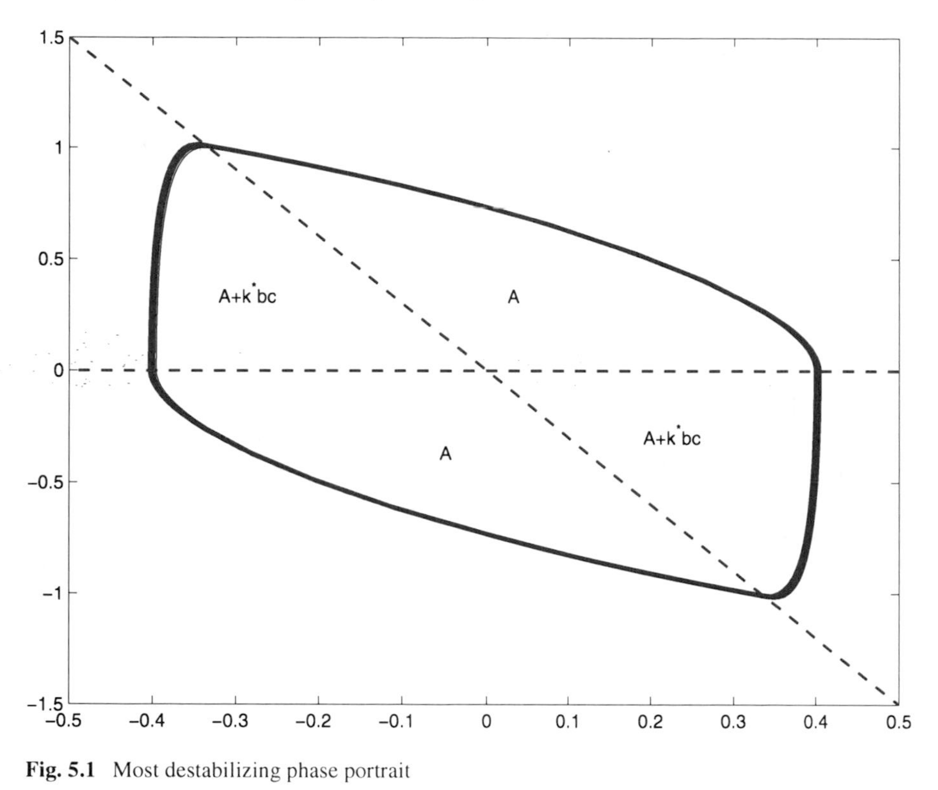

Fig. 5.1 Most destabilizing phase portrait

Example 5.6 Assume that

$$\alpha_1 = \alpha_2 = 1, \qquad c_1 = -3, \qquad c_2 = -1.$$

It is clear that $A + kbc$ is stable for any $k \geq 0$, and we have $k_1^* = +\infty$. Solving equation (5.8) gives $k = 1 \pm \sqrt{7}/2$. Further verification shows that $k_2^* = 1 + \sqrt{7}/2 \approx 2.3229$. Applying the above computational procedure (with $k_1^* = 1e + 10$ and $\epsilon = 1e - 8$), we obtain $k^* = 36.5031$. Figure 5.1 depicts the most destabilizing phase trajectory of the switched system with A and $A + k^*bc$ being the subsystems. It is clear that the trajectory is periodic. The switching surfaces (the dashed lines) are also depicted in the figure.

5.2 Adaptive Control via Supervisory Switching

For a control system with relatively small uncertainties or disturbances, it is possible to design a robust (single) controller that renders the uncertain system working well. However, when the uncertainties are of large scale, a robust controller might not exist, and usually the methodology of adaptive control has to be sought. By the well-known Astrom–Wittenmark textbook [12], an adaptive controller is "a controller that can modify its behavior in response to changes in the dynamics of the plant and the character of the disturbances". The classical adaptive control theory seeks a parameterized controller where the parameter is taking value over a continuum, and a controller is chosen based on the online estimate of the parameter and the principle of certainty equivalence. While effective when the parameter enters linearly and the parameter set is convex, the classical approach faces severe limitations when the unknown parameter enters the plant in nonlinear/nonconvex ways. To overcome the inherent difficulties, an alternative approach known as supervisory adaptive control appeared since the 1980s. A primary difference between this approach and the classical approach lies in the mechanism of controller selection, which is logic-based switching in the supervisory control rather than continuous tuning in the classical approach. The key idea behind the approach is to incorporate, besides the plant and the parameterized controller, a "high-level" supervisor that coordinates the switching among the candidate controllers in the way that an optimal controller is finally switched on so that a satisfactory performance is achieved. For this, a monitoring signal is designed based on the measured input/output data, and the switching signal is chosen to minimize the monitoring signal. In this way, suitable design of the monitoring signal is a crucial step toward the supervisory switching adaptive control diagram.

5.2.1 *Preliminaries*

Consider a single-input single-output plant with an unknown (but fixed) parameter described by

$$\hat{y}(s) = g(\lambda, s)\hat{u}(s), \tag{5.9}$$

where $\lambda \in \Lambda$, which is a compact subset of an Euclidean space, $g(\lambda, s) = \alpha_\lambda(s)/\beta_\lambda(s)$ is a strictly proper transfer function for each λ, β_λ is monic, α_λ and β_λ are coprime, and $g(\lambda, s)$ depends continuously on λ in the sense that the coefficients of polynomials α_λ and β_λ depend continuously on λ. The plant with transfer function $g(\lambda, s)$ is denoted by P_λ. We denote by λ^* the (unknown) true parameter, and P^{λ^*} (or P in short) the true plant.

We are to address the problem of adaptive stabilization within the supervisory adaptive control scheme. To this end, we need a feasibility assumption as follows.

Assumption 5.1 For each $\lambda \in \Lambda$, a stabilizing controller C_λ exists for process P_λ.

Note that, when controller C_λ stabilizes plant P_λ, it also stabilizes any plant P_μ when μ is sufficiently close to λ. This, together with the facts of continuous dependence of λ in P_λ and compactness of Λ, implies the existence of a finite set of parameters $\{\lambda_1, \dots, \lambda_k\}$ and a related set of stabilizing controllers $\{\mathrm{C}_1, \dots, \mathrm{C}_k\}$ such that $\bigcup_{i=1}^k \Lambda^i = \Lambda$, where $\Lambda^i = \{\lambda \in \Lambda : \mathrm{C}_i \text{ stabilizes } \mathrm{P}_\lambda\}$.

Next, we present a constructive way of determining a finite set of stabilizing controllers. For a plant P and a controller C, let $\mathbf{T}(\mathrm{P}, \mathrm{C})$ be the closed-loop generalized sensitivity transfer function matrix defined by [254],

$$\mathbf{T}(\mathrm{P}, \mathrm{C}) = \frac{1}{1 - \mathrm{CP}} \begin{bmatrix} -\mathrm{PC} & \mathrm{P} \\ -\mathrm{C} & 1 \end{bmatrix}.$$

For each $\lambda \in \Lambda$, we can find a stabilizing controller C_λ for plant P_λ such that

$$\left\| \mathbf{T}(\mathrm{P}_\lambda, \mathrm{C}_\lambda) \right\|_\infty = \frac{1}{\sqrt{1 - \| \frac{\alpha_\lambda}{\beta_\lambda} \|_{\mathrm{H}}^2}}, \tag{5.10}$$

where $\| \cdot \|_\infty$ denotes the $\mathcal{H}_\infty$ norm, and $\| \cdot \|_{\mathrm{H}}$ the Hankel norm. On the other hand, controller C_λ stabilizes any plant with transfer function $\alpha(s)/\beta(s)$ when

$$\left\| \begin{matrix} \alpha_\lambda - \alpha \\ \beta_\lambda - \beta \end{matrix} \right\|_\infty \left\| \mathbf{T}(\mathrm{P}_\lambda, \mathrm{C}_\lambda) \right\|_\infty < 1.$$

Fix sufficiently small positive real numbers ϵ and δ.

Procedure for Calculating a Finite Set of Stabilizing Controllers

(1) Grid the parametric set Λ into $\Upsilon = \{\lambda_1, \ldots, \lambda_k\}$ uniformly distributed with radius δ.
(2) For each $\lambda \in \Upsilon$, calculate controller C_λ that satisfies (5.10) and estimate the parametric subset

$$\Lambda^\lambda \subseteq \left\{ \mu \in \mathbf{R}^p : \left\| \begin{matrix} \alpha_\mu - \alpha_\lambda \\ \beta_\mu - \beta_\lambda \end{matrix} \right\|_\infty \le (1 - \epsilon) \sqrt{1 - \left\| \begin{matrix} \alpha_\lambda \\ \beta_\lambda \end{matrix} \right\|_{\mathrm{H}}^2} \right\}.$$

(3) Verify the coverage $\Lambda \subseteq \bigcup_{\lambda \in \Upsilon} \Lambda^\lambda$. If yes, go to Step (4). Otherwise, set $\delta := \delta/2$, and go back to Step (1).
(4) Prune Υ as long as coverage is kept.

In the above procedure, the major computational load is the estimate of the parametric subset in Step (2) and the coverage verification in Step (3). Due to the continuous dependence of the process on the parameter, the parametric subset admits nonempty interior; hence it is possible to approximate the set by means of a (convex) polyhedron. In this case, the coverage verification can be made by means of numerical softwares such as MATLAB GBT Toolbox [252].

In the sequel, we assume that the procedure is carried out successfully, which returns a finite parametric set $\{\lambda_1, \ldots, \lambda_k\}$, the corresponding controller set $\{\mathrm{C}_{\lambda_1}, \ldots, \mathrm{C}_{\lambda_k}\}$, and stabilizing region set $\{\Lambda^{\lambda_1}, \ldots, \Lambda^{\lambda_k}\}$ with $\Lambda \subseteq \bigcup_{i=1}^k \Lambda^{\lambda_i}$. Each process P_{λ_i} is said to be a *nominal model* that represents a family of systems $\mathrm{P}_\lambda : \lambda \in \Lambda^{\lambda_i}$. The stabilizing controller set $\{\mathrm{C}_{\lambda_1}, \ldots, \mathrm{C}_{\lambda_k}\}$ is said to be the *candidate controllers* for the plant. For notational convenience, redenote P_{λ_i} by P_i, Λ^{λ_i} by Λ^i, and C_{λ_i} by C_i for $i = 1, \ldots, k$.

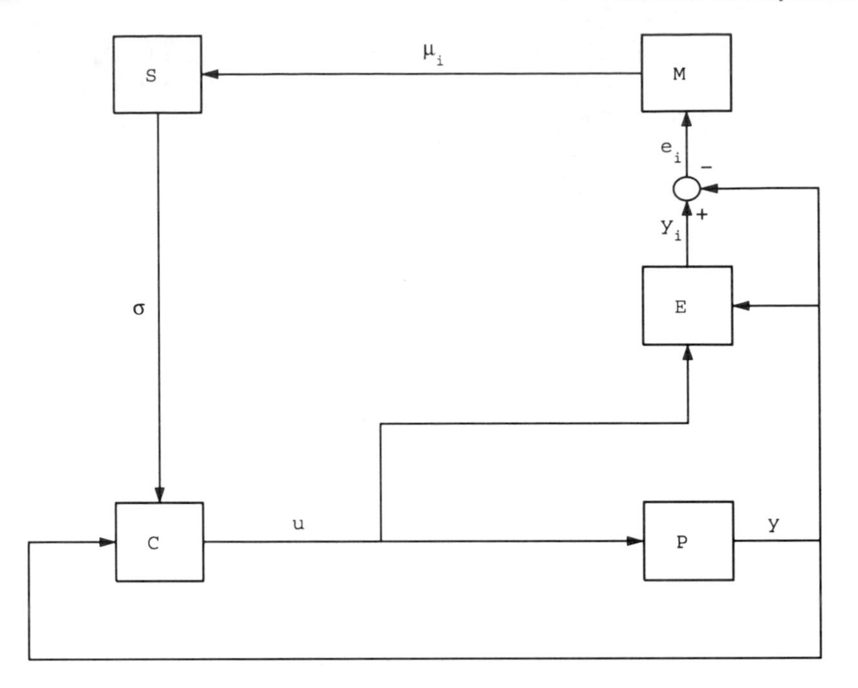

Fig. 5.2 Supervisory switching diagram

5.2.2 Estimator-based Supervisory Switching

As we are to find a supervisor that coordinates the switching among the candidate controllers such that the closed-loop system is stable, we need to design a switching law based on measured input/output data. Heuristically, if the plant parameter λ^* is known to belong to a parametric subset Λ^{i^*}, setting the switching signal to constant i^* could achieve the desired performance. However, as the plant parameter is unknown, we have to introduce a "monitoring signal" based on measured data and then design a monitoring-signal-driven switching law. For this, we incorporate a multiestimator whose outputs are signals y_i, $i \in \{1, \dots, k\}$. When the right (i^*th) controller is switched into the loop y_{i^*} would asymptotically converge to y, the plant output. The estimation errors $e_i = y_i - y$ are a key in producing the monitoring signal in a suitable manner.

To summarize, the supervisory switching scheme consists of (1) a multiestimator E with input (u, y) and outputs y_i, $i \in \{1, \dots, k\}$; (2) a monitoring signal generator M with inputs e_i and outputs μ_i called monitoring signals; and (3) a switching logic S with inputs μ_i and output σ. The system diagram is shown in Fig. 5.2.

First, for every $i \in \{1, \dots, k\}$, let

$$\begin{aligned} \dot{x}_C &= A_i^C x_C + b_i^C y, \\ u &= h_i x_C + r_i y \end{aligned} \tag{5.11}$$

be a realization of controller C_i, all sharing the same state x_C, where A_i^C is stable, and

$$\begin{aligned} \dot{x}_E &= A^E x_E + b^E y + d^E u, \\ y_i &= c_i x_E \end{aligned} \tag{5.12}$$

be a realization of the estimators, all sharing the same state x_E. Note that A^E, b^E, and d^E can be chosen to be parameter-independent with A^E stable. By defining the composite state $x = \begin{bmatrix} x_E \\ x_C \end{bmatrix}$, we have the system equations

$$\begin{aligned} \dot{x} &= A_\sigma x + d_\sigma e_\sigma, \\ y &= [c_{i^*}\ 0]x - e_{i^*}, \\ u &= f_\sigma x + g_\sigma e_{i^*}, \end{aligned} \tag{5.13}$$

where $\sigma \in \{1, \dots, k\}$ is the switching signal, $e_\sigma = y_\sigma - y$ is the estimation error, and

$$\begin{aligned} A_i &= \begin{bmatrix} A^E + (b^E + d^E r_i)c_i & d^E h_i \\ b_i^C c_i & A_i^C \end{bmatrix}, \\ d_i &= -\begin{bmatrix} b^E + d^E r_i \\ b_i^C \end{bmatrix}, \\ f_i &= [r_i c_i \ \ h_i], \\ g_i &= -r_i, \quad i = 1, \dots, k. \end{aligned}$$

As C_i stabilizes P_i for each $i = 1, \dots, k$, there exists a positive real number λ_0 such that all eigenvalues of A_i's are on the left of the vertical line $s = -\lambda_0$ on the complex plane. On the other hand, as C_{i^*} robustly stabilizes the plant set $\{P_\lambda : \lambda \in \Lambda^{i^*}\}$, the error e_{i^*} converges exponentially when the "right" controller is switched onto the loop. This means the existence of a positive real number λ_1 such that

$$|e_{i^*}(t)| \le \vartheta_1 e^{-\lambda t} \tag{5.14}$$

for any $\lambda < \lambda_1$, where ϑ_1 is a constant relying on the initial condition. This further implies that

$$\int_0^t e^{2\lambda\tau} e_{i^*}^2(\tau)\, d\tau \le \vartheta_2 \tag{5.15}$$

for any $\lambda < \lambda_1$, where ϑ_2 is a constant relying on the initial condition. Fix a $\lambda < \lambda_0$ such that both (5.14) and (5.15) hold.

Then, define the monitoring signal to be

$$\mu_i(t) = \int_0^t e_i^2(\tau)\, d\tau + \epsilon_i, \tag{5.16}$$

where ϵ_i, $i = 1, \dots, k$ are positive real numbers. Note that the monitoring signal is a solution of the differential equation

$$\dot{\mu}_i(t) = e_i^2(t), \qquad \mu_i(0) = \epsilon_i.$$

It is clear that the monitoring signal is the $\mathcal{L}^2$ performance w.r.t. the (virtue) output error and is strictly increasing and positively lower bounded. Constants ϵ_i, $i = 1, \dots, k$, are decided by the designer according to the prior experience which measure the possibility that the ith controller stabilizes the true plant, or otherwise arbitrarily chosen.

Next, fix a hysteresis factor $h > 0$ and define the switching signal recursively by

$$\begin{aligned} t_{j+1} &= \min\bigl\{t > t_j : (1+h)\min\bigl\{\mu_1(t), \dots, \mu_k(t)\bigr\} \le \mu_{\sigma(t_j)}(t)\bigr\}, \\ \sigma(t_{j+1}) &= \arg\min\bigl\{\mu_1(t_{j+1}), \dots, \mu_k(t_{j+1})\bigr\} \end{aligned} \tag{5.17}$$

for $j = 0, 1, 2, \dots$, where initially $t_0 = 0$ and $\sigma(t_0) = \arg\min\{\epsilon_1, \dots, \epsilon_k\}$. When there are two or more indices achieving the minimum, just choose one arbitrarily. It is clear that the switching law is scale-independent in the sense that, when the monitoring signal is scaled by $\Theta(t)$ with Θ a positive function, the switching signal will keep unchanged.

Finally, we briefly analyze the system performance for the supervised uncertain system. For this, we need a technical lemma as follows.

Lemma 5.7 *For any $t > 0$, the number of switches N_t within $[0, t)$ is upper bounded by*

$$N_t \le 1 + k + \frac{k}{\ln(1+h)} \ln\biggl(\frac{\mu_{i^*}(t)}{\min_j \epsilon_j}\biggr). \tag{5.18}$$

Proof Let $(t_0, i_0), \dots, (t_{N_t}, i_{N_t})$ be the switching sequence in $[0, t)$. Denote $t_{N_t+1} = t$. From the switching law we have

$$\mu_{i_l}(t) \le (1+h)\mu_j(t) \quad \forall l = 0, \dots, N_t,\ j = 1, \dots, k,\ t \in [t_l, t_{l+1}], \tag{5.19}$$

and

$$\mu_{i_l}(t_l) \le \mu_j(t_l) \quad \forall l = 0, \dots, N_t.$$

In particular, it follows from (5.19) that

$$\mu_{i_l}(t_{l+1}) \le (1+h)\mu_{i_{l+1}}(t_{l+1}) \quad \forall l = 0, \dots, N_t - 1.$$

Let s be the most frequently appeared index in the set $\{i_0, \dots, i_{N_t}\}$. It is clear that there exist a natural number $\nu \ge \lceil (N_t - 1)/k \rceil$ and integers $0 \le \kappa_1 < \kappa_2 < \cdots < \kappa_\nu < N_t$ such that $i_{\kappa_j} = s$ for $j = 1, \dots, \nu$. When $\nu \le 1$, inequality (5.18) is trivially true. Suppose that $\nu \geq 2$; then we have

$$(1+h)\mu_s(t_{\kappa_j}) \leq (1+h)\mu_{i_{\kappa_j+1}}(t_{\kappa_j}) \leq (1+h)\mu_{i_{\kappa_j+1}}(t_{\kappa_j+1})$$
$$= \mu_s(t_{\kappa_j+1}) \leq \mu_s(t_{\kappa_{j+1}}) \quad \forall j = 1, \dots, \nu - 1,$$

where the relationship $t_{\kappa_j+1} \leq t_{\kappa_{j+1}}$ and the monotonicity of μ_j's have been utilized. As a result, we have

$$\mu_{i^*}(t) \geq \mu_{i^*}(t_{\kappa_\nu}) \geq \mu_s(t_{\kappa_\nu}) \geq (1+h)^{\nu-1}\mu_s(t_{\kappa_1})$$
$$\geq (1+h)^{\nu-1}\mu_s(t_0) \geq (1+h)^{\nu-1}\min_j \epsilon_j,$$

which implies that

$$\nu \leq 1 + \frac{1}{\ln(1+h)} \ln\left(\frac{\mu_{i^*}(t)}{\min_j \epsilon_j}\right).$$

As $N_t \leq k\nu + 1$, inequality (5.18) follows. □

It follows from the lemma and inequality (5.15) that

$$N_t \leq 1 + k + \frac{k}{\ln(1+h)} \ln\left(\frac{\vartheta_2 + \epsilon_{i^*}}{\min_i \epsilon_i}\right),$$

which is a finite number independent of t. As a result, there exist an index j^* and a time T^* such that $\sigma(t) = j^*$ for $t \geq T^*$. By the switching law, we have

$$\mu_{j^*}(t) \leq (1+h)\mu_{i^*}(t) \leq (1+h)(\vartheta_2 + \epsilon_{i^*}),$$

which means that $e_{j^*} \in \mathcal{L}^2$. As A_{j^*} is stable and $e_{i^*} \in \mathcal{L}^2$, it follows from relationship (5.13) that $y \to 0$, and all other signals are bounded.

To summarize, we have the following conclusion.

Theorem 5.8 *All the signals in the closed-loop system remain bounded for any initial conditions, and $y(t) \to 0$ as $t \to +\infty$.*

Remark 5.9 As in the traditional adaptive scheme, the identified parameter j^* is not necessarily the "true parameter" i^*. In general, as we do not exclude the possibility that more than one candidate controller stabilizes the true plant, the identified parameter may be initial-condition dependent. Another observation is that, though the supervisor can coordinate the switching to a final (stabilizing) controller in a finite time, it cannot prevent a destabilizing controller from connecting to the loop more than one time. This means that the switching frequency may be high during the identifying process, which may worsen the transient performance as switching itself is usually undesirable.

5.2.3 An Example

Let the set of plants be

$$\mathrm{P}_\lambda: \quad g(\lambda, s) = \frac{\lambda(s-1)}{(s+1)(s-2)}, \qquad \lambda \in \Lambda = [1, 40]. \tag{5.20}$$

It is clear that all the plants are unstable and nonminimum phase systems. The large parameter uncertainty excludes the possibility to design a single, fixed-parameter linear controller that regulates the system in a satisfactory way.

It follows from the standard frequency domain theory that, for any fixed $\lambda \in \Lambda$, a stabilizing controller can be chosen to be

$$\mathrm{C}_\lambda = \lambda^{-1} \frac{448s^2 + 450s - 18}{31s(s-9)}.$$

As the unknown parameter is a scalar multiplicative gain in the system, we take $\lambda_j = \gamma^{j-1}$ for $j = 1, \ldots, k$, where $\gamma > 1$ should be chosen such that Λ^{λ_j} covers the interval $[\frac{\lambda_j}{\sqrt{\gamma}}, \sqrt{\gamma}\lambda_j]$ for any $j = 1, \ldots, k$, which is confirmative if

$$\left\| \begin{matrix} \alpha_\gamma - \alpha_1 \\ \beta_\gamma - \beta_1 \end{matrix} \right\|_\infty \left\| \mathbf{T}(\mathrm{P}_1, \mathrm{C}_1) \right\|_\infty < 1.$$

Simple calculation shows that $\gamma = 1.2$ satisfies the requirement. As a result, we can choose $k = 21$ and

$$\lambda_j = 1.2^{j-1}, \quad j = 1, \ldots, k.$$

This corresponds to 21 candidate controllers C_j with $\mathrm{C}_j = 1.2^{1-j} \frac{448s^2+450s-18}{31s(s-9)}$, $j = 1, \ldots, 21$. For any $j = 1, \ldots, 21$, the controller C_j stabilizes the set of plants P_λ with $\lambda \in [\frac{1}{\sqrt{1.2}}\lambda_j, \sqrt{1.2}\lambda_j]$.

Next, we construct a supervisor that coordinates the switching among the candidate controllers. For this, note that the jth estimator can be chosen as

$$\dot{x}^E = \begin{bmatrix} 0 & 1 \\ -1 & -2 \end{bmatrix} x^E + \begin{bmatrix} 1 \\ -3 \end{bmatrix} y + \begin{bmatrix} 3 \\ -3 \end{bmatrix} u,$$
$$y_j = [\lambda_j \quad 0] x^E,$$

and the jth controller can be realized by

$$\dot{x}^C = \begin{bmatrix} 9 & 1 \\ 0 & 0 \end{bmatrix} x^C + \begin{bmatrix} \frac{144.5806}{\lambda_j} \\ -\frac{0.5806}{\lambda_j} \end{bmatrix} y,$$
$$u = [1 \quad 0] x^C + 14.4516 y.$$

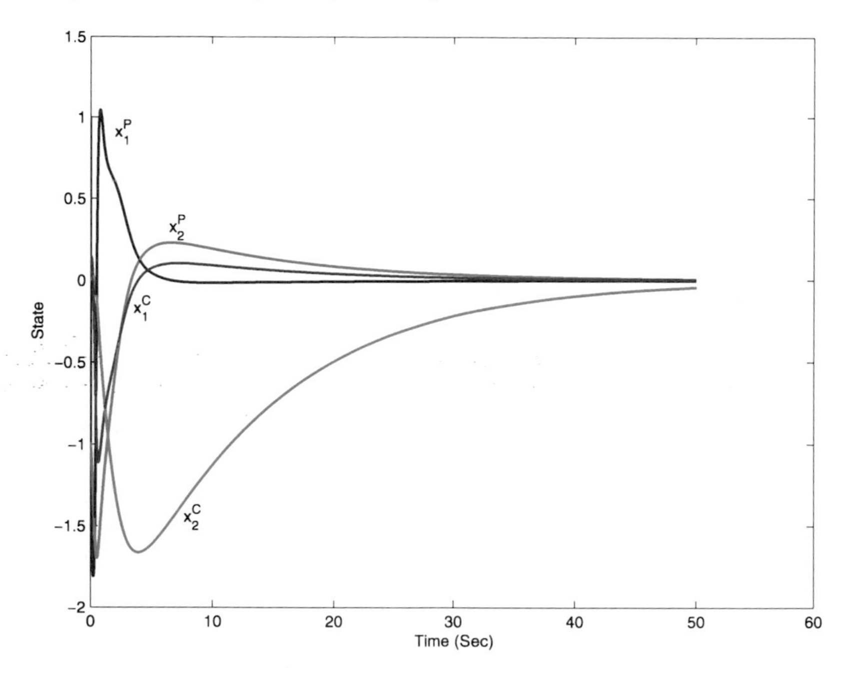

Fig. 5.3 The closed-loop state trajectories

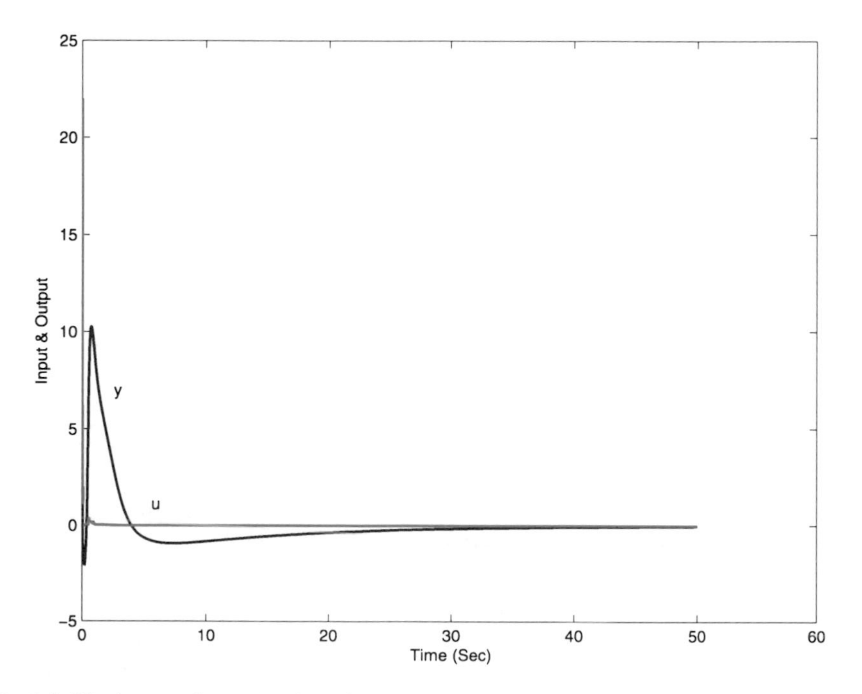

Fig. 5.4 The input and output trajectories

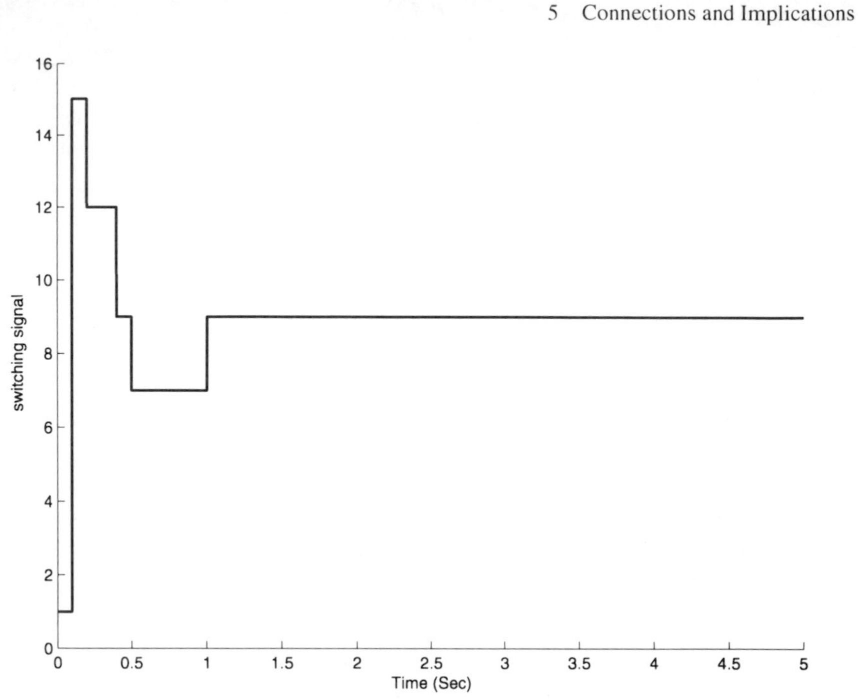

Fig. 5.5 The switching signal

To proceed with simulation, assume that the true parameter is 4 and that the true plant can be realized by

$$\dot{x}^P = \begin{bmatrix} 0 & 1 \\ -1 & -2 \end{bmatrix} x^P + \begin{bmatrix} 1 \\ -3 \end{bmatrix} y + \begin{bmatrix} 3 \\ -3 \end{bmatrix} u,$$

$$y = [4 \quad 0] x^P.$$

In addition, the factor h in the switching law (5.17) is set to be $h = 0.1$.

Let the initial state of the true plant be $x^P(0) = [1, -1]^T$, and both the candidate controller and the estimator be initially at the origin. Suppose that the first candidate controller is connected into the loop initially, that is, $\sigma(0) = 1$. Figures 5.3 and 5.4 depict the closed-loop state and input–output trajectories, respectively. These exhibit satisfactory closed-loop performance, though the convergence rate is quite low due to the fact that A_{i^*} has an eigenvalue at -0.0825, which is very close to the imaginary axis. The switching signal is shown in Fig. 5.5, which identifies the "right" controllers via 5 switches in 1 sec. It is interesting to note that, before the final identification of the right controller, the controller had been connected to the loop but disconnected after a while. This indicates that the switching mechanism fails to identify the right controller even if it is connected to the loop. Furthermore, we set the right controller initially, i.e., $\sigma(0) = i^* = 9$, and carry out the simulation again. The switching signal in Fig. 5.6 exhibits that the finally identified controller

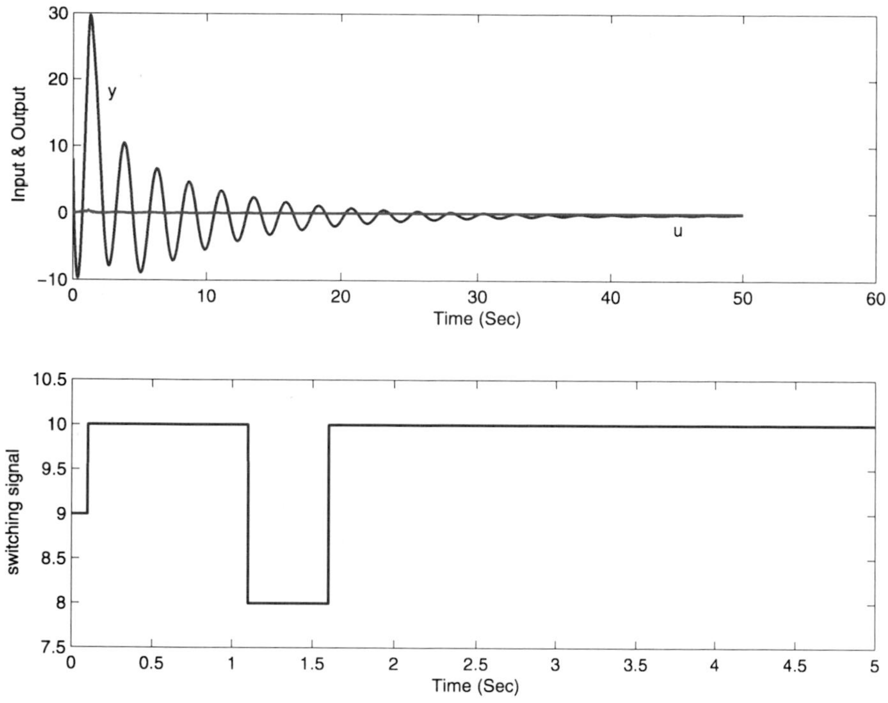

Fig. 5.6 The input–output and switching signals

is not the "right" controller. This is not surprising due to the fact that the identified controller is a stabilizing controller for the true plant.

5.3 Stability Analysis of Fuzzy Systems via Piecewise Switching

Fuzzy systems and fuzzy logic control have gained much attention due to their wide applications to many areas including machine intelligence, signal processing, and management, to list a few. As an alternative of the conventional control design methodologies, fuzzy control is powerful in dealing with nonlinear dynamical systems. However, the development of systematic methods for analysis and control of fuzzy systems has proven to be very hard. In particular, quadratic stability analysis leads usually to conservativeness, and nonquadratic analysis may result in intractable computational burden.

In this section, we examine a special yet typical class of fuzzy systems, which was known to be linear Takagi–Sugeno systems or the T–S fuzzy model in short. A forced-free T–S system is described by a set of rules in the form

$$\begin{aligned} \mathrm{R}^l: &\text{ IF } z_1 \text{ is } F_1^l \text{ AND } \cdots \text{ AND } z_s \text{ is } F_s^l, \\ &\text{THEN } x(t+1) = A_l x(t) + a_l, \end{aligned} \tag{5.21}$$

where R^l denotes the lth fuzzy inference rule, $l \in M = \{1, \dots, m\}$, F_i^l, $l = 1, \dots, m$, $i = 1, \dots, s$, are fuzzy sets, $x(t) \in \mathbf{R}^n$ is the state, z_i, $i = 1, \dots, s$, are measurable variables of the system, and (A_l, a_l) is the lth local model. By the standard fuzzy inference, the inferred system can be written as

$$x(t+1) = A\big(\mu(x)\big)x(t) + a\big(\mu(x)\big), \tag{5.22}$$

where

$$A(\mu) = \sum_{l=1}^{m} \mu_l A_l, \qquad a(\mu) = \sum_{l=1}^{m} \mu_l a_l$$

with $\mu_l = \frac{\prod_{i=1}^{s} F_i^l}{\sum_{j=1}^{m} \prod_{i=1}^{s} F_i^j}$. It is clear that $\mu_l \geq 0$ for $l \in M$ and that $\sum_{l \in M} \mu_l = 1$. Therefore, $A(\mu)$ is always a convex combination of the matrices $A_1, \dots, A_m$. When the affine term vanishes, and $A_1, \dots, A_m$ admit a common quadratic Lyapunov function, the fuzzy system is globally exponentially stable. While simple, the idea may lead to very conservative criteria. Indeed, as the membership functions are state dependent, the common Lyapunov function approach does not utilize this useful information. To tackle this problem, we introduce a general framework of piecewise switched linear systems and conduct stability analysis for the systems. The approach is then applied to stability of T–S systems.

5.3.1 Piecewise Switched Linear Systems

Piecewise switched linear systems are piecewise dynamical systems with switched linear systems as local dynamics. Such a system may represent, for instance, a nonlinear process where approximate linearization is taken at different operating points, and the approximate error and other perturbations are taken into account as parameter/structural uncertainties. From a system framework's perspective, piecewise switched linear systems extend both switched linear systems and piecewise linear systems.

A discrete-time piecewise switched linear system can be described by

$$x(t+1) = A_{i,\sigma_i} x(t) + a_{i,\sigma_i} \quad \text{for } x(t) \in \Omega_i, \tag{5.23}$$

where $x(t) \in \mathbf{R}^n$ is the state, $\Omega_1, \dots, \Omega_k$ are convex polyhedra with the union being the state space, $\sigma_i \in \{1, \dots, m_i\}$, $i = 1, \dots, k$, and $A_{i,j} \in \mathbf{R}^{n \times n}$ and $a_{i,j} \in \mathbf{R}^n$, $j = 1, \dots, m_i$, $i = 1, \dots, k$, are real constant matrices and vectors, respectively. For treatment convenience, the system can be restated to be

$$\bar{x}(t+1) = \bar{A}_{i,\sigma_i} \bar{x}(t) \quad \text{for } x(t) \in \Omega_i,$$

where $\bar{A}_{i,j} = \left[\begin{smallmatrix} A_{i,j} & a_{i,j} \\ 0 & 1 \end{smallmatrix}\right]$, $j = 1, \dots, m_i$, $i = 1, \dots, k$, and $\bar{x} = \left[\begin{smallmatrix} x \\ 1 \end{smallmatrix}\right]$.

We assume that the system is well defined in the sense that a unique solution exists and extends to infinity in time for any initial condition and switching signal.

While several stabilities can be defined for the system under various switching mechanisms, here we focus on the (guaranteed) global exponential stability which means that any state trajectory exponentially converges to the origin under arbitrary switching.

Take the Lyapunov candidate as

$$V(x) = x^T P_i(x)x + 2q_i(x)x + r_i(x) = \bar{x}^T \bar{P}_i(x)\bar{x}, \quad x \in \Omega_i, \tag{5.24}$$

where $\bar{P}_i = \left[\begin{smallmatrix} P_i & q_i^T \\ q_i & r_i \end{smallmatrix}\right]$. When $0 \in \Omega_i$, we require that $q_i = 0$ and $r_i = 0$, which means that $V(x) = x^T P_i(x)x$ for $x \in \Omega_i$. Assume that $V(x)$ is continuous in each Ω_i, $i = 1, \dots, k$. Note that we do not impose the continuity of $V(x)$ over the cell boundaries.

Proposition 5.10 *The piecewise switched linear system is globally exponentially stable if there is a Lyapunov candidate V as in* (5.24) *satisfying*

(1) *for any $i = 1, \dots, k$, there exist positive real numbers α_i and β_i such that*

$$\alpha_i x^T x \le V(x) \le \beta_i x^T x \quad \forall x \in \Omega_i \tag{5.25}$$

(2) *for any $i = 1, \dots, k$, there exists a positive real number γ_i such that*

$$\begin{gathered} \bar{x}^T \bar{A}_{i,j}^T \bar{P}_l(A_{i,j}x)\bar{A}_{i,j}\bar{x} - \bar{x}^T \bar{P}_i(x)\bar{x} \le -\gamma_i x^T x \\ \forall x \in \Omega_i, \; j = 1, \dots, m_i, \end{gathered} \tag{5.26}$$

where $l = \arg\{\mu : A_{i,j}x \in \Omega_\mu\}$

Proof The proof is straightforward. Indeed, let

$$\alpha = \min_i \alpha_i, \qquad \beta = \max_i \beta_i, \qquad \gamma = \min_i \gamma_i.$$

Taking an arbitrarily given state trajectory $x(t)$, $t = 0, 1, 2, \dots$, it can be seen from Item (2) that

$$V\big(x(t+1)\big) \le \left(1 - \frac{\gamma}{\beta}\right) V\big(x(t)\big) \quad \forall t = 0, 1, 2, \dots,$$

which, together with Item (1), implies that

$$\big|x(t)\big| \le \sqrt{\frac{\beta}{\alpha}} \left(1 - \frac{\gamma}{\beta}\right)^{t/2} \big|x(0)\big| \quad \forall t = 0, 1, 2, \dots.$$

□

Remark 5.11 Note that $\bar{x}^T \bar{P}_i(x)\bar{x}$ is a common Lyapunov function for $\bar{A}_{i,j}$, $j = 1, \dots, m_i$ over the cell Ω_i. When $\bar{P}_i(x) = \bar{P}_i$ is a constant matrix, the subsystems of the ith local model admit a common quadratic Lyapunov function. When the subsystems do not admit such a common Lyapunov function, we could seek piecewise quadratic ones instead. To verify the proposition, we need to examine one-step cell transitions of the state, that is, to determine the indices of the cells that $A_{i,j}x$ belong to for any $x \in \Omega_i$, $j = 1, \ldots, m_i$, $i = 1, \ldots, k$.

To search a qualified Lyapunov candidate that satisfies inequality (5.25), we take the parameterization in terms of polyhedral cell bounding, as presented in Sect. 3.3.

Proposition 5.12 *Suppose that $\bar{E}_i = [E_i, e_i]$, $i = 1, \ldots, k$, are polyhedral cell bounding matrices. The piecewise switched linear system is globally exponentially stable if there exist symmetric matrices $\bar{P}_1, \ldots, \bar{P}_k$ satisfying*

(1) *for any $i = 1, \ldots, k$, there exists a matrix W_i with positive real entries such that*

$$\bar{P}_i - \bar{E}_i^T W_i \bar{E}_i > 0 \tag{5.27}$$

(2) *for any $i = 1, \ldots, k$, there exists a positive real number γ_i such that*

$$\begin{aligned} &\bar{x}^T \bar{A}_{i,j}^T \bar{P}_l \bar{A}_{i,j} \bar{x} - \bar{x}^T \bar{P}_i \bar{x} \le -\gamma_i x^T x \\ &\forall x \in \Omega_i, \ j = 1, \ldots, m_i, \end{aligned} \tag{5.28}$$

where $l = \arg\{\mu : A_{i,j}x \in \Omega_\mu\}$

The proposition could be proven by simply combining Proposition 5.10 with Theorem 3.23. The details are omitted.

Remark 5.13 The above criteria are much less conservative than the existence of a common quadratic Lyapunov function, and they could be verified by means of linear matrix inequalities. However, to find qualified piecewise quadratic Lyapunov functions, we usually need to further partition the cells, which makes the computation inefficient for higher-dimensional systems.

5.3.2 Stability Analysis of T–S Fuzzy Systems

A T–S fuzzy system (5.22) can be aggregated into a piecewise switched linear system in a clear manner. Indeed, suppose that $\Omega_1, \ldots, \Omega_k$ partition the state space. Define

$$\zeta_i = \{ j : \exists x \in \Omega_i \text{ s.t. } \mu_j(x) > 0 \}, \quad i = 1, \ldots, k,$$

that is, ζ_i is the set of indices of the normalized membership functions that do not vanish in the region Ω_i. Let m_i be the cardinality of the set ζ_i, and denote the corresponding set of subsystem matrices by $A_{i,1}, \ldots, A_{i,m_i}$ and the affine terms by $a_{i,1}, \ldots, a_{i,m_i}$, respectively. The corresponding piecewise switched linear system (5.23) is said to be the *aggregated system* w.r.t. partition $\{\Omega_1, \ldots, \Omega_k\}$.

Lemma 5.14 *Suppose that $\Omega_1, \ldots, \Omega_k$ partition the state space. T–S fuzzy system* (5.22) *is globally exponentially stable if its aggregated system is globally exponentially stable.*

While simple and straightforward, the lemma provides a general approach for verifying the stability of a T–S fuzzy system. By appropriately choosing the state partition, the approach may lead to less conservative stability criteria. For this, we need to find a minimum partition $\{\Omega_1^*, \ldots, \Omega_{k^*}^*\}$ in the sense that, for any l and j, either $\mu_l(x) \neq 0 \ \forall x \in \Omega_j^*$ or $\mu_l(x) \equiv 0$ in Ω_j^*. A minimum partition can be calculated from the supporting sets of the normalized membership functions. In fact, denote $\Upsilon_i = \{x : \mu_i(x) > 0\}$ for $i = 1, \ldots, m$ and by Υ_i^c the complement to Υ_i. Then, each Ω_i^* is exactly an intersection of some Υ_i's and Υ_i^c's. In particular, in the regions where $\mu_l = 1$ for some l, all other membership functions are equal to zero. We will call such a region an *operating regime*. In between operating regimes, there are regions where $0 < \mu_l < 1$, and these regions are called *interpolation regimes*.

Example 5.15 Suppose that a T–S fuzzy system is with local modes

$$A_1 = \begin{bmatrix} -0.9425 & 0.4308 \\ 0.2132 & 0.2615 \end{bmatrix}, \qquad a_1 = \begin{bmatrix} -0.4111 \\ 0.6656 \end{bmatrix},$$

$$A_2 = \begin{bmatrix} -0.2821 & -0.6661 \\ 0.7458 & 0.9664 \end{bmatrix}, \qquad a_2 = \begin{bmatrix} 0 \\ 0 \end{bmatrix},$$

$$A_3 = \begin{bmatrix} 0.1133 & 0.9112 \\ 0.2925 & -1.5132 \end{bmatrix}, \qquad a_3 = \begin{bmatrix} -1.1257 \\ 0.9847 \end{bmatrix},$$

and normalized membership functions are scheduled by variable x_1 as

$$\mu_1(x_1) = \begin{cases} 1 & \text{if } x_1 \leq -5, \\ (x_1+3)/2 & \text{if } -5 < x_1 \leq -3, \\ 0 & \text{otherwise}, \end{cases}$$

$$\mu_2(x_1) = \begin{cases} 0 & \text{if } x_1 \leq -5, \\ (5+x_1)/2 & \text{if } -5 < x_1 \leq -3, \\ 1 & \text{if } -3 < x_1 \leq 1, \\ 2-x_1 & \text{if } 1 \leq x_1 < 2, \\ 0 & \text{otherwise}, \end{cases}$$

$$\mu_3(x_1) = \begin{cases} 0 & \text{if } x_1 \leq 1, \\ x_1 - 1 & \text{if } 1 < x_1 \leq 2, \\ 1 & \text{otherwise}. \end{cases}$$

By partitioning the state space into five regions as shown in Fig. 5.7, the fuzzy systems can be aggregated into a piecewise switched linear system, where Ω_1, Ω_3,

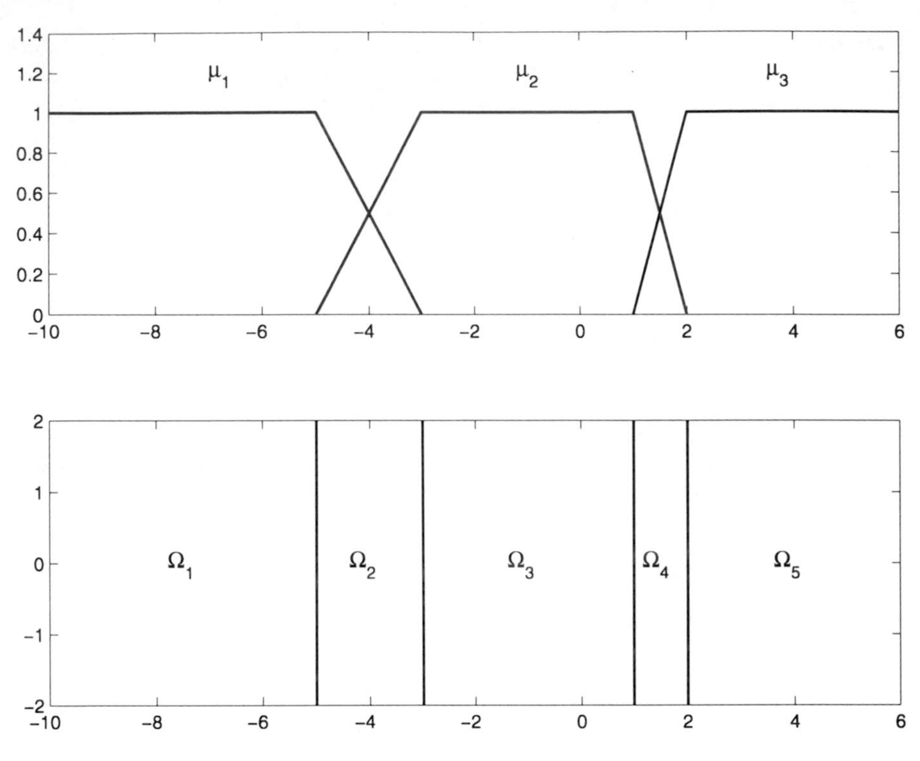

Fig. 5.7 The membership functions and state partitions

Ω_5 are related to single local modes (A_1, a_1), A_2, and (A_3, a_3), respectively, while Ω_2 and Ω_4 are related to mode pairs $\{(A_1, a_1), A_2\}$ and $\{A_2, (A_3, a_3)\}$, respectively.

By searching piecewise quadratic Lyapunov functions in form (5.24) that verify Proposition 5.10, we obtain

$$P_1 = P_2 = \begin{bmatrix} 1.0934 & 0.0216 \\ 0.0216 & 0.8491 \end{bmatrix}, \qquad q_1 = q_2 = [0.6570,\ 0.7234],$$

$$r_1 = r_2 = 1.3152,$$

$$P_3 = \begin{bmatrix} 3.4437 & 1.3694 \\ 1.3694 & 2.3554 \end{bmatrix}, \qquad q_3 = [0,\ 0], \qquad r_3 = 0,$$

$$P_4 = P_5 = \begin{bmatrix} 5.9608 & 2.4491 \\ 2.4491 & 1.8491 \end{bmatrix}, \qquad q_4 = q_5 = [-1.7496,\ 0.5635],$$

$$r_4 = r_5 = 3.7524.$$

By Lemma 5.14, the fuzzy system is globally asymptotically stable. The level sets of the Lyapunov functions and the sample phase portrait are depicted in Fig. 5.8. It is clear that the Lyapunov function is not continuous over the region boundaries.

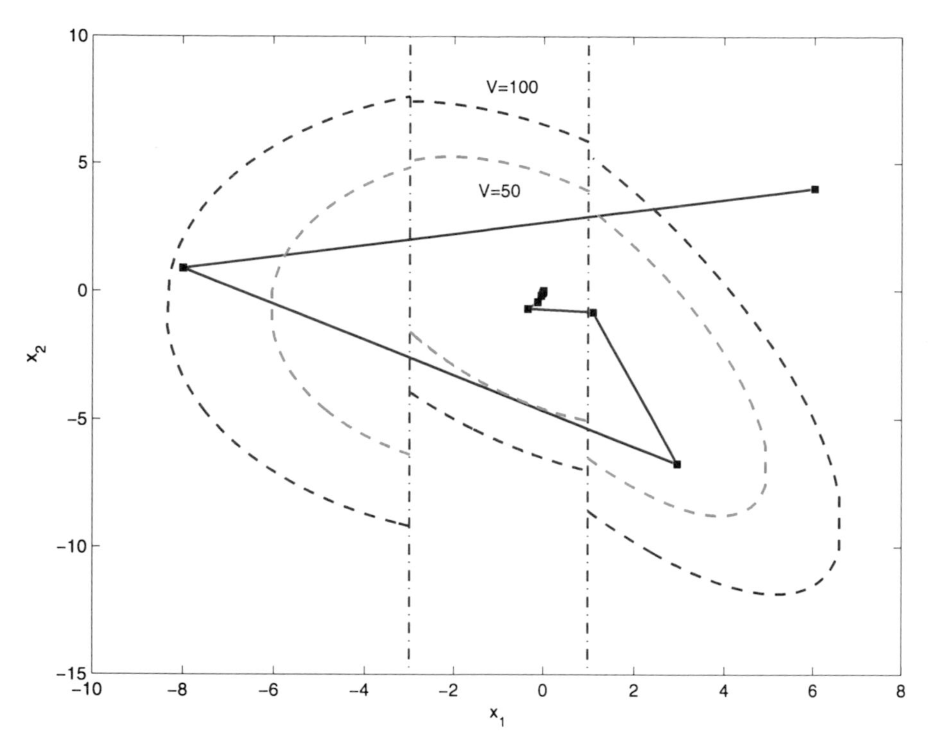

Fig. 5.8 A sample phase portrait with level sets

5.4 Consensus of Multiagent Systems with Proximity Graphs

5.4.1 Introduction

A multiagent system is a collection of physical or virtual entities, each with the abilities of acting and perceiving the environment (possibly in a partial manner) and communicating with others. A typical example is a school of migrating birds that stay together, coordinate turns, and avoid collision with obstacles. A bird is capable of adjusting its speed and flying toward the center of the flock within its sight. The motion of an individual bird can be mathematically described by

$$\dot{x}_i(t) = \sum_{j \in \mathcal{N}_i(t)} \big(x_j(t) - x_i(t)\big),$$

where i is the bird's label, x_i is the physical state (w.r.t. a fixed coordinate), and $\mathcal{N}_i$ is a neighbor set of the bird, i.e., the set of birds within its sight. Let r_i be the sight radius of the bird. Then, the neighbor set is

$$\mathcal{N}_i = \big\{ j \neq i : |x_j - x_i| \leq r_i \big\}.$$

Suppose that there are m agents in the flocking system and that the agents are with a uniform radius of perception, i.e., $r_i = r$ for all $i = 1, \dots, m$. The multiagent system is thus represented by

$$\dot{x}_i(t) = \sum_{|x_j(t)-x_i(t)|\le r} \big(x_j(t) - x_i(t)\big), \quad i = 1, \dots, m, \tag{5.29}$$

where $x_i \in \mathbf{R}^n$, $i = 1, \dots, m$. Note that the system is exactly a piecewise linear system where the switching surfaces are the hyperplanes

$$H_{i,j} = \big\{\big[x_1^T, \dots, x_m^T\big]^T \in \mathbf{R}^{mn} : |x_i - x_j| = r\big\}, \quad i \neq j$$

and their intersections. For any fixed $x = [x_1^T, \dots, x_m^T]^T \in \mathbf{R}^{mn}$, there corresponds a sequence of neighbor sets $\mathcal{N}_1^x, \dots, \mathcal{N}_m^x$ with

$$\mathcal{N}_i^x = \big\{j \neq i : |x_j - x_i| \le r\big\}, \quad i = 1, \dots, m.$$

Define

$$\Omega(x) = \big\{y \in \mathbf{R}^{mn} : \mathcal{N}_i^y = \mathcal{N}_i^x,\ i = 1, \dots, m\big\}$$

and

$$\Lambda(x) = \begin{bmatrix} \Lambda_{11}(x) & \dots & \Lambda_{1m}(x) \\ & \ddots & \\ \Lambda_{m1}(x) & \dots & \Lambda_{mm}(x) \end{bmatrix},$$

where

$$\Lambda_{ij}(x) = \begin{cases} I_n & \text{if } j \in \mathcal{N}_i^x, \\ -|\mathcal{N}_i^x| I_n & \text{if } j = i, \\ 0_{n\times n} & \text{otherwise,} \end{cases}$$

and $|\mathcal{N}_i^x|$ is the cardinality of the set $\mathcal{N}_i^x$. It is clear that we can reexpress the system equations as

$$\dot{x} = \Lambda(x)x = \big(A(x) \otimes I_n\big)x, \tag{5.30}$$

where

$$A(x) = \begin{bmatrix} A_{11}(x) & \dots & A_{1m}(x) \\ & \ddots & \\ A_{m1}(x) & \dots & A_{mm}(x) \end{bmatrix}, \quad A_{ij}(x) = \begin{cases} 1 & \text{if } j \in \mathcal{N}_i^x, \\ -|\mathcal{N}_i^x| & \text{if } j = i, \\ 0 & \text{otherwise.} \end{cases}$$

Note the following equivalencies:

$$\big(\mathcal{N}_1^x, \dots, \mathcal{N}_m^x\big) = \big(\mathcal{N}_1^y, \dots, \mathcal{N}_m^y\big) \iff \Omega(x) = \Omega(y) \iff A(x) = A(y).$$

As $\mathcal{N}_i^x \subset \{1, \dots, m\}$ for any $i = 1, \dots, m$ and $x \in \mathbf{R}^{mn}$, the value space of sequence $(\mathcal{N}_1^x, \dots, \mathcal{N}_m^x)$ admits a finite number of elements (up to $\prod_{i=0}^{m-1} 2^i$ elements). As a result, there exist a finite number of polyhedra $\Omega_1, \dots, \Omega_N$ that partition the state space in the sense that $\bigcup_{j=1}^N \Omega_j = \mathbf{R}^{mn}$ and $\Omega_k \cap \Omega_j^o = \emptyset$ for all $k \neq j$ and that any $\Omega(x)$ belongs to the partitioned polyhedra set, i.e., $\Omega(x) \in \{\Omega_1, \dots, \Omega_N\}$. Let A^i be the matrix $A(x)$ corresponding to Ω_i. The multiagent system can be rewritten as

$$\dot{x} = (A^j \otimes I_n)x, \quad x \in \Omega_j, \tag{5.31}$$

which is exactly a piecewise linear system.

5.4.2 A Consensus Criterion

For multiagent system (5.31), an interesting problem is to find conditions under which the agreement can be achieved in a certain sense. More precisely, for a system motion $x(\cdot)$, let $V_t = (G, \Upsilon_t^x)$ be the dynamic graph, where $G = \{1, \dots, m\}$ is the set of nodes, and $\Upsilon_t^x \subseteq G \times G$ is the set of edges at t,

$$\Upsilon_t^x = \{(i, j) : j \neq i, \ |x_j(t) - x_i(t)| \leq r\}.$$

For a node i, the set of neighbors at t is

$$\mathcal{N}_i^x(t) = \{j \neq i : (i, j) \in \Upsilon_t^x\} = \{j \neq i : |x_j(t) - x_i(t)| \leq r\}.$$

Nodes i and j are said to be *agree at time t* if $x_i(t) = x_j(t)$. The motion is said to be *in consensus at t* if any two nodes agree at t. The motion is said to be *in asymptotic consensus* if consensus is achieved asymptotically, i.e., $\lim_{t\to+\infty}(x_i(t) - x_j(t)) = 0$ for all $i, j \in G$. In particular, when the limit $\lim_{t\to+\infty} x_i(t)$ exists, the consensus limit is said to be the *group decision value*. If it happens that the limit is equal to the algebraic average of the initial state, i.e., $\lim_{t\to+\infty} x_i(t) = \frac{1}{m}\sum_{i=1}^m x_i(t_0)$, then the average consensus is achieved.

The multiagent system (5.30) is equivalently represented by its dynamic graph in the following sense. Let $B(t) = [b_{ij}(t)]_{m\times m}$ be the adjacency matrix of graph V_t, that is, $b_{ij}(t) = 1$ if $(i, j) \in \Upsilon_t$, and $b_{ij}(t) = 0$ otherwise. Let $C(t) = \operatorname{diag}(c_1(t), \dots, c_m(t))$ be the valency matrix with $c_i(t) = \sum_{j=1}^m b_{ij}(t)$, $i = 1, \dots, m$. Furthermore, define the graph's dynamic Laplacian matrix to be $L(t) = C(t) - B(t)$. Note that V_t is an undirected graph for any t and that the adjacency matrix is a symmetric matrix, so is the Laplacian matrix. Correspondingly, a motion $x(\cdot)$ of the multiagent system satisfies

$$\dot{x}(t) = -(L(t) \otimes I_n)x(t). \tag{5.32}$$

The system is a piecewise linear system with the system matrices all being graph Laplacian matrices. As a result, the properties of the Laplacian matrices play an important role in analyzing the system behavior. Fortunately, the well-developed algebraic graph theory (see, e.g, [28, 89]) provides a rigorous tool for this.

Fix a (undirected) graph $V = (G, \Upsilon)$, and suppose that its Laplacian matrix is L. It is clear that L is symmetric; hence all its eigenvalues are real. We denote the spectra to be

$$\lambda_1 \leq \lambda_2 \leq \cdots \leq \lambda_m.$$

An observation is that the matrix is row-stochastic, that is, the sum of elements in each row is zero. As an implication, zero is an eigenvalue with eigenvector $\mathbf{1} = [1, \ldots, 1]^T$. This further implies that $\mathbf{1}^T x(t)$ is a constant independent of the time. As the matrix is row-dominated by its diagonal entries, it is in fact semi-definite positive. Combining the above analysis gives $\lambda_1 = 0$. λ_2 is called the *algebraic connectivity*, and the smallest nontrivial eigenvalue is called the *spectral gap*.

Lemma 5.16 (See [28]) *The multiplicity of zero as an eigenvalue of L is exactly the number of largest connected components of the graph.*

As a corollary, the algebraic connectivity is the spectral gap if and only if the graph is connected. In this case, rank $L = m - 1$.

With the above preparations, we are ready to prove the main result of this subsection.

Theorem 5.17 *Multiagent system* (5.30) *achieves asymptotic average consensus if the associated dynamic graph is always connected.*

Proof Note that, without loss of generality, we can assume that the system is one-dimensional. In fact, by denoting $y^j = [x_1^j, \ldots, x_m^j]^T$, $j = 1, \ldots, n$, with x_i^j being the jth entry of vector x_i, we can rewrite system (5.32) into

$$\dot{y}^j = -L(t)y^j, \quad j = 1, \ldots, n.$$

The consensus of the multidimensional system is thus converted to that of a set of one-dimensional systems.

Let $\alpha = \sum_{j=1}^m x_j(0)/m$ be the average of the initial states, and $\delta = [x_1 - \alpha, \ldots, x_m - \alpha]^T$ be the group disagreement vector. It follows from $L(t)\mathbf{1} = 0$ that $\dot{\delta} = -L(t)\delta$. Define the Lyapunov-like function

$$V(\delta) = \delta^T \delta.$$

Simple calculation gives

$$\frac{dV}{dt} = -2\delta^T L(t)\delta \leq -2\lambda_2\big(L(t)\big)V \leq -2\lambda_2^* V,$$

where the fact that $\delta^T \mathbf{1} = 0$ has been used, and λ_2^* is the smallest algebraic connectivity of the Laplacian matrices $L(t)$. Note that this number is positive since $L(t) \in \{-A_1, \ldots, -A_m\}$. As a result, $\delta \to 0$ as $t \to +\infty$, and the consensus is thus established. □

Remark 5.18 The theorem provides an elegant criterion on consensus of the multiagent system (5.30) in terms of connectivity of the associated dynamic graph. This exhibits the power of the algebraic graph theory in the analysis of multiagent consensus. In particular, for one-dimensional systems, the sufficient condition is also necessary due to the fact that any disconnected agent at a time will be isolated for any forward time. However, graph connectivity is not necessary for two- or higher-dimensional systems.

5.4.3 A Verifiable Criterion

The criterion in Theorem 5.17 is not verifiable since it requires graph connectivity at all times. As multiagent system (5.30) is piecewise linear with autonomous switching, the connectivity relies totally on the initial agent locations. By applying the idea of graph transition analysis introduced in Sect. 3.3.4, we present here a verifiable consensus criterion in terms of initial network connectivity.

Let $\kappa_i(t)$ denote the degree of node i at time t, i.e., $\kappa_i(t) = |\mathcal{N}_i(t)|$. The dynamic network is said to be *initially connected* if the static graph $(G, \Upsilon(0))$ is connected, and *throughout connected* if the graph $(G, \Upsilon(t))$ is connected for any time $t \in \mathcal{T}_0$.

Define the Lyapunov-like function

$$V(x) = \max_{i,j=1,\dots,m}^{i\neq j} \left\{(x_i - x_j)^T(x_i - x_j) \colon |x_i - x_j| \le r\right\}, \tag{5.33}$$

which is the largest square distance of the connected edges. When no edge exists in the graph, let $V(x) = +\infty$. It is clear that $V(x)$ is always positive unless all x_i's coincide with each other. The function is continuous as long as the interconnection relationship keeps unchanged, that is, when neither new edge is added nor old edge is disconnected.

Let $x(t)$ be the motion of the multiagent system.

The following lemma states a monotone condition for $V(x(t))$ that will be used later.

Lemma 5.19 *Suppose that*

$$\begin{aligned} &\big(x_i(t) - x_j(t)\big)^T\bigg(\sum_{k\in\mathcal{N}_i(t)}\big(x_k(t) - x_i(t)\big) - \sum_{k\in\mathcal{N}_j(t)}\big(x_k(t) - x_j(t)\big)\bigg) \le 0 \\ &\quad \forall (i,j)\ s.t.\ \big|x_i(t) - x_j(t)\big| = \max\big\{\big|x_k(t) - x_l(t)\big| \colon (k,l) \in \Upsilon(t)\big\} \\ &\quad \forall t \in \mathcal{T}_0. \end{aligned} \tag{5.34}$$

Then, $V(x(t))$ is nonincreasing as long as the interconnection relationship keeps unchanged.

Proof Under condition (5.34), function V is nonincreasing as its derivative is nonpositive. As a result, the maximum edge length does not increase when no new edge is added. This leads directly to the conclusion. □

Remark 5.20 The monotonicity of $V(x(t))$ means that, while a connected edge can generally become longer in length, the maximum edge length is always decreasing. A simple yet useful observation is that, if $V(x(t))$ is nonincreasing as long as the interconnection relationship keeps unchanged, it is possible that one or more new edges will appear, but no existing edge will be disconnected. In this case, $V(x(t))$ is always right-continuous, and the number of discontinuities is finite (in fact, upper bounded by $m(m-2)$). Therefore, if $V(x(t))$ is always nonincreasing except at the discontinuous instants, then any initially connected network will keep connected for all the forward time.

Next, with the aid of Lemma 5.19 and Remark 5.20, we are able to establish the main result of this subsection as follows.

Theorem 5.21 *Suppose that the multiagent system is initially connected and satisfies*

$$\kappa_i + \kappa_j \geq \frac{3}{2}m - 2 \quad \forall (i,j) \in \Upsilon \tag{5.35}$$

and

$$\kappa_i + \kappa_j \geq \frac{3}{2}m - 4 \quad \forall (i,j) \notin \Upsilon,\ i \neq j, \tag{5.36}$$

at the initial time. Then, the network is throughout connected, and the asymptotic average consensus is achieved.

Proof First, pick up any node pair (i,j) with $i \neq j$, let ν_i^j be the number of nodes (excluding i and j) that connect to both i and j, and ι_i^j be the number of nodes (excluding i and j) that connect to i but disconnect to j. It is clear that, among the $m-2$ nodes except for i and j, there are $\kappa_i - 1$ that connect to i, and there are $\kappa_j - 1$ that connect with j. According to the Pigeonhole Principle, the number of common neighbors of i and j, ν_i^j, satisfies the inequality

$$\nu_i^j \geq (\kappa_i - 1) + (\kappa_j - 1) - (m-2) = \kappa_i + \kappa_j - m.$$

Similarly, we have

$$\iota_i^j = (\kappa_i - 1) - \nu_i^j \leq m - \kappa_j - 1$$

and

$$\iota_j^i = (\kappa_j - 1) - \nu_i^j \leq m - \kappa_i - 1.$$

Second, observe that, under the conditions of the theorem, we have

$$\nu_i^j + m \geq \kappa_i + \kappa_j \geq 3m - (\kappa_i + \kappa_j) - 4 \quad \forall (i,j) \in \Upsilon \tag{5.37}$$

when the network has the same connectivity as in the initial time. Note that inequality (5.37) further implies that

$$\begin{aligned} \nu_i^j + 2 &\geq 2m - (\kappa_i + \kappa_j) - 2 \\ &= (m - \kappa_i - 1) + (m - \kappa_j - 1) \\ &\geq \iota_i^j + \iota_j^i \quad \forall (i,j) \in \Upsilon. \end{aligned} \tag{5.38}$$

Third, pick any two connected nodes i and j with

$$|x_i - x_j| = \max\{|x_k - x_l| \colon (k,l) \in \Upsilon\}.$$

It can be seen that

$$\begin{aligned} &\sum_{k \in \mathcal{N}_i} (x_k - x_i) - \sum_{k \in \mathcal{N}_j} (x_k - x_j) \\ &\quad = (2 + \nu_i^j)(x_j - x_i) + \sum_{k \in \psi_i^j} (x_k - x_i) - \sum_{k \in \psi_j^i} (x_k - x_j), \end{aligned}$$

where ψ_i^j is the set of nodes (excluding i and j) that connect to i but disconnect to j. It follows that

$$\begin{aligned} &(x_i - x_j)^T \Big(\sum_{k \in \mathcal{N}_i} (x_k - x_i) - \sum_{k \in \mathcal{N}_j} (x_k - x_j) \Big) \\ &\quad \leq -(2 + m_i^j - \iota_i^j - \iota_j^i)|x_i - x_j|^2 \leq 0 \end{aligned} \tag{5.39}$$

when the network has the same connectivity as that in the initial time.

Finally, suppose that the multiagent system is initially connected and relationship (5.35) initially holds. Then, it follows from (5.39) and Lemma 5.19 that each edge will keep connected until new edges add into the network graph. A further implication is that each node degree is nondecreasing and inequality (5.35) still holds for old edges. On the other hand, inequality (5.36) guarantees that each newly added edge also satisfies (5.35). The above reasonale shows that relationships between (5.35) and (5.36) hold for all forward time, which further implies throughout connectivity under the assumption of initial network connectivity. It follows from Theorem 5.17 that the system achieves asymptotic average consensus. □

Remark 5.22 The theorem presents an easily verifiable sufficient condition, which only relies on initial network graph, for consensus of the multiagent system under the linear feedback protocol. It is interesting to note that the set of initial configurations under the theorem condition is in fact invariant and attractive. To be more precise, define the region

$$\Big\{x : \forall i \neq j,\ \#\big\{k \notin (i,j) : |x_i - x_k| \leq r\big\} + \#\big\{k \notin (i,j) : |x_j - x_k| \leq r\big\} \geq \frac{3}{2}m - 4\Big\},$$

where $\#\{\cdot\}$ denotes the cardinality of a set. The region is attractive in that, once the multiagent system state enters into the region, it will never leave the region for any future time. Moreover, any consensus state trajectory must enter into and stay inside the region due to the fact that the consensus equilibrium is an interior point of the region.

Corollary 5.23 *A sufficient condition for throughout connectivity is the initial connectivity and that each initial node degree is not less than* $\frac{3}{4}m - 1$.

The corollary reveals an important fact: if the node degree exceeds a certain threshold (three fourth of the graph order), no edge will be disconnected, and the throughout connectivity reduces to initial connectivity. While conservative, this result is simple and verifiable, and it readily applies to a phase of any consensus process.

Next, we apply the approach to the multiagent networks with five or less nodes.

The case of two-node networks is trivial, and consensus is equivalent to initial connectedness. For three-node networks, if the network is initially connected, then the consensus follows from Theorem 5.21. Otherwise, reaching consensus is equivalent to the existence of an initial edge, and the (least) distance between the initially isolated state to the edge is equal to or less than r.

The case of four-node networks is more interesting, to which the proposed approach applies either directly or indirectly. Figure 5.9 demonstrates the types of connected networks, where (a)–(e) verifies Theorem 5.21. For a system with initial interaction (f), it is clear that the sum of any two node degrees is less than or equal to four, which verifies inequality (5.38). This means that the connection will keep unchanged unless a new edge appears. On the other hand, the appearance of any new edge leads the network to satisfy (5.35) and (5.36). Therefore, any multiagent system with network (f) reaches consensus under the linear control protocol.

The above analysis is summarized into the following proposition.

Proposition 5.24 *For a multiagent network with four or less nodes, initial connectedness is equivalent to throughout connectedness.*

Finally, let us examine the multiagent networks with five nodes, which exhibit more diverse and complex dynamics. When the size of initial graph is fourteen or more, the network is always initially connected. In particular, if the size is sixteen or

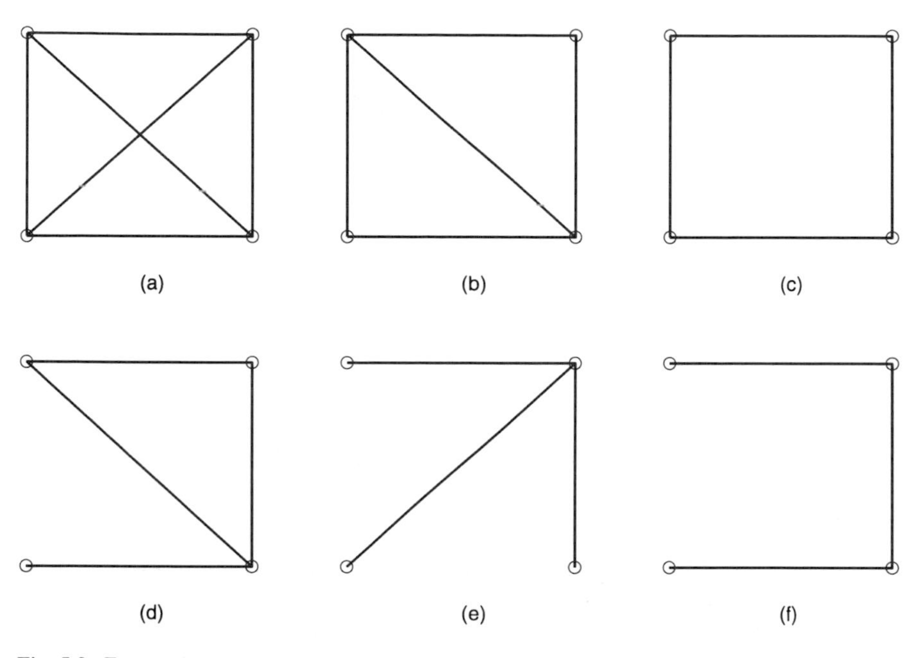

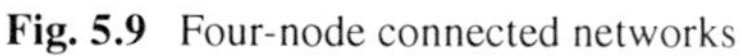

Fig. 5.9 Four-node connected networks

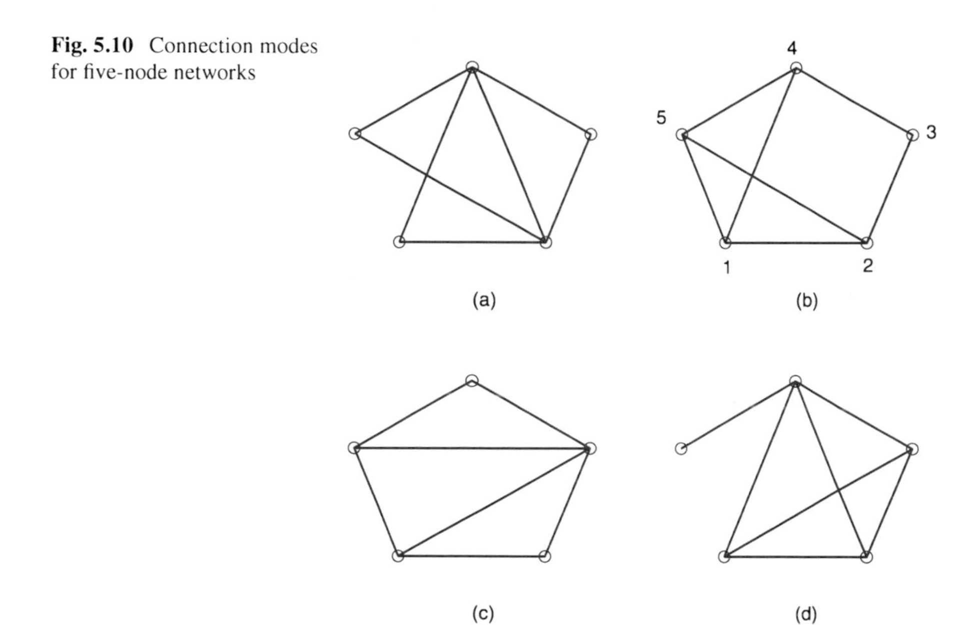

Fig. 5.10 Connection modes for five-node networks

more, then Theorem 5.21 applies, and the network is throughout connected. When the size is exactly fourteen, there are four connection types, which are depicted in Fig. 5.10. It can be easily verified that Theorem 5.21 applies to case (a).

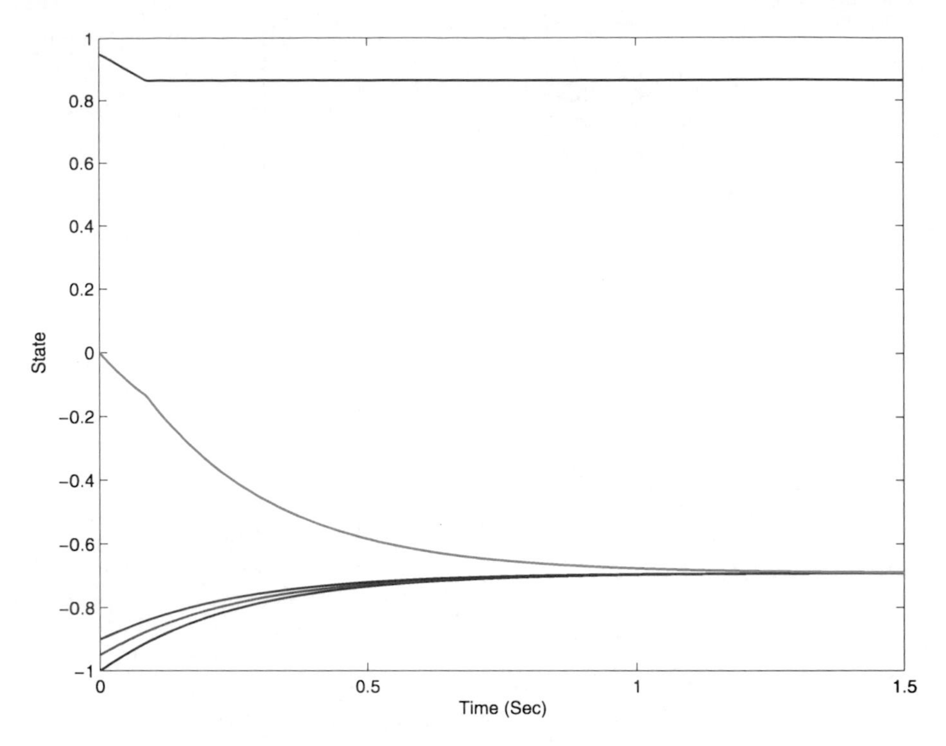

Fig. 5.11 Initial connectivity with no consensus

For case (b), the theorem is not applicable, but an analysis can be conducted based on Lemma 5.19. In fact, by (5.38), any edge other than (2, 3) and (3, 4) will keep connected if the interconnection keep unchanged. For edge (2, 3), calculate

$$\begin{aligned}\frac{d}{dt}|x_3 - x_2|^2 &= 2(x_3 - x_2)^T\big[(x_4 - x_3) + (x_2 - x_3)\\ &\quad + (x_2 - x_1) + (x_2 - x_3) + (x_2 - x_5)\big]\\ &= -6\big|(x_3 - x_2)\big|^2 + 2(x_3 - x_2)^T\big[(x_2 - x_1) + (x_4 - x_5)\big],\end{aligned}$$

which means that the length of edge (2, 3) will strictly decrease when the edge length is the largest among all the edges, and the edge will keep connected if the interconnection keep unchanged. The same property holds for edge (3, 4). On the other hand, the introduction of any new edge leads the network to verify Theorem 5.21. As a result, the network is throughout connected in this case.

System with initial network (c) is throughout connected due to the facts that inequality (5.38) holds for each edge and Theorem 5.21 applies when new edges appear.

Network (d) is most interesting as disconnectedness does might happen. To see this, we take a one-dimensional system with

$$x_1(0) = -1, \qquad x_2(0) = -0.95, \qquad x_3(0) = -0.9,$$

$$x_4(0) = 0, \qquad x_5(0) = 0.95, \qquad r = 1.$$

It can be verified that the system admits network (d) as initial graph. Figure 5.11 shows the state trajectory, where node 5 becomes isolated at time 0.083 sec. It can be seen that, as neighbors of x_4, x_5 imposes x_4 going up, but x_1, x_2, and x_3 impose x_4 going down. As a result of the "competition", x_4 goes down and becomes invisible from x_5. This makes x_5 isolated.

5.5 Stabilizing Design of Controllable Switched Linear Systems

5.5.1 Problem Formulation

A practical motivation for studying hybrid dynamical systems stems from the fact that the hybrid control scheme provides an effective approach for controlling highly nonlinear complex dynamical systems. Indeed, while linear design techniques are widely used in control system synthesis, as far as nonlinear dynamics are concerned, in practice a linear controller is only valid around a specific operating point. It is thus a common practice to design more than one linear controller, each at a different operating region and a switching mechanism that coordinates the switching among them.

Consider the switched linear control system given by

$$\dot{x}(t) = A_{\sigma(t)}x(t) + B_{\sigma(t)}u_{\sigma(t)}(t), \tag{5.40}$$

where $x(t) \in \mathbf{R}^n$ is the continuous state, $\sigma(t) \in \{1, \dots, m\}$ is the switching signal, $u_i(t) \in \mathbf{R}^{p_i}$ is the control input, $A_i \in \mathbf{R}^{n\times n}$ and $B_i \in \mathbf{R}^{n\times p_i}$, $i \in \{1, \dots, m\}$, are real constant matrices. Denote the continuous state by $\phi(t; 0, x, u, \sigma)$.

Suppose that the system is completely controllable, that is, for any state $x \in \mathbf{R}^n$, there exist a time $t_f \geq 0$, a switching law $\sigma : [0, t_f] \mapsto \{1, \dots, m\}$, and inputs $u_i : [0, t_f] \mapsto \mathbf{R}^{p_i}$, $i \in \{1, \dots, m\}$, such that solution of the continuous state at t_f is the origin, i.e., $\phi(t_f; 0, x, u, \sigma) = 0$.

The problem of exponential stabilization is to find a control law and a switching law such that the switched system is exponentially stable. To be more precise, find a pathwise state-feedback switching law and a state-feedback control law that steer the controllable switched system exponentially stable.

5.5.2 Multilinear Feedback Design

In this subsection, we investigate the possibility of solving the stabilization problem with a feedback control scheme. The main idea is to associate with each subsystem a set of candidate linear controllers such that the extended switched system is stabilizable. We thus can apply the design approach presented in Sect. 4.4 for calculating a pathwise state-feedback switching law for the extended system.

Suppose that F_i^j, $j=1,\dots,k_i$, are linear gain matrices associated with the ith subsystem. Then, with the linear controller

$$u_i = F_i x,\quad F_i \in \{F_i^1,\dots,F_i^{k_i}\},$$

the switched system is extended to

$$\dot{x} = C_\varsigma x, \tag{5.41}$$

where $\varsigma \in \{1,\dots,\sum_{i=1}^m k_i\}$, $C_l = A_i + B_i F_i^j$ for $i=1,\dots,m,\ j=1,\dots,k_i$, and $l=\sum_{\mu=1}^{i-1} k_\mu + j$.

Definition 5.25 Switched system (5.40) is *multifeedback exponentially stabilizable* if there exist natural numbers $k_1,\dots,k_m$ and gain matrices F_i^j, $i=1,\dots,m$, $j=1,\dots,k_i$, such that the extended forced-free switched system (5.41) is exponentially stabilizable. In this case, the gain matrix set $\{F_i^j : i=1,\dots,m,\ j=1,\dots,k_i\}$ is a solution of the exponential stabilization problem.

To proceed, we introduce the concept of unit sphere switched contractility.

Definition 5.26 Switched system (5.40) is *unit sphere switched contractive*, if for any state x with unit norm, there exist a time $T>0$, an input u, and a switching signal σ such that $|\phi(T;0,x,u,\sigma)|<1$.

Lemma 5.27 *The switched linear system is multifeedback exponentially stabilizable iff it is unit sphere switched contractive w.r.t. any given norm.*

Proof It is clear that multifeedback exponential stabilizability implies unit sphere switched contractility. Thus, we need only to prove the converse.

For any arbitrarily given but fixed state x with unit norm, let T_x, u, and σ be such that $|\phi(T_x;0,x,u,\sigma)|=\rho_x<1$. Without loss of generality, we assume that state curve $\{y(t)=\phi(t;0,x,u,\sigma): t\in[0,T_x]\}$ does not pass by the origin, otherwise redefine T_x to be

$$\min\{t : |\phi(t;0,x,u,\sigma)|=\rho_x\}.$$

Let $i_t = \arg\max_{j=1}^n |y_j(t)|$. It can be seen that i_t is piecewise constant and y_{i_t} is nonzero and piecewise continuous w.r.t. t. For any $t\in[0,T_x]$, denote $z=y(t)$, and define a $p_i\times n$ matrix F^z by

$$F^z(j,k) = \begin{cases} (u_{\sigma(t)}(t))_j/z_k & \text{if } k=i_t,\\ 0 & \text{otherwise.}\end{cases}$$

It is clear that F^z is well defined and piecewise continuous w.r.t. z. In addition, $u_{\sigma(t)}(t) = F^z x(t)$. Therefore, we have

$$\dot{y}(t) = \left(A_{\sigma(t)} + B_{\sigma(t)} F^{y(t)}\right) y(t).$$

According to the continuous dependence of initial state, there exist a natural number N, a time sequence $0 = t_0 < t_1 < t_2 < \cdots < t_N < T_x$, and a constant matrix sequence $F_0^x, F_1^x, \ldots, F_N^x$ such that solution of the linear time-varying system

$$\dot{w} = \left(A_{\sigma(t)} + B_{\sigma(t)} F_k^x\right) w(t), \quad t \in [t_k, t_{k+1}),\ k = 0, 1, \ldots, N,$$
$$w(0) = x, \quad t_{N+1} = T_x,$$

satisfies $|w(T_x) - y(T_x)| \le \frac{1-\rho_x}{2}$. Assume without loss of generality that the switching times of σ in $[0, T_x]$ are in the set $\{t_1, \ldots, t_N\}$, otherwise just incorporate the times into the set. As

$$w(T_x) = \exp\left(\left(A_{\sigma(t_0)} + B_{\sigma(t_0)} F_0^x\right)(t_1 - t_0)\right) \cdots$$
$$\times \exp\left(\left(A_{\sigma(t_N)} + B_{\sigma(t_N)} F_N^x\right)(t_{N+1} - t_N)\right) x \stackrel{\text{def}}{=} \Phi_x x,$$

we have $|\Phi_x x| \le \frac{1+\rho_x}{2} < 1$.

To summarize, for any state on the unit sphere, there exist a finite number of gain matrices such that the associated state trajectory is norm contractive. By the standard arguments as in [234, Theorem 3.9], there exist multilinear feedback controllers for each input channel such that the extended forced-free switched linear system is exponentially stable. This completes the proof. □

It is clear that complete controllability implies unit sphere switched contractility. According to Lemma 5.27, any controllable switched system is multifeedback exponentially stabilizable.

Next, we develop a design procedure to compute a set of multifeedback gains such that the extended switched system is exponentially stabilizable.

Sample the switched system into a discrete-time switched linear system

$$z_{k+1} = C_\varrho^\tau x + D_\varrho^\tau u, \tag{5.42}$$

where $\tau > 0$ is the sampling period, $\varrho \in \{1, \ldots, m\}$ is the switching signal, and $C_i^\tau = e^{A_i \tau}$, $D_i^\tau = \int_0^\tau \exp(A_i t)\,dt B_i$, $i = 1, \ldots, m$. It has been known that the sampled system is controllable under almost all period τ (see, e.g., [234, Lemma 4.53]). Pick up such a τ. From the controllability of the sampled system, we can find a natural number k and an index sequence $i_1, \ldots, i_k$ such that

$$\operatorname{rank}\left[D_{i_k}^\tau, C_{i_k}^\tau D_{i_{k-1}}^\tau, \ldots, C_{i_k}^\tau \cdots C_{i_2}^\tau D_{i_1}^\tau\right] = n. \tag{5.43}$$

Fix a Schur stable matrix E. By perturbation analysis, for almost all matrices $\tilde{E}^\tau$, the equation

$$E + \tilde{E}^\tau = \left(C_{i_k}^\tau + D_{i_k}^\tau F_k^\tau\right) \cdots \left(C_{i_1}^\tau + D_{i_1}^\tau F_1^\tau\right)$$

admits at least one solution $(F_1^\tau, \dots, F_k^\tau)$ which satisfies the relationships

$$E + \tilde{E}^\tau - C_{i_k}^\tau \cdots C_{i_1}^\tau = \left[D_{i_k}^\tau, C_{i_k}^\tau D_{i_{k-1}}^\tau, \dots, C_{i_k}^\tau \cdots C_{i_2}^\tau D_{i_1}^\tau\right] \begin{bmatrix} \bar{F}_k^\tau \\ \vdots \\ \bar{F}_1^\tau \end{bmatrix} \tag{5.44}$$

and

$$\begin{aligned} \bar{F}_1^\tau &= F_1^\tau, \\ \bar{F}_2^\tau &= F_2^\tau (C_{i_1}^\tau + D_{i_1}^\tau F_1^\tau), \\ &\vdots \\ \bar{F}_k^\tau &= F_k^\tau (C_{i_{k-1}}^\tau + D_{i_{k-1}}^\tau F_{k-1}^\tau) \cdots (C_{i_1}^\tau + D_{i_1}^\tau F_1^\tau). \end{aligned} \tag{5.45}$$

On the other hand, the nonsingularity of E implies the nonsingularity of $(C_{i_j}^\tau + D_{i_j}^\tau F_j^\tau)$, which in turn implies the solvability of F_j^τ by means of $\bar{F}_j^\tau$ in (5.45). A useful observation here is that, even when $k \geq n$, it is always possible to solve (5.44) with at most n nonzero F_j^τ's.

Proposition 5.28 *Suppose that* $\lim_{\tau \to 0} \tilde{E}^\tau = 0$. *For sufficiently small* τ, *any solution* $(F_1^\tau, \dots, F_k^\tau)$ *of* (5.44) *and* (5.45) *is a solution of multifeedback stabilization problem of the continuous-time switched linear system.*

The proof is simply based on the observation that, as $\tau \to 0+$, we have

$$e^{(A_{i_k} + B_{i_k} F_k^\tau)\tau} \cdots e^{(A_{i_1} + B_{i_1} F_1^\tau)\tau} \to E.$$

As a result, the switched system with the multifeedback controllers is stabilizable by means of periodic switching signals (see, e.g., [234, Corollary 3.12]).

While the extended switched system is consistently stabilizable by a periodic switching signal, the periodic switching may not be practically applaudable. Indeed, the switched system with such a switching is possibly with poor transient performance and small stability margin. For this, we use the state-feedback pathwise switching mechanism instead. The computational procedure for finding the state-feedback pathwise switching law can be found in Sect. 4.4.2.

To summarize, to achieve stability within the multilinear feedback control scheme, we need to implement the following steps.

(1) Compute a set of multilinear feedback control gain matrices.
  (1.1) Sample the switched system as in (5.42).
  (1.2) Solve (5.44) and (5.45).
(2) Compute a state-feedback pathwise switching law.

Example 5.29 Consider system (5.40) with $n = 3$, $m = 2$, and

$$\begin{aligned} A_1 &= \begin{bmatrix} 0 & 0 & 0 \\ 1 & 1 & 0 \\ 0 & 0 & 1 \end{bmatrix}, & B_1 &= \begin{bmatrix} 1 \\ 0 \\ 0 \end{bmatrix}, \\ A_2 &= \begin{bmatrix} 0 & 0 & 0 \\ 0 & 1 & 0 \\ 1 & 0 & 1 \end{bmatrix}, & B_2 &= \begin{bmatrix} 0 \\ 0 \\ 0 \end{bmatrix}. \end{aligned} \tag{5.46}$$

It can be verified that the system is completely controllable. However, it has been shown that the system cannot be stabilized by means of (single) linear feedback controllers [234, Example 5.23].

We are to design a multilinear feedback control law and a state-feedback pathwise switching law that achieve exponential stability. For this, the first step is to sample the continuous-time system into a discrete-time controllable switched system. Choose the sampling rate to be $\tau = 0.2$ sec, and let

$$E = \begin{bmatrix} 0.1 & 0 & 0 \\ 0 & 0.1 & -0.1 \\ -0.2 & 0 & 0.1 \end{bmatrix} \quad \text{and} \quad \tilde{E} = 0_{3\times 3}.$$

By solving (5.44) and (5.45) for the sampled-data system, we obtain the feedback gain matrices

$$\begin{aligned} F_1 &= [-8.0276, 0, -32.1765], \\ F_2 &= [-7.3529, -35.5013, 1.8918], \\ F_3 &= [-2.6700, 21.9002, -4.0833]. \end{aligned}$$

The extended forced-free system has four subsystems with matrices $C_i = A_1 + B_1 F_i$, $i = 1, 2, 3$, and $C_4 = A_2$, respectively.

The next step is to compute a state-feedback pathwise switching law. For this, apply the computational algorithm for stabilizing switching design presented in Sect. 4.4.2. Choose $\tau = 0.3$ sec (note that this sampling rate is not equal to that of multilinear feedback design). It can be verified that the four switching paths

$$\vartheta_1 = (1), \qquad \vartheta_2 = (2), \qquad \vartheta_3 = (3, 2, 4, 1), \qquad \vartheta_4 = (3, 2, 4, 1, 3, 2, 4, 1)$$

are piecewise contractive w.r.t. the ℓ_1-norm. By implementing the multilinear feedback control law and the state-feedback pathwise switching law, we obtain an extended switched system that is exponentially stable. Figure 5.12 depicts sample state/input/switching trajectories of the switched system with initial state $x_0 = [-1\ 1.5\ 1]^T$. The switching signal is taken from the extended system to accommodate the multilinear feedbacks (for instance, when the first subsystem is activated with the third linear feedback controller, the switching signal is equal to 3).

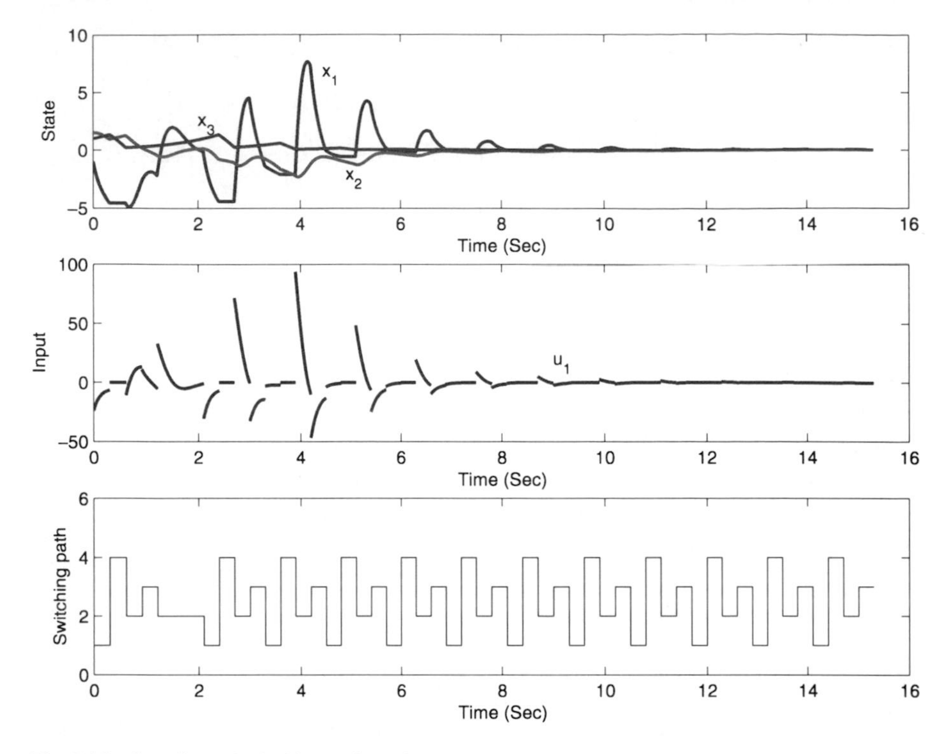

Fig. 5.12 State/input/switching trajectories

5.6 Notes and References

Absolute stability of Lur'e systems has long been a stimulating motivation for nonlinear control theory. The problem is notoriously difficult, and it is still among the most famous open problems in the control society. Nevertheless, remarkable progress has been made so far, especially for lower-dimensional systems. In particular, Pyatnitskiy and Rapoport presented a necessary and sufficient condition for absolute stability of second- and third-order systems [194, 196]. Technically, they characterized the "most destabilizing" phase portrait using variational calculus, and this amounts to solving a nonlinear equation with three unknowns that no efficient numerical verification is available in general. The idea of constructing the so-called most destabilizing nonlinearity is clearly related to finding a worst possible switching strategy that makes the extreme switched linear system diverging with a largest rate. Along this line, several new characterizations were carried out [44, 112, 162, 163] that solve the problem of absolute stability for planar Lur'e systems in a computational point of view. Other related and interesting progress can be found in [165, 207].

The algebraic criterion, Theorem 5.1 in Sect. 5.1.1, was borrowed from [17, 18, 44]. The application to the problem of absolute stability in Sect. 5.1.2 is straightforward.

Switching-based adaptive control can be traced back to the pioneering work by Fu and Barmish [84], who proposed an adaptive control scheme with multiple controllers and a sequential switching logic that disconnects the "wrong" controllers by evaluating the performance index. Since the 1990s, the adaptive stabilization via hybrid control has witnessed a great progress, mostly in the framework of "logic-based supervisory control" developed by Morse and his coworkers [107, 177, 178]. While the design of candidate controllers/estimators is somewhat standard [6], the design of switching logics could be quite involved [109, 111, 178, 183]. It has been shown that a properly designed switching logic can not only stabilize a system with large-scale unknown parameters, but also improve the performances, for instance, transient response [182] and robustness [9, 179]. For more recent progress, the reader is referred to [19, 214] and the references therein.

The switching adaptive scheme presented in Sect. 5.2 is a simplified and combined version taken from [6, 109]. In particular, Sect. 5.2.1 was adopted from [6], and Sect. 5.2.2 was adopted from [109]. Lemma 5.18 was taken from [108], and the numerical example in Sect. 5.2.3 was taken from [6].

A fuzzy system is a dynamical system based on fuzzy logic, which was first initiated by Zadeh in the 1960s [272, 273]. Several types of fuzzy models have been proposed, among which the T–S fuzzy systems (known also as Type-3 fuzzy systems) have gained more and more attention and recognition [215, 244]. A T–S fuzzy system uses fuzzy rules to describe the global nonlinear dynamics by means of a set of local linear modes that are smoothly connected by fuzzy membership functions. The clear two-level "execution-supervisor" structure makes it powerful in approximating general nonlinear systems and flexible in high-level design [51, 248]. Note that if we take the IF-THEN fuzzy logic as a switching logic, then a T–S system is closely related to the corresponding switched system. Therefore, stability analysis for fuzzy systems can be benefited from stability analysis for switched systems. Much earlier effort has been paid to present stability criteria based on common quadratic Lyapunov functions [247], and later it was extended to the piecewise quadratic Lyapunov approach that may lead to less conservative criteria [76, 127]. Most criteria can be expressed in terms of linear matrix inequalities (LMIs), which permits efficient numerical solvers [46]. Despite a remarkable progress made so far, stability analysis is still a challenging problem to be solved. The reader is referred to the surveys [77, 248] and the references therein for more details.

The framework and stability criteria in Sect. 5.3 were mainly adopted from [76, 127]. The piecewise switched linear system is a straightforward extension of a piecewise system with a set of switched linear systems as subsystems. Proposition 5.12 simply combines the piecewise quadratic Lyapunov function approach with transition analysis among the partitions.

The consensus or agreement problem for multiagent systems is a typical problem in distributed computational intelligence, which is to achieve a common value asymptotically by means of proper cooperative control among the agents in a distributed manner. Numerous linear/nonlinear protocols have been proposed to solve the problem in the literature. In particular, with the aid of the Laplacian matrix theory, the problem was tackled in an elegant manner by means of linear protocols for multiagent systems with either static or dynamic connections [75, 153, 187]. A multiagent system with a linear protocol is exactly a piecewise linear system, and the consensus analysis can be conducted based on stability of piecewise linear systems. It was established that, under the linear feedback protocol and the assumption that the network is always connected, the system admits a common quadratic Lyapunov function, and asymptotic consensus is achieved. This was summarized in Sect. 5.4.2, where a first-order differential equation was used for simplification.

While the connectedness assumption in Theorem 5.17 can be further relaxed in various manners (see, e.g., [113, 120, 175]), the verification of the connectivity assumptions are not tractable for dynamic network interconnections. To tackle this issue, one way is to seek for nonlinear protocols as in [121, 275]. Alternatively, by combining the transition analysis presented in Sect. 3.3.4 and a novel Lyapunov-like function method, we can estimate a set of initial configurations that is in fact invariant and attractive, and hence the requirement of throughout connectivity can be removed. This was first presented in [237], from which Sect. 5.4.3 was adopted.

For switched linear control systems, the problem of stabilization by means of simultaneous switching/control design is both important and challenging. A primary reason for this lies in the fact that a nontrivial gap exists between controllability and feedback stabilizability. To be more precise, it was revealed that controllability does not imply feedback stabilizability if one linear feedback controller is attached to a subsystem [234]. However, when more than one controller is allowed to connect with a subsystem, controllability does imply stabilizability [242, 268, 269]. Therefore, a natural way to address the problem of stabilization, as adopted here in Sect. 5.5.2, is first to introduce multiple controllers that not necessarily connect with the subsystems in a one-to-one manner and then to design a state-feedback path-wise switching law for the (extended) switched system. The notion of unit sphere switched contractility given in Definition 5.26 and its connection with stabilizability were borrowed from [228]. The design procedure of calculating multifeedback gains was a combination of that presented in [242, 268]. For more recent progress on the problem of stabilization, the reader is referred to [238, 243, 279].

References

1. Agrachev AA, Liberzon D. Lie-algebraic stability criteria for switched systems. SIAM J Control Optim. 2001;40(1):253–69.
2. Aguiar AP, Hespanha JP, Pascoal AM. Switched seesaw control for the stabilization of underactuated vehicles. Automatica. 2007;43(12):1997–2008.
3. Aizerman MA. On a problem concerning stability "in large" of the dynamic systems. Usp Mat Nauk. 1949;4(4):186–8.
4. Aizerman MA, Gantmakher FR. Absolyutnaya ustoichivost reguliruemykh sistem (Absolute stability of the controlled systems). Moscow: Akad Nauk SSSR; 1963.
5. Al'pin YA, Ikramov KD. Reducibility theorems for pairs of matrices as rational criteria. Linear Algebra Appl. 2000;313(1–3):155–61.
6. Anderson BDO, Brinsmead TS, De Bruyne F, Hespanha JP, Liberzon D, Morse AS. Multiple model adaptive control, part 1: finite controller coverings. Int J Robust Nonlinear Control. 2000;10(11–12):909–29.
7. Ando T, Shih M-H. Simultaneous contractibility. SIAM J Matrix Anal Appl. 1998;19(2): 487–98.
8. Angeli D. A note on stability of arbitrarily switched homogeneous systems. Preprint; 1999.
9. Angeli D, Mosca E. Lyapunov-based switching supervisory control of nonlinear uncertain systems. IEEE Trans Autom Control. 2002;47(3):500–5.
10. Arnold L, Wihstutz V, editors. Lyapunov exponents. New York: Springer; 1986.
11. Asarin E, Bournez O, Dang T, Maler O. Approximate reachability analysis of piecewise-linear dynamical systems. In: Lynch N, Krogh BH, editors. Hybrid systems: computation and control. Berlin: Springer; 2000. p. 20–31.
12. Astrom KJ, Wittenmark B. Adaptive control. 2nd ed. Eaglewood Cliffs: Prentice Hall; 1995.
13. Aubin JP, Cellina A. Differential inclusions. Berlin: Springer; 1984.
14. Auslander J, Seibert P. Prolongations and stability in dynamical systems. Ann Inst Fourier (Grenoble). 1964;14:237–68.
15. Bacciotti A. Stabilization by means of state space depending switching rules. Syst Control Lett. 2004;53(3–4):195–201.
16. Bacciotti A, Mazzi L. A discussion about stabilizing periodic and near-periodic switching signals. In: Proc IFAC NOLCOS; 2010. p. 250–5.
17. Balde M, Boscain U. Stability of planar switched systems: the nondiagonalizable case. Commun Pure Appl Anal. 2008;7:1–21.
18. Balde M, Boscain U, Mason P. A note on stability conditions for planar switched systems. Int J Control. 2009;82(10):1882–8.
19. Baldi S, Battistelli G, Mosca E, Tesi P. Multi-model unfalsified adaptive switching supervisory control. Automatica. 2010;46(2):249–59.
20. Baotic M, Christophersen FJ, Morari M. Constrained optimal control of hybrid systems with a linear performance index. IEEE Trans Autom Control. 2006;51(12):1903–19.

Z. Sun, S.S. Ge, *Stability Theory of Switched Dynamical Systems*,
Communications and Control Engineering,
DOI 10.1007/978-0-85729-256-8,

21. Barvinok A. A course in convexity. Providence: Am Math Soc; 2002.
22. Bell JP. A gap result for the norms of semigroups of matrices. Linear Algebra Appl. 2005;402(1–3):101–10.
23. Bemporad A, Morari M. Control of systems integrating logic, dynamics, and constraints. Automatica. 1999;35(3):407–27.
24. Bengea SC, DeCarlo RA. Optimal control of switching systems. Automatica. 2005;41(1):11–27.
25. Berger M, Wang Y. Bounded semigroups of matrices. Linear Algebra Appl. 1992;166:21–7.
26. Bertsekas DP. Dynamic programming and optimal control, vol II. 3rd ed. Nashua: Athena Scientific; 2010.
27. Bharucha BH. On the stability of randomly varying systems. PhD dissertation, Dept Elec Eng, Univ Calif Berkeley; 1961.
28. Biggs N. Algebraic graph theory. Cambridge: Cambridge University Press; 1993.
29. Biswas P, Grieder P, Lofberg J, Morari M. A survey on stability analysis of discrete-time piecewise affine systems. In: Proc IFAC World Congress, Prague, Czech Republic; 2005.
30. Blanchini F. Nonquadratic Lyapunov function for robust control. Automatica. 1995; 31(3):451–61.
31. Blanchini F. Set invariance in control. Automatica. 1999;35(11):1747–67.
32. Blanchini F. The gain scheduling and the robust state feedback stabilization problems. IEEE Trans Autom Control. 2000;45(11):2061–70.
33. Blanchini F, Miani S. On the transient estimate for linear systems with time-varying uncertain parameters. IEEE Trans Circuits Syst I, Fundam Theory Appl. 1996;43(7):592–6.
34. Blanchini F, Miani S. A new class of universal Lyapunov functions for the control of uncertain linear systems. IEEE Trans Autom Control. 1999;44(3):641–7.
35. Blanchini F, Savorgnan C. Stabilizability of switched linear systems does not imply the existence of convex Lyapunov functions. In: Proc IEEE CDC; 2006. p. 119–24.
36. Blanchini F, Savorgnan C. Stabilizability of switched linear systems does not imply the existence of convex Lyapunov functions. Automatica. 2008;44(4):1166–70.
37. Blondel VD, Nesterov Yu. Computationally efficient approximations of the joint spectral radius. SIAM J Matrix Anal Appl. 2005;27(1):256–72.
38. Blondel VD, Nesterov Yu, Theys J. On the accuracy of the ellipsoidal norm approximation of the joint spectral radius. Linear Algebra Appl. 2005;394:91–107.
39. Blondel VD, Theys J, Vladimirov AA. An elementary counterexample to the finiteness conjecture. SIAM J Matrix Anal Appl. 2003;24(4):963–70.
40. Blondel VD, Tsitsiklis JN. Complexity of stability and controllability of elementary hybrid systems. Automatica. 1999;35(3):479–89.
41. Blondel VD, Tsitsiklis JN. A survey of computational complexity results in systems and control. Automatica. 2000;36(9):1249–74.
42. Bolzern P, Colaneria P, De Nicolaob G. On almost sure stability of continuous-time Markov jump linear systems. Automatica. 2006;42(6):983–8.
43. Bolzern P, Colaneria P, De Nicolaob G. Markov jump linear systems with switching transition rates: mean square stability with dwell-time. Automatica. 2010;46(6):1081–8.
44. Boscain U. Stability of planar switched systems: the linear single input case. SIAM J Control Optim. 2002;41:89–112.
45. Bousch T, Mairesse J. Asymptotic height optimization for topical IFS, Tetris heaps and the finiteness conjecture. J Am Math Soc. 2002;15(1):77–111.
46. Boyd S, El Ghaoui L, Feron E, Balakrishnan V. Linear matrix inequalities in systems and control theory. Philadelphia: SIAM; 1994.
47. Brayton RK, Tong CH. Stability of dynamic systems: a constructive approach. IEEE Trans Circuits Syst. 1979;26(4):224–34.
48. Brayton RK, Tong CH. Constructive stability and asymptotic stability of dynamic systems. IEEE Trans Circuits Syst. 1980;27(11):1121–30.
49. Camlibel MK, Heemels WPMH, Schumacher JM. Algebraic necessary and sufficient conditions for the controllability of conewise linear systems. IEEE Trans Autom Control. 2008;53(3):762–74.

50. Camlibel MK, Pang JS, Shen J. Conewise linear systems: non-Zenoness and observability. SIAM J Control Optim. 2006;45(5):1769–800.
51. Cao SG, Rees NW, Feng G. Universal fuzzy controllers for a class of nonlinear systems. Fuzzy Sets Syst. 2001;122(1):117–23.
52. Cardim R, Teixeira MCM, Assuncao E, Covacic MR. Variable-structure control design of switched systems with an application to a DC–DC power converter. IEEE Trans Ind Electron. 2009;56(9):3505–13.
53. Cheng DZ, Guo L, Huang J. On quadratic Lyapunov functions. IEEE Trans Autom Control. 2003;48(5):885–90.
54. Cheng DZ, Guo L, Lin YD, Wang Y. Stabilization of switched linear systems. IEEE Trans Autom Control. 2005;50(5):661–6.
55. Cheng DZ, Lin YD, Wang Y. Accessibility of switched linear systems. IEEE Trans Autom Control. 2006;51(9):1486–91.
56. Chesi G, Colaneri P, Geromel JC, Middleton R, Shorten R. On the minimum dwell time for linear switching systems. In: Proc ACC; 2010. p. 2487–92.
57. Chesi G, Garulli A, Tesi A, Vicino A. Homogeneous polynomial forms for robustness analysis of uncertain systems. Berlin: Springer; 2009.
58. Clarke FH, Ledyaev YS, Sontag ED, Subbotin AI. Asymptotic controllability implies feedback stabilization. IEEE Trans Autom Control. 1997;42(10):1394–407.
59. Costa OLV, Fragoso MD, Marquez RP. Discrete-time Markov jump linear systems. London: Springer; 2005.
60. Daafouz J, Riedinger P, Iung C. Stability analysis and control synthesis for switched systems: a switched Lyapunov function approach. IEEE Trans Autom Control. 2002;47(11):1883–7.
61. Dai XP, Huang Y, Xiao MQ. Almost sure stability of discrete-time switched linear systems: a topological point of view. SIAM J Control Optim. 2008;47(4):2137–56.
62. Dai XP, Huang Y, Xiao MQ. Criteria of stability for continuous-time switched systems by using Liao-type exponents. SIAM J Control Optim. 2010;48(5):3271–96.
63. Daubechies I, Lagarias JC. Corrigendum/addendum to: sets of matrices all infinite products of which converge. Linear Algebra Appl. 2001;327(1–3):69–83.
64. Dayawansa WP, Martin CF. A converse Lyapunov theorem for a class of dynamical systems which undergo switching. IEEE Trans Autom Control. 1999;44(4):751–60.
65. De Jong H, Gouze J-L, Hernandez C, Page M, Sari T, Geiselmann J. Qualitative simulation of genetic regulatory networks using piecewise-linear models. Bull Math Biol. 2004;66: 301–40.
66. De Persis C, De Santis R, Morse AS. Switched nonlinear systems with state-dependent dwell-time. Syst Control Lett. 2003;50(4):291–302.
67. De Santis E, Di Benedetto MD, Pola G. A structural approach to detectability for a class of hybrid systems. Automatica. 2009;45(5):1202–6.
68. Drenick R, Shaw L. Optimal control of linear plants with random parameters. IEEE Trans Autom Control. 1964;9(3):236–44.
69. Elsner L. The generalized spectral-radius theorem: an analytic-geometric proof. Linear Algebra Appl. 1995;220:151–8.
70. Fang Y. Stability analysis of linear control systems with uncertain parameters. PhD dissertation, Dept Syst Contr Ind Eng, Case Western Reserve Univ; 1994.
71. Fang Y. A new general sufficient condition for almost sure stability of jump linear systems. IEEE Trans Autom Control. 1997;42:378–82.
72. Fang Y, Loparo KA. Stochastic stability of jump linear systems. IEEE Trans Autom Control. 2002;47(7):1204–8.
73. Fang Y, Loparo KA. On the relationship between the sample path and moment Lyapunov exponents for jump linear systems. IEEE Trans Autom Control. 2002;47(9):1556–60.
74. Fang Y, Loparo KA. Stabilization of continuous-time jump linear systems. IEEE Trans Autom Control. 2002;47(10):1590–603.
75. Fax JA, Murray RM. Information flow and cooperative control of vehicle formations. IEEE Trans Autom Control. 2004;49(9):1465–76.

76. Feng G. Stability analysis of discrete time fuzzy dynamic systems based on piecewise Lyapunov functions. IEEE Trans Fuzzy Syst. 2004;12(1):22–8.
77. Feng G. A survey on analysis and design of model-based fuzzy control systems. IEEE Trans Fuzzy Syst. 2006;14(5):676–97.
78. Feng G, Ma J. Quadratic stabilization of uncertain discrete-time fuzzy dynamic systems. IEEE Trans Circuits Syst I, Fundam Theory Appl. 2001;48(11):1337–44.
79. Feng X, Loparo KA, Ji Y, Chizeck HJ. Stochastic stability properties of jump linear systems. IEEE Trans Autom Control. 1992;37(1):38–53.
80. Feron E. Quadratic stabilizability of switched systems via state and output feedback. Massachusetts Inst Tech, Tech Rep CICS-P-468; 1996.
81. Feuer A, Goodwin GC, Salgado M. Potential benefits of hybrid control for linear time invariant plants. In: Proc ACC; 1997. p. 2790–4.
82. Filippov AF. Stability for differential equations with discontinuous and many-valued right-hand sides. Differ Uravn (Minsk). 1979;15:1018–27.
83. Filippov AF. Differential equations with discontinuous right-hand side. Moscow: Nauka; 1985.
84. Fu M, Barmish B. Adaptive stabilization of linear systems via switching control. IEEE Trans Autom Control. 1986;31(12):1097–103.
85. Gaines FJ, Thompson RC. Sets of nearly triangular matrices. Duke Math J. 1968;35(3):441–54.
86. Ge SS, Sun Z. Switched controllability via bumpless transfer input and constrained switching. IEEE Trans Autom Control. 2008;53(7):1702–6.
87. Geromel JC, Colaneri P. Stability and stabilization of continuous-time switched linear systems. SIAM J Control Optim. 2006;45(5):1915–30.
88. Geromel JC, Colaneri P, Bolzern P. Dynamic output feedback control of switched linear systems. IEEE Trans Autom Control. 2008;53(3):720–33.
89. Godsil C, Royle GF. Algebraic graph theory. Berlin: Springer; 2001.
90. Goebel R, Sanfelice RG, Teel AR. Invariance principles for switching systems via hybrid systems techniques. Syst Control Lett. 2008;57(12):980–6.
91. Goebel R, Sanfelice RG, Teel AR. Hybrid dynamical systems. IEEE Control Syst Mag. 2009;29(2):28–93.
92. Goncalves JM, Megretski A, Dahleh A. Global stability of relay feedback systems. IEEE Trans Autom Control. 2001;46(4):550–62.
93. Goncalves JM, Megretski A, Dahleh A. Global analysis of piecewise linear systems using impact maps and surface Lyapunov functions. IEEE Trans Autom Control. 2003;48(12):2089–106.
94. Griggs WM, King CK, Shorten RN, Mason O, Wulff K. Quadratic Lyapunov functions for systems with state-dependent switching. Linear Algebra Appl. 2010;433(1):52–63.
95. Gripenberg G. Computing the joint spectral radius. Linear Algebra Appl. 1996;234:43–60.
96. Guglielmi N, Zennaro M. On the limit products of a family of matrices. Linear Algebra Appl. 2003;362:11–27.
97. Guo YQ, Wang YY, Xie LH, Zheng JC. Stability analysis and design of reset systems: theory and an application. Automatica. 2009;45(2):492–7.
98. Gurvits L. Stabilities and controllabilities of switched systems (with applications to the quantum systems). In: Proc 15th int symp math theory network syst; 2002.
99. Han TT, Ge SS, Lee TH. Persistent dwell-time switched nonlinear systems: variation paradigm and gauge design. IEEE Trans Autom Control. 2010;55(2):321–37.
100. Hedlund S, Johansson M. PWLTOOL: a MATLAB toolbox for analysis of piecewise linear systems. Dept Automat Contr, Lund Inst Tech; 1999.
101. Heemels WP, Brogliato B. The complementarity class of hybrid dynamical systems. Eur J Control. 2003;9(2–3):322–60.
102. Heemels WP, De Schutter B, Bemporad A. Equivalence of hybrid dynamical models. Automatica. 2001;37(7):1085–91.
103. Hegselmann R, Krause U. Opinion dynamics and bounded confidence: models, analysis, and simulation. J Artif Soc Soc Simul. 2002;5(3):2–34.

104. Hespanha JP. Uniform stability of switched linear systems: extensions of LaSalle's invariance principle. IEEE Trans Autom Control. 2004;49(4):470–82.
105. Hespanha JP, Liberzon D, Angeli D, Sontag ED. Nonlinear norm-observability notions and stability of switched systems. IEEE Trans Autom Control. 2005;50(2):154–68.
106. Hespanha JP, Liberzon D, Morse AS. Logic-based switching control of a nonholonomic system with parametric modeling uncertainty. Syst Control Lett. 1999;38(3):167–77.
107. Hespanha JP, Liberzon D, Morse AS. Overcoming the limitations of adaptive control by means of logic-based switching. Syst Control Lett. 2003;49(1):49–65.
108. Hespanha JP, Liberzon D, Morse AS. Hysteresis-based switching algorithms for supervisory control of uncertain systems. Automatica. 2003;39:263–72.
109. Hespanha JP, Liberzon D, Morse AS, Anderson BDO, Brinsmead TS, De Bruyne F. Multiple model adaptive control, part 2: Switching. Int J Robust Nonlinear Control. 2001;11(5):479–96.
110. Hespanha JP, Morse AS. Stability of switched systems with average dwell-time. In: Proc IEEE CDC; 1999. p. 2655–60.
111. Hockerman-Frommer J, Kulkarni SR, Ramadge PJ. Controller switching based on output prediction errors. IEEE Trans Autom Control. 1998;43(5):596–607.
112. Holcman D, Margaliot M. Stability analysis of switched homogeneous systems in the plane. SIAM J Control Optim. 2003;41(5):1609–25.
113. Hong YG, Gao LX, Cheng D, Hu JP. Lyapunov-based approach to multiagent systems with switching jointly connected interconnection. IEEE Trans Autom Control. 2007;52(5):943–8.
114. Hu TS, Lin ZL. Composite quadratic Lyapunov functions for constrained control systems. IEEE Trans Autom Control. 2003;48(3):440–50.
115. Hu TS, Ma LQ, Lin ZL. Stabilization of switched systems via composite quadratic functions. IEEE Trans Autom Control. 2008;53(11):2571–85.
116. Imura J. Well-posedness analysis of switch-driven piecewise affine systems. IEEE Trans Autom Control. 2003;48(11):1926–35.
117. Ingalls B, Sontag ED, Wang Y. An infinite-time relaxation theorem for differential inclusions. Proc Am Math Soc. 2003;131(2):487–99.
118. Ishii H, Francis BA. Stabilizing a linear system by switching control with dwell time. IEEE Trans Autom Control. 2002;47(12):1962–73.
119. Iwatania Y, Hara S. Stability tests and stabilization for piecewise linear systems based on poles and zeros of subsystems. Automatica. 2006;42(10):1685–95.
120. Jadbabaie A, Lin J, Morse AS. Coordination of groups of mobile agents using nearest neighbor rules. IEEE Trans Autom Control. 2003;48(6):988–1001.
121. Ji M, Egerstedt M. Distributed coordination control of multiagent systems while preserving connectedness. IEEE Trans Robot. 2007;23(4):693–703.
122. Ji ZJ, Wang L, Guo XX. Design of switching sequences for controllability realization of switched linear systems. Automatica. 2007;43(4):662–8.
123. Jiang SX, Voulgaris PG. Performance optimization of switched systems: a model matching approach. IEEE Trans Autom Control. 2009;54(9):2058–71.
124. Jiang ZP, Wang Y. A converse Lyapunov theorem for discrete-time systems with disturbances. Syst Control Lett. 2002;45(1):49–58.
125. Johansson M. Piecewise linear control systems. New York: Springer; 2003.
126. Johansson M, Rantzer A. Computation of piecewise quadratic Lyapunov functions for hybrid systems. IEEE Trans Autom Control. 1998;43:555–9.
127. Johansson M, Rantzer A, Arzen K-E. Piecewise quadratic stability of fuzzy systems. IEEE Trans Fuzzy Syst. 1999;7(6):713–22.
128. John F. Extremum problems with inequalities as subsidiary conditions. In: Studies and essays presented to R. Courant on his 60th birthday. New York: Interscience; 1948. p. 187–204.
129. Kellet CM, Teel A. Weak converse Lyapunov theorems and control-Lyapunov functions. SIAM J Control Optim. 2004;42:1934–59.
130. Khalil HK. Nonlinear systems. 3rd ed. Upper Saddle River: Prentice Hall; 2002.
131. Khas'minskii RZ. Necessary and sufficient condition for the asymptotic stability of linear stochastic systems. Theory Probab Appl. 1967;12(1):144–7.

132. Kolmanovsky I, McClamroch NH. Developments in nonholonomic control problems. IEEE Control Syst Mag. 1995;15(6):20–36.
133. Kozin F. On relations between moment properties and almost sure Lyapunov stability for linear stochastic systems. J Math Anal Appl. 1965;10:324–53.
134. Kozin F. A survey of stability of stochastic systems. Automatica. 1969;5(1):95–112.
135. Krasovskii NN. Stability of motion. Stanford: Stanford Univ Press; 1963.
136. Kushner HJ. Stochastic stability and control. New York: Academic Press; 1967.
137. Laffey TJ. Simultaneous triangularization of matrices—low rank case and the nonderogatory case. Linear Multilinear Algebra. 1978;6(1):269–305.
138. Lagarias JC, Wang Y. The finiteness conjecture for the generalized spectral radius of a set of matrices. Linear Algebra Appl. 1995;214:17–42.
139. Lasota A, Strauss A. Asymptotic behavior for differential equations which cannot be locally linearized. J Differ Equ. 1971;10(1):152–72.
140. Lee JW, Dullerud GE. Uniform stabilization of discrete-time switched and Markovian jump linear systems. Automatica. 2006;42(2):205–18.
141. Lee JW, Khargonekar PP. Optimal output regulation for discrete-time switched and Markovian jump linear systems. SIAM J Control Optim. 2008;47(1):40–72.
142. Lee JW, Khargonekar PP. Detectability and stabilizability of discrete-time switched linear systems. IEEE Trans Autom Control. 2009;54(3):424–37.
143. Leenaerts DMW. On linear dynamic complementary systems. IEEE Trans Circuits Syst I, Fundam Theory Appl. 1999;46(8):1022–6.
144. Li ZG, Soh YC, Wen CY. Sufficient conditions for almost sure stability of jump linear systems. IEEE Trans Autom Control. 2000;45(7):1325–9.
145. Li ZG, Soh YC, Wen CY. Switched and impulsive systems: analysis, design and applications. Berlin: Springer; 2005.
146. Liberzon D. Switching in systems and control. Boston: Birkhäuser; 2003.
147. Liberzon D, Hespanha JP, Morse AS. Stability of switched systems: a Lie-algebraic condition. Syst Control Lett. 1999;37(3):117–22.
148. Liberzon MR. Essays on the absolute stability theory. Autom Remote Control. 2006; 67(10):1610–44.
149. Lin H, Antsaklis PJ. Stability and stabilizability of switched linear systems: a short survey of recent results. In: Proc IEEE ISIC; 2005. p. 24–9.
150. Lin H, Antsaklis PJ. Switching stabilizability for continuous-time uncertain switched linear systems. IEEE Trans Autom Control. 2007;52(4):633–46.
151. Lin H, Antsaklis PJ. Stability and stabilizability of switched linear systems: a survey of recent results. IEEE Trans Autom Control. 2009;54(2):308–22.
152. Lin Y, Sontag ED, Wang Y. A smooth converse Lyapunov theorem for robust stability. SIAM J Control Optim. 1996;34(1):124–60.
153. Lin Z, Broucke M, Francis B. Local control strategies for groups of mobile autonomous agents. IEEE Trans Autom Control. 2004;49(4):622–9.
154. Loewy R. On ranges of real Lyapunov transformations. Linear Algebra Appl. 1976;13(1):79–89.
155. Lu L, Lin ZL. A switching anti-windup design using multiple Lyapunov functions. IEEE Trans Autom Control. 2010;55(1):142–8.
156. Lur YY. A note on a gap result for norms of semigroups of matrices. Linear Algebra Appl. 2006;419(2–3):368–72.
157. Lur'e AI. Nekotorye nelineinye zadachi teorii avtomaticheskogo regulirovaniya (Some nonlinear problems of the automatic control theory). Moscow: Gostekhizdat; 1951.
158. Maesumi M. An efficient lower bound for the generalized spectral radius of a set of matrices. Linear Algebra Appl. 1996;240:1–7.
159. Mancilla-Aguilar JL, Garcia RA. A converse Lyapunov theorem for nonlinear switched systems. Syst Control Lett. 2000;41(1):67–71.
160. Marcelo D, Fragoso MD, Costa OLV. A unified approach for stochastic and mean square stability of continuous-time linear systems with Markovian jumping parameters and additive disturbances. SIAM J Control Optim. 2005;44(4):1165–91.

161. Margaliot M. Stability analysis of switched systems using variational principles: an introduction. Automatica. 2006;42(12):2059–77.
162. Margaliot M, Gitizadeh R. The problem of absolute stability: a dynamic programming approach. Automatica. 2004;40(7):1247–52.
163. Margaliot M, Langholz G. Necessary and sufficient conditions for absolute stability: the case of second-order systems. IEEE Trans Circuits Syst I, Fundam Theory Appl. 2003;50(2):227–34.
164. Margaliot M, Liberzon D. Lie-algebraic stability conditions for nonlinear switched systems and differential inclusions. Syst Control Lett. 2006;55(1):8–16.
165. Margaliot M, Yfoulis C. Absolute stability of third-order systems: a numerical algorithm. Automatica. 2006;42(10):1705–11.
166. Mariton M. Almost sure and moment stability of jump linear systems. Syst Control Lett. 1988;11(5):393–7.
167. Mariton M. Jump linear systems in automatic control. New York: Marcel Dekker; 1990.
168. Mason P, Boscain U, Chitour Y. On the minimal degree of a common Lyapunov function for planar switched systems. In: Proc IEEE CDC; 2004. p. 2786–91.
169. McClamroch NH, Kolmanovsky I. Performance benefits of hybrid control design for linear and nonlinear systems. Proc IEEE. 2000;88(7):1083–96.
170. Molchanov AP, Pyatnitskiy YeS. Lyapunov functions defining the necessary and sufficient conditions for absolute stability of the nonlinear control systems, I. Avtom Telemeh. 1986;3:63–73.
171. Molchanov AP, Pyatnitskiy YeS. Lyapunov functions defining the necessary and sufficient conditions for absolute stability of the nonlinear control systems, II. Avtom Telemeh. 1986;4:5–15.
172. Molchanov AP, Pyatnitskiy YeS. Lyapunov functions defining the necessary and sufficient conditions for absolute stability of the nonlinear control systems, III. Avtom Telemeh. 1986;5:38–49.
173. Molchanov AP, Pyatnitskiy YeS. Criteria of asymptotic stability of differential and difference inclusions encountered in control theory. Syst Control Lett. 1989;13(1):59–64.
174. Morari M, Baotic M, Borrelli F. Hybrid systems modeling and control. Eur J Control. 2003;9(2–3):177–89.
175. Moreau L. Stability of continuous-time distributed consensus algorithms. In: Proc IEEE CDC; 2004. p. 3998–4003.
176. Mori Y, Mori T, Kuroe Y. A solution to the common Lyapunov function problem for continuous-time systems. In: Proc IEEE CDC; 1997. p. 3530–1.
177. Morse AS. Logic-based switching and control. In: Francis BA, Tannenbaum AR, editors. Feedback control, nonlinear systems, and complexity. New York: Springer; 1995. p. 173–95.
178. Morse AS. Supervisory control of families of linear set-point controllers, part 1: Exact matching. IEEE Trans Autom Control. 1996;41(10):1413–31.
179. Morse AS. Supervisory control of families of linear set-point controllers, part 2: Robustness. IEEE Trans Autom Control. 1997;42(11):1500–15.
180. Morse AS. Lecture notes on logically switched dynamical systems. In: Agrachev AA, Morse AS, Sontag ED, Sussmann HJ, Utkin VI, editors. Nonlinear and optimal control theory. Berlin: Springer; 2008. p. 61–161.
181. Mount DM. Bioinformatics: sequence and genome analysis. 2nd ed. New York: Cold Spring Harbor Laboratory Press; 2004.
182. Narendra KS, Balakrishnan J. A common Lyapunov function for stable LTI systems with commuting A-matrices. IEEE Trans Autom Control. 1994;39(12):2469–71.
183. Narendra KS, Balakrishnan J. Adaptive control using multiple models. IEEE Trans Autom Control. 1997;42(2):171–87.
184. Nesterov Y. Squared functional systems and optimization problems, high performance optimization. Appl Optim. 2000;33:405–40.
185. Notredame C. Recent progresses in multiple sequence alignment: a survey. Pharmacogenomics. 2002;3(1):131–44.

186. Oktem H. A survey on piecewise-linear models of regulatory dynamical systems. Nonlinear Anal. 2005;63(3):336–49.
187. Olfati-Saber R, Murray RM. Consensus problems in networks of agents with switching topology and time-delays. IEEE Trans Autom Control. 2004;49(9):1520–33.
188. Opoiytsev VI. Conversion of principle of contractive maps. Usp Mat Nauk. 1976;31:169–98.
189. Opoiytsev VI. A multiplicative ergodic theorem: Lyapunov characteristic numbers for dynamical systems. Trans Mosc Math Soc. 1968;19:197–231.
190. Parrilo PA, Jadbabaie A. Approximation of the joint spectral radius using sum of squares. Linear Algebra Appl. 2008;428:2385–402.
191. Pearson WR, Lipman DJ. Improved tools for biological sequence comparison. Proc Natl Acad Sci USA. 1988;85:2444–8.
192. Peng YP. Feedback stabilization and performance optimization of switched systems. PhD dissertation, South China Univ Tech; 2010.
193. Protasov Yu. The geometric approach for computing the joint spectral radius. In: Proc IEEE CDC; 2005. p. 3001–6.
194. Pyatnitskiy ES, Rapoport LB. Criteria of asymptotic stability of differential inclusions and periodic motions of time-varying nonlinear control systems. IEEE Trans Circuits Syst I, Fundam Theory Appl. 1996;43(3):219–29.
195. Radjavi H, Rosenthal P. Simultaneous triangularization. New York: Springer; 1999.
196. Rapoport LB. Asymptotic stability and periodic motions of selector-linear differential inclusions. In: Garofalo F, Glielmo L, editors. Robust control via variable structure and Lyapunov techniques. New York: Springer; 1996. p. 269–85.
197. Riedinger P, Sigalotti M, Daafouz J. On the algebraic characterization of invariant sets of switched linear systems. Automatica. 2010;46(6):1047–52.
198. Rifford L. Existence of Lipschitz and semiconcave control-Lyapunov functions. SIAM J Control Optim. 2000;39(4):1043–64.
199. Rota GC, Strang X. A note on the joint spectral radius. Indag Math. 1960;22:379–81.
200. Santarelli KR, Dahleh MA. Optimal controller synthesis for a class of LTI systems via switched feedback. Syst Control Lett. 2010;59(3–4):258–64.
201. Seatzu C, Corona D, Giua A, Bemporad A. Optimal control of continuous-time switched affine systems. IEEE Trans Autom Control. 2006;51(5):726–41.
202. Shen JL. Observability analysis of conewise linear systems via directional derivative and positive invariance techniques. Automatica. 2010;46(5):843–51.
203. Shor NZ. Class of global minimum bounds of polynomial functions. Cybernetics. 1987;23(6):731–4.
204. Shorten RN, Narendra KS. On the existence of a common quadratic Lyapunov functions for linear stable switching systems. In: Proc Yale Workshop Adapt Learn Syst; 1998.
205. Shorten RN, Narendra KS. Necessary and sufficient conditions for the existence of a common quadratic Lyapunov function for two stable second order linear time-invariant systems. In: Proc ACC; 1999. p. 1410–4.
206. Shorten RN, Narendra KS. Necessary and sufficient conditions for the existence of a common quadratic Lyapunov function for a finite number of stable second order linear time-invariant systems. Int J Adapt Control Signal Process. 2003;16:709–28.
207. Shorten RN, Narendra KS, Mason O. A result on common quadratic Lyapunov functions. IEEE Trans Autom Control. 2003;48(1):110–3.
208. Shorten RN, Wirth F, Mason O, Wulff K, King C. Stability criteria for switched and hybrid systems. SIAM Rev. 2007;49(4):545–92.
209. Solmaz S, Shorten RN, Wulff K, Cairbre F. A design methodology for switched discrete time linear systems with applications to automotive roll dynamics control. Automatica. 2008;44(9):2358–63.
210. Sontag ED. Nonlinear regulation: the piecewise linear approach. IEEE Trans Autom Control. 1981;26(2):346–58.
211. Sontag ED. Smooth stabilization implies coprime factorization. IEEE Trans Autom Control. 1989;34(4):435–43.

212. Sontag ED. Interconnected automata and linear systems: a theoretical framework in discrete-time. In: Alur R, Henzinger TA, Sontag ED, editors. Hybrid systems III—Verification and control. Berlin: Springer; 1996. p. 436–48.
213. Stanford DP, Urbano JM. Some convergence properties of matrix sets. SIAM J Matrix Anal Appl. 1994;14(4):1132–40.
214. Stefanovic M, Safonov M. Safe adaptive switching control: stability and convergence. IEEE Trans Autom Control. 2008;53(9):2012–21.
215. Sugeno M. On stability of fuzzy systems expressed by fuzzy rules with singleton consequents. IEEE Trans Fuzzy Syst. 1999;7(2):201–24.
216. Sun Z. Stabilizability and insensitiveness of switched systems. IEEE Trans Autom Control. 2004;49(7):1133–7.
217. Sun Z. A modified stabilizing law for switched linear systems. Int J Control. 2004;77(4):389–98.
218. Sun Z. A general robustness theorem for switched linear systems. In: Proc IEEE ISIC; 2005. p. 8–11.
219. Sun Z. Combined stabilizing strategies for switched linear systems. IEEE Trans Autom Control. 2006;51(4):666–74.
220. Sun Z. Stabilization and optimal switching of switched linear systems. Automatica. 2006;42(5):783–8.
221. Sun Z. Guaranteed stability of switched linear systems revisited. In: Proc IEEE ICCA 2007; 2007. p. 18–23.
222. Sun Z. Converse Lyapunov theorem for switched stability of switched linear systems. In: Proc Chinese contr conf; 2007. p. 678–80.
223. Sun Z. Matrix measure approach for stability of switched linear systems. In: IFAC NOLCOS; 2007. p. 557–60.
224. Sun Z. A note on marginal stability of switched systems. IEEE Trans Autom Control. 2008;53(2):625–31.
225. Sun Z. Stabilizing switching design for switched linear systems: a state-feedback path-wise switching approach. Automatica. 2009;45(7):1708–14.
226. Sun Z. A graphic approach for stability of piecewise linear systems. In: Proc Chinese conf dec contr; 2009. p. 1016–9.
227. Sun Z. The problem of slow switching for switched linear systems. In: Proc ICCAS-SICE; 2009. p. 4843–6.
228. Sun Z. Robust switching of switched linear systems. In: Proc IFAC NOLCOS; 2010. p. 256–9.
229. Sun Z. Stability and contractivity of conewise linear systems. In: Proc IEEE MSC; 2010. p. 2094–8.
230. Sun Z. Stability of piecewise linear systems revisited. Annu Rev Control. 2010;34(2): 221–31.
231. Sun Z. Switching distance and robust switching for switched linear systems. Automatica. 2010; submitted.
232. Sun Z, Ge SS. Dynamic output feedback stabilization of a class of switched linear systems. IEEE Trans Circuits Syst I, Fundam Theory Appl. 2003;50(8):1111–5.
233. Sun Z, Ge SS. Analysis and synthesis of switched linear control systems. Automatica. 2005;41(2):181–95.
234. Sun Z, Ge SS. Switched linear systems: control and design. London: Springer; 2005.
235. Sun Z, Ge SS. On stability of switched linear systems with perturbed switching paths. J Control Theory Appl. 2006;4(1):18–25.
236. Sun Z, Ge SS, Lee TH. Reachability and controllability criteria for switched linear systems. Automatica. 2002;38(5):775–86.
237. Sun Z, Huang J. A note on connectivity of multi-agent systems with proximity graphs and linear feedback protocol. Automatica. 2009;45(9):1953–6.
238. Sun Z, Peng Y. Stabilizing design for switched linear control systems: a constructive approach. Trans Instrum Meas. 2010;32(6):706–35.

239. Sun Z, Shorten RN. On convergence rates of simultaneously triangularizable switched linear systems. IEEE Trans Autom Control. 2005;50(8):1224–8.
240. Sworder D. Control of a linear system with a Markov property. IEEE Trans Autom Control. 1965;10(3):294–300.
241. Sworder D. Feedback control of a class of linear systems with jump parameters. IEEE Trans Autom Control. 1969;14(1):9–14.
242. Szabo Z, Bokor J, Balas G. Generalized piecewise linear feedback stabilizability of controlled linear switched systems. In: Proc IEEE CDC; 2008. p. 3410–4.
243. Szabo Z, Bokor J, Balas G. Controllability and stabilizability of linear switched systems. In: Proc ECC; 2009.
244. Takagi T, Sugeno M. Fuzzy identification of systems and its applications to modeling and control. IEEE Trans Syst Man Cybern. 1985;15(1):116–32.
245. AH Tan. Direction-dependent systems—a survey. Automatica. 2009;45(12):2729–43.
246. Tan PV, Millerioux G, Daafouz J. Left invertibility, flatness and identifiability of switched linear dynamical systems: a framework for cryptographic applications. Int J Control. 2010;83(1):145–53.
247. Tanaka K, Sugeno M. Stability analysis and design of fuzzy control systems. Fuzzy Sets Syst. 1992;12(2):135–56.
248. Tanaka K, Wang HO. Fuzzy control systems design and analysis: a linear matrix inequality approach. New York: Wiley; 2001.
249. Theys J. Joint spectral radius: theory and approximations. PhD dissertation, Dept Math Eng, Univ Louvain; 2005.
250. Tokarzewski J. Stability of periodically switched linear systems and the switching frequency. Int J Syst Sci. 1987;18(4):697–726.
251. Tsitsiklis JN, Blondel VD. The Lyapunov exponent and joint spectral radius of pairs of matrices are hard, when not impossible, to compute and to approximate. Math Control Signals Syst. 1997;10(1):31–40.
252. Veres SM. The geometric bounding toolbox, user's manual & reference. UK: SysBrain; 2001.
253. Vidyasagar M. Nonlinear systems analysis. 2nd ed. Eaglewood Cliffs: Prentice Hall; 1993.
254. Vinnicombe G. Frequency domain uncertainty and the graph topology. IEEE Trans Autom Control. 1993;38:1371–83.
255. Vladimirov A, Elsner L, Beyn WJ. Stability and paracontractivity of discrete linear inclusions. Linear Algebra Appl. 2000;312(1–3):125–34.
256. Vu L, Liberzon D. Common Lyapunov functions for families of commuting nonlinear systems. Syst Control Lett. 2005;54(5):405–16.
257. Wang W, Nesic D. Input-to-state stability and averaging of linear fast switching systems. IEEE Trans Autom Control. 2010;55(5):1274–9.
258. Wicks MA, Peleties V, DeCarlo RA. Construction of piecewise Lyapunov functions for stabilizing switched systems. In: Proc IEEE CDC; 1994. p. 3492–7.
259. Wicks MA, Peleties P, DeCarlo RA. Switched controller synthesis for the quadratic stabilization of a pair of unstable linear systems. Eur J Control. 1998;4(2):140–7.
260. Wielandt H. Losung der Aufgabe 338 (When are irreducible components of a semigroup of matrices bounded?). Jahresber Dtsch Math-Ver. 1954;57:4–5.
261. Wirth F. A converse Lyapunov theorem for linear parameter-varying and linear switching systems. SIAM J Control Optim. 2005;44(1):210–39.
262. Witsenhausen HS. A class of hybrid-state continuous-time dynamic systems. IEEE Trans Autom Control. 1966;11(2):161–7.
263. Wu AG, Feng G, Duan GR, Gao HJ. A stabilizing slow-switching law for switched discrete-time linear systems. In: Proc IEEE MSC; 2010. p. 2099–104.
264. Wu J, Sun Z. New slow-switching laws for switched linear systems. In: Proc IEEE ICCA; 2010. p. 2274–7.
265. Xia X. Well-posedness of piecewise-linear systems with multiple modes and multiple criteria. IEEE Trans Autom Control. 2002;47(10):1716–20.

266. Xie G, Wang L. Controllability and stabilizability of switched linear-systems. Syst Control Lett. 2003;48(2):135–55.
267. Xie G, Wang L. Periodical stabilization of switched linear systems. J Comput Appl Math. 2005;18(1):176–87.
268. Xie G, Wang L. Controllability implies stabilizability for discrete-time switched linear systems. In: Hybrid systems: computation and control. Berlin: Springer; 2005. p. 667–82.
269. Xie G, Wang L. Periodic stabilizability of switched linear control systems. Automatica. 2009;45(9):2141–8.
270. Xu X, Antsaklis PJ. Optimal control of switched systems based on parameterization of the switching instants. IEEE Trans Autom Control. 2004;49(1):2–16.
271. Yakubovich VA, Leonov GA, Gelig AK. Ustoichivost Nelineinykh Sistem s Needinstvennym Sostoyaniem Ravnovesiya (Stability of Nonlinear Systems with Nonunique Equilibrium State). Moscow: Nauka; 1978.
272. Zadeh LA. Fuzzy sets. Inf Control. 1965;8(3):338–53.
273. Zadeh LA. Fuzzy algorithm. Inf Control. 1968;12:94–102.
274. Zahreddine Z. Matrix measure and application to stability of matrices and interval dynamical systems. Int J Math Math Sci. 2003;2:75–85.
275. Zavlanos MM, Jadbabaie A, Pappas GJ. Flocking while preserving network connectivity. In: Proc IEEE CDC; 2007. p. 2919–24.
276. Zhai G, Lin H, Antsaklis PJ. Quadratic stabilizability of switched linear systems with polytopic uncertainties. Int J Control. 2003;76(7):747–53.
277. Zhai G, Yasuda K. Stability analysis for a class of switched systems. Trans Soc Instrum Control Eng. 2000;36(5):409–15.
278. Zhang LX, Gao HJ. Asynchronously switched control of switched linear systems with average dwell time. Automatica. 2010;46(5):953–8.
279. Zhang W, Abate A, Hu JH, Vitus MP. Exponential stabilization of discrete-time switched linear systems. Automatica. 2009;45(11):2526–36.
280. Zhang W, Hu JH. On optimal quadratic regulation for discrete-time switched linear systems. In: Egerstedt M, Mishra B, editors. Hybrid systems: computation and control. Berlin: Springer; 2008. p. 584–97.
281. Zhao J, Hill DJ. Dissipativity theory for switched systems. IEEE Trans Autom Control. 2008;53(4):941–53.
282. Zhao J, Hill DJ. Passivity and stability of switched systems: a multiple storage function method. Syst Control Lett. 2008;57(2):158–64.
283. Zhao J, Hill DJ, Liu T. Synchronization of complex dynamical networks with switching topology: a switched system point of view. Automatica. 2009;45(11):2502–11.

Index

Z. Sun, S.S. Ge, *Stability Theory of Switched Dynamical Systems*,
Communications and Control Engineering,
DOI 10.1007/978-0-85729-256-8, © Springer-Verlag London Limited 2011